Cloud Computing Advancements in Design, Implementation, and Technologies

Shadi Aljawarneh
Isra University, Jordan

Information Science
REFERENCE

Managing Director:	Lindsay Johnston
Senior Editorial Director:	Heather A. Probst
Book Production Manager:	Sean Woznicki
Development Manager:	Joel Gamon
Assistant Acquisitions Editor:	Kayla Wolfe
Typesetter:	Deanna Jo Zombro
Cover Design:	Nick Newcomer

Published in the United States of America by
Information Science Reference (an imprint of IGI Global)
701 E. Chocolate Avenue
Hershey PA 17033
Tel: 717-533-8845
Fax: 717-533-8661
E-mail: cust@igi-global.com
Web site: http://www.igi-global.com

Library of Congress Cataloging-in-Publication Data

Cloud computing advancements in design, implementation, and technologies /
Shadli Aljawarneh, editor.
    p. cm.
  Includes bibliographical references and index.
  Summary: "This book outlines advancements in the state-of-the-art standards
and practices of cloud computing, in an effort to identify emerging trends
that will ultimately define the future of the cloud"--Provided by publisher.
  ISBN 978-1-4666-1879-4 (hardcover) -- ISBN 978-1-4666-1880-0 (ebook) -- ISBN
978-1-4666-1881-7 (print & perpetual access) 1. Cloud computing. I.
Aljawarneh, Shadli.
  QA76.585.C57 2013
  004.6782--dc23
                        2012005394

British Cataloguing in Publication Data
A Cataloguing in Publication record for this book is available from the British Library.

The views expressed in this book are those of the authors, but not necessarily of the publisher.

Table of Contents

Detailed Table of Contents

Junaid Arshad, University of Leeds, UK
Paul Townend, University of Leeds, UK
Jie Xu, University of Leeds, UK

Cloud computing is an emerging computing paradigm which introduces novel opportunities to establish large scale, flexible computing infrastructures. However, security underpins extensive adoption of Cloud computing. This paper presents efforts to address one of the significant issues with respect to security of Clouds i.e. intrusion detection and severity analysis. An abstract model for integrated intrusion detection and severity analysis for Clouds is proposed to facilitate minimal intrusion response time while preserving the overall security of the Cloud infrastructures. In order to assess the effectiveness of the proposed model, detailed architectural evaluation using Architectural Trade-off Analysis Model (ATAM) is used. A set of recommendations which can be used as a set of best practice guidelines while implementing the proposed architecture is discussed.

Ismael Solis Moreno, University of Leeds, UK
Jie Xu, University of Leeds, UK

Due to all the pollutants generated during its production and the steady increases in its rates, energy consumption is causing serious environmental and economic problems. In this context, the growing use and adoption of ICTs is being highlighted not only as one as the principal problem sources but also as one of the principal areas that could help in the problem's reduction. Cloud computing is an emerging model for distributed utility computing and is being considered as an attractive opportunity for saving energy through central management of computational resources. To be successful, the design of energy-efficient mechanisms must start playing a mayor role. This paper argues the importance of energy-efficient mechanisms within cloud data centers and remarks on the significance of the "energy-performance" relationship in boosting the adoption of these mechanisms in real scenarios. It provides an analysis of the current approaches and the outline of key opportunities that need to be addressed to improve the "energy-performance" relationship in this promising model.

Luis M. Vaquero, Telefónica Investigación y Desarrollo, Spain

Juan Cáceres, Telefónica, Spain

Daniel Morán, Universidad Nacional de Educación a Distancia, Spain

This paper presents a brief overview of the available literature on distributed systems scalability that serves as a justification for presenting some of the most prominent challenges that current Cloud systems need to face in order to deliver their pledged easy-to-use scalability. Through illustrative comparisons and examples, this paper aims to make the reader's acquaintance with this long needed problem in distributed systems: user-oriented service-level scalability. Scalability issues are analyzed from the Infrastructure as a Service (IaaS) and the Platform as a Service (PaaS) point of view, as they deal with different functions and abstraction levels. Next generation Cloud provisioning models rely on advanced monitoring and automatic scaling decision capabilities to ensure quality of service (QoS), security and economic sustainability.

Zhiwei Xu, Chinese Academy of Sciences, China

Bo Yan, Chinese Academy of Sciences, China

Yongqiang Zou, Tencent Research, China

As a main subfield of cloud computing applications, internet services require large-scale data computing. Their workloads can be divided into two classes: customer-facing query-processing interactive tasks that serve hundreds of millions of users within a short response time and backend data analysis batch tasks that involve petabytes of data. Hadoop, an open source software suite, is used by many Internet services as the main data computing platform. Hadoop is also used by academia as a research platform and an optimization target. This paper presents five research directions for optimizing Hadoop; improving performance, utilization, power efficiency, availability, and different consistency constraints. The survey covers both backend analysis and customer-facing workloads. A total of 15 innovative techniques and systems are analyzed and compared, focusing on main research issues, innovative techniques, and optimized results.

Hong Cai, IBM China Software Development Lab, China

Berthold Reinwald, IBM Almaden Research Center, USA

Ning Wang, IBM China Software Development Lab, China

Chang Jie Guo, IBM China Research Lab, China

SaaS (Software as a Service) provides new business opportunities for application providers to serve more customers in a scalable and cost-effective way. SaaS also raises new challenges and one of them is multi-tenancy. Multi-tenancy is the requirement of deploying only one shared application to serve multiple customers (i.e. tenant) instead of deploying one dedicated application for each customer. This paper describes the authors' practice of developing and deploying multi-tenant technologies. This paper targets a technology that could quickly enable existing Java EE (Enterprise Edition) applications to be multi-tenancy enabled thus having the benefit of quick time to market. This paper describes the overall framework of multi-tenant SaaS platform, how to migrate an existing Java EE application, how to provision the multi-tenant application, and how to onboard the tenants. The paper also shows experiments which compare the economics of multi-tenant SaaS deployment versus traditional application deployment (one application for one tenant) with precise data.

#### Chapter 6

Cloud Computing promises novel and valuable capabilities for computer users and is explored in all possible areas of information technology dependant fields. However, the literature suffers from hype and divergent definitions and viewpoints. Cloud powered higher education can gain significant flexibility and agility. Higher education policy makers must assume activist roles in the shift towards cloud computing. Classroom experiences show it is a better tool for teaching and collaboration. As it is an emerging service technology, there is a need for standardization of services and customized implementation. Its evolution can change the facets of rural education. It is important as a possible means of driving down the capital and total costs of IT. This paper examines and discusses the concept of Cloud Computing from the perspectives of diverse technologists, cloud standards, services available today, the status of cloud particularly in higher education, and future implications.

#### Chapter 7

Even though public cloud providers already exist and offer computing and storage services, cloud computing is still a buzzword for scientists in various fields such as engineering, finance, social sciences, etc. These technologies are currently mature enough to leave the experimental laboratory in order to be used in real-life scenarios. To this end, the authors consider that the prime example use case of cloud computing is a web hosting service. This paper presents the architectural approach as well as the technical solution for applying elastic web hosting onto a private cloud infrastructure using only free software. Through several available software applications and tools, anyone can build their own private cloud on top of a local infrastructure and benefit from the dynamicity and scalability provided by the cloud approach.

#### Chapter 8

Over a decade ago, cloud computing became an important topic for small and large businesses alike. The new concept promises scalability, security, cost reduction, portability, and availability. While addressing this issue over the past several years, there have been intensive discussions about the importance of cloud computing technologies. Therefore, this paper reviews the transition from traditional computing to cloud computing and the benefit for businesses, cloud computing architecture, cloud computing services classification, and deployment models. Furthermore, this paper discusses the security policies and types of internal risks that a small business might encounter implementing cloud computing technologies. It addresses initiatives towards employing certain types of security policies in small businesses implementing cloud computing technologies to encourage small business to migrate to cloud computing by portraying what is needed to secure their infrastructure using traditional security policies without the complexity used in large corporations.

This paper demonstrates financial enterprise portability, which involves moving entire application services from desktops to clouds and between different clouds, and is transparent to users who can work as if on their familiar systems. To demonstrate portability, reviews for several financial models are studied, where Monte Carlo Methods (MCM) and Black Scholes Model (BSM) are chosen. A special technique in MCM, Least Square Methods, is used to reduce errors while performing accurate calculations. Simulations for MCM are performed on different types of Clouds. Benchmark and experimental results are presented for discussion. 3D Black Scholes are used to explain the impacts and added values for risk analysis. Implications for banking are also discussed, as well as ways to track risks in order to improve accuracy. A conceptual Cloud platform is used to explain the contributions in Financial Software as a Service (FSaaS) and the IBM Fined Grained Security Framework. This study demonstrates portability, speed, accuracy, and reliability of applications in the clouds, while demonstrating portability for FSaaS and the Cloud Computing Business Framework (CCBF).

Information security is a key challenge in the Cloud because the data will be virtualized across different host machines, hosted on the Web. Cloud provides a channel to the service or platform in which it operates. However, the owners of data will be worried because their data and software are not under their control. In addition, the data owner may not recognize where data is geographically located at any particular time. So there is still a question mark over how data will be more secure if the owner does not control its data and software. Indeed, due to shortage of control over the Cloud infrastructure, use of ad-hoc security tools is not sufficient to protect the data in the Cloud; this paper discusses this security. Furthermore, a vision and strategy is proposed to mitigate or avoid the security threats in the Cloud. This broad vision is based on software engineering principles to secure the Cloud applications and services. In this vision, security is built into all phases of Service Development Life Cycle (SDLC), Platform Development Life Cycle (PDLC) or Infrastructure Development Life Cycle (IDLC).

#### Chapter 11

Cloud Computing (CC) is revolutionizing the methodology by which IT services are being utilized. It is being introduced and marketed with many attractive promises that are enticing to many companies and managers, such as reduced capital costs and relief from managing complex information technology infrastructure. However, along with desirable benefits come risks and security concerns that must be considered and addressed correctly. Thus, security issues must be considered as a major issue when considering Cloud Computing. This paper discusses Cloud Computing and its related concepts; highlights and categorizes many of the security issues introduced by the "cloud"; surveys the risks, threats, and vulnerabilities; and makes the necessary recommendations that can help promote the benefits and mitigate the risks associated with Cloud Computing.

Thermal management of integrated circuit (IC) and system-in-package (SIP) has gained importance as the power density and requirement for IC design have increased and need exists to analyse the heat dissipation performance characteristics of IC under use. In this paper, the authors examine the thermal characteristics of materials of IC. The authors leverage Cloud Computing architecture to remotely compute the dissipation performance parameters. Understanding thermal dissipation performance, which explains the thermal management of IC, is important for chip performance, as well as power and energy consumption in a chip or SIP. Using architectural understanding of Software as a Service (SaaS), the authors develop an efficient, fast, and secure simulation technique by leveraging control volume method (CVM) of linearization of relevant equations. Three chips are kept in tandem to make it a multi-chip module (MCM) to realise it as a smaller and lighter package. The findings of the study are presented for different dimensions of chips inside the package.

The term computer anti-forensics (CAF) generally refers to a set of tactical and technical measures intended to circumvent the efforts and objectives of the field of computer and network forensics (CF). Many scientific techniques, procedures, and technological tools have evolved and effectively applied in the field of CF to assist scientists and investigators in acquiring and analyzing digital evidence for the purpose of solving cases that involve the use or misuse of computer systems. CAF has emerged as a CF counterpart that plants obstacles throughout the path of computer investigations. The purpose of this paper is to highlight the challenges introduced by anti-forensics, explore various CAF mechanisms, tools, and techniques, provide a coherent classification for them, and discuss their effectiveness. Moreover, the authors discuss the challenges in implementing effective countermeasures against these techniques. A set of recommendations are presented with future research opportunities.

Cloud Computing provides on-demand access to a shared pool of configurable computing resources. The major issue lies in managing extremely large agile data centers which are generally over provisioned to handle unexpected workload surges. This paper focuses on green computing by introducing Power-Aware Meta Scheduler, which provides right fit infrastructure for launching virtual machines onto host. The major challenge of the scheduler is to make a wise decision in transitioning state of the processor cores by exploiting various power saving states inherent in the recent microprocessor technology. This is done by dynamically predicting the utilization of the cloud data center. The authors have extended existing cloudsim toolkit to model power aware resource provisioning, which includes generation of dynamic workload patterns, workload prediction and adaptive provisioning, dynamic lifecycle management of random workload, and implementation of power aware allocation policies and chip aware VM scheduler. The experimental results show that the appropriate usage of different power saving states guarantees significant energy conservation in handling stochastic nature of workload without compromising the performance, both when the data center is in low as well as moderate utilization.

Chapter 15

Promise Mvelase, CSIR Meraka Institute, South Africa
Nomusa Dlodlo, CSIR Meraka Institute, South Africa
Quentin Williams, CSIR Meraka Institute, South Africa
Matthew Adigun, University of Zululand, South Africa

Small, Medium, and Micro enterprises (SMMEs) usually do not have adequate funds to acquire ICT infrastructure and often use cloud computing. In this paper, the authors discuss the implementation of virtual enterprises (VE) to enable SMMEs to respond quickly to customers' demands and market opportunities. The virtual enterprise model is based on the ability to create temporary co-operations and realize the value of a short term business opportunity that the partners cannot fully capture on their own. The model of virtual enterprise is made possible through virtualisation technology, which is a building block of cloud computing. To achieve a common goal, enterprises integrate resources, organisational models, and process models. Through the virtual business operating environment offered by cloud computing, the SMMEs are able to increase productivity and gain competitive advantage due to the cost benefit incurred. In this paper, the authors propose a virtual enterprise enabled cloud enterprise architecture based on the concept of virtual enterprise at both business and technology levels. The business level comprises of organisational models, process models, skills, and competences whereas the technology level comprises of IT resources.

Chapter 16

P. Sasikala, Makhanlal Chaturvedi National University of Journalism and Communication,
India

Opportunities for improving IT efficiency and performance through centralization of resources have increased dramatically in the past few years with the maturation of technologies, such as service oriented architecture, virtualization, grid computing, and management automation. A natural outcome of this is what has become increasingly referred to as cloud computing, where a consumer of computational capabilities sets up or makes use of computing in the cloud network in a self service manner. Cloud computing is evolving, and enterprises are setting up cloud-like, centralized shared infrastructures with automated capacity adjustment that internal departmental customers utilize in a self service manner. Cloud computing promises to speed application deployment, increase innovation, and lower costs all while increasing business agility. This paper discusses the various architectural strategies for clean and green cloud computing. It suggests a variety of ways to take advantage of cloud applications and help identify key issues to figure out the best approach for research and business.

Chapter 17

Jeffrey Chang, London South Bank University, UK

Cloud computing is hailed as the next-revolution of computing services. Although there is no precise definition, cloud computing refers to a scalable network infrastructure where consumers receive IT services such as software and data storage through the Internet like a utility on a subscription basis. With an increasing number of data centres hosted by large companies such as Amazon, Google, and Microsoft cloud computing offers potential benefits including cost savings, simpler IT, and reduced energy consumption. Central government and local authorities, like commercial organisations, are considering cloud-based services. However, concerns are raised over issues such as security, access, data protection, and ownership. This paper develops a framework for analysing the likely impact of cloud computing on local government and suggests an agenda for research in this emerging area.

Chapter 18

Sanjay P. Ahuja, University of North Florida, USA
Alan C. Rolli, University of North Florida, USA

Cloud computing as a computational model has gathered tremendous traction. It is not completely clear what this term represents though it generally is thought to include a pay-as-you model for computation and storage. This paper explains what Cloud Computing is and contrasts it with Grid computing. It describes the major cloud services offered, discusses architectural details, and gives details about the infrastructure of Cloud Computing. This paper surveys the state of cloud computing and associated research and discusses the probable directions to the future of this evolving field of computing.

Chapter 19

P. Sasikala, Makhanlal Chaturvedi National University of Journalism and Communication, India

With the popularization and improvement of social and industrial IT development, information appears to explosively increase, and people put much higher expectations on the services of computing, communication and network. Today's public communication network is developing in the direction that networks are widely interconnected using communication network infrastructure as backbone and Internet protocols; at the same time, cloud computing, a computing paradigm in the ascendant, provides new service modes. Communication technology has the trend of developing towards computing technology and applications, and computing technology and applications have the trend of stepping towards service orientation architecture. Communication technology and information technology truly comes to a convergence. Telecom operators are planning to be providers of comprehensive information services in succession. To adopt cloud computing technology not only facilitates the upgrade of their communication network technology, service platform and supporting systems, but also facilitates the construction of the infrastructure and operating capacity of providing comprehensive information services. In this paper, the development processes of public communication network and computing are reviewed along with some new concepts for cloud computing.

Ghalem Belalem, University of Oran (Es Senia), Africa
Said Limam, University of Oran (Es Senia), Africa

Cloud computing refers to both the applications delivered as services over the Internet and the hardware and systems software in the datacenters that provide those services. Failures of any type are common in current datacenters, partly due to the number of nodes. Fault tolerance has become a major task for computer engineers and software developers because the occurrence of faults increases the cost of using resources and to meet the user expectations, the most fundamental user expectation is, of course, that his or her application correctly finishes independent of faults in the node. This paper proposes a fault tolerant architecture to Cloud Computing that uses an adaptive Checkpoint mechanism to assure that a task running can correctly finish in spite of faults in the nodes in which it is running. The proposed fault tolerant architecture is simultaneously transparent and scalable.

Preface

"Do not worry about your difficulties in Mathematics.
I can assure you mine are still greater." Albert Einstein

This publication summarizes the 2011 volume contents of *International Journal of Cloud Applications and Computing* (IJCAC) that addresses current trends in Cloud Computing, such as Cloud services, applications, and technologies. To illustrate the role of applications and services in the growth of Cloud computing industries, a number of examples focusing on the learning, government, and security are used.

This book is intended for researchers and practitioners who are interested in issues that arise from using technologies of cloud computing advancements. In addition, this book is also targeted to anyone who wants to learn more about the cloud computing advancements in design and implementation. Cloud computing has become a hot topic in recent years and people at varying levels in any organization need to understand cloud computing in different ways.

OVERVIEW OF THE BOOK

This book presents the current state of cloud computing advancements in design, implementation, and technology. Issues in the field and their influence in the science of cloud computing and applications are summarized as follows.

Issue 1

Junaid Arshad *et al.* present efforts to address one of the significant issues with respect to security of Clouds i.e. intrusion detection and severity analysis. An abstract model for integrated intrusion detection and severity analysis for Clouds is proposed to facilitate minimal intrusion response time while preserving the overall security of the Cloud infrastructures.

Ismael Solis Moreno *et al.* argue the importance of energy-efficient mechanisms within cloud data centers and remark on the significance of the "energy-performance" relationship in boosting the adoption of these mechanisms in real scenarios. The chapter provides an analysis of the current approaches and an outline of key opportunities that need to be addressed to improve the "energy-performance" relationship in this promising model.

Luis M. Vaquero *et al.* aim to make the reader's acquaintance with this long needed problem in distributed systems: user-oriented service-level scalability. Scalability issues are analyzed from the Infrastructure as a Service (IaaS) and the Platform as a Service (PaaS) point of view, as they deal with different functions and abstraction levels.

Zhiwei Xu, *et al.* present five research directions for optimizing Hadoop-improving performance, utilization, power efficiency, availability, and different consistency constraints. The survey covers both back-end analysis and customer-facing workloads.

Hong Cai *et al.* describe the authors' practice of developing and deploying multi-tenant technologies. This paper targets a technology that could quickly enable existing Java EE (Enterprise Edition) applications to be multi-tenancy enabled thus having the benefit of quick time to market.

Issue 2

P. Sasikala examines and discusses the concept of Cloud Computing from the perspectives of diverse technologists, cloud standards, services available today, the status of cloud particularly in higher education, and future implications.

Roland Kübert and Gregory Katsaros present the architectural approach as well as the technical solution for applying elastic web hosting onto a private cloud infrastructure using only free software. Through several available software applications and tools, anyone can build their own private cloud on top of a local infrastructure and benefit from the dynamicity and scalability provided by the cloud approach.

Louay Karadsheh *et al.* review the transition from traditional computing to cloud computing and the benefit for businesses, cloud computing architecture, cloud computing services classification, and deployment models. They also discuss the security policies and types of internal risks that a small business might encounter implementing cloud computing technologies.

Victor Chang *et al.* demonstrate portability, speed, accuracy, and reliability of applications in the clouds, while demonstrating portability for Financial Software as a Service (FSaaS) and the Cloud Computing Business Framework (CCBF).

Shadi Aljawarneh discusses this security issue. Furthermore, a vision and strategy is proposed to mitigate or avoid the security threats in the Cloud. This broad vision is based on software engineering principles to secure the Cloud applications and services.

Issue 3

Kamal Dahbur and others explain Cloud Computing and many of its related concepts, highlighting and categorizing many of the security issues introduced by the Cloud.

S.K. Maharana and others develop an efficient, faster, and secure simulation technique by leveraging control volume method (CVM) of linearization of relevant equations for the study.

Kamal Dahbur and others highlight the challenges introduced by anti-forensics, explore the various CAF mechanisms, tools, and techniques, provide a coherent classification for them, and discuss their effectiveness thoroughly. Moreover, this paper also highlights the challenges seen in implementing effective countermeasures against these techniques. Finally, a set of recommendations is presented with future research opportunities.

R.Jeyarani and others focus on green computing by introducing Power-Aware Meta Scheduler, which provides the right fit infrastructure for launching virtual machines onto a host. The experimental results show that the appropriate usage of different power saving states guarantees significant energy conservation in handling stochastic nature of workload without compromising the performance, both when the data center is in low as well as moderate utilization.

Promise Mvelase and others focus on the implementation of virtual enterprises (VE), to enable SMMEs to respond quickly to customers' demands and market opportunities. The virtual enterprise model is based on the ability to create temporary co-operations and to realize the value of a short term business opportunity that the partners cannot (or can, but only to a lesser extent) capture on their own.

Issue 4

P. Sasikala discusses the various architectural strategies for clean and green cloud computing. The chapter analyzes the various environments, infrastructures, and resources available in cloud computing.

Jeffrey Chang attempts to develop a framework for analyzing the likely impact of cloud computing on local government and suggested an agenda for research in this emerging area.

P. Sasikala reviews the development processes of public communication network and computing along with some new concepts for cloud computing. So the following works should be done pressingly: (1) turn the computing resources into network elements, form a optimal architecture of network, platform, terminal, and service; (2) formulate the compatible and interoperable technical standard of computing resources; (3) integrate the research, manufacture, network constructing, application development, and service supply to build a more open business model.

Ghalem Belalem *et al.* propose a fault tolerant architecture to Cloud Computing that uses an adaptive checkpoint mechanism to assure that a task running can correctly finish in spite of faults in the nodes in which it is running.

FUTURE TRENDS IN CLOUD COMPUTING ADVANCEMENTS IN DESIGN, IMPLEMENTATION, AND TECHNOLOGIES

Cloud computing analysts have stated "the global market for cloud computing will grow from $40.7 billion in 2011 to more than $241 billion in 2020" (Forrester Research, 2011).

It should be noted on *mobility*: "Over 80% of the Fortune 100 is deploying or piloting tablets with sales expected to increase by 123%" (Ellis, 2012). Whereas, as regards *storage*: "Nearly 40% increase expected by 2012 in backing up to/storing data in the cloud. Each day, AWS adds the equivalent server capacity to power Amazon when it was a global, $2.76B enterprise (circa 2000)" (Varia, 2009).

According to Economist Intelligence Survey (October 2011), the cloud landscape is evolving. A few points from 2010:

(i)   Peak of market hype on cloud computing -- driven by cost savings and IT efficiency gains.
(ii) Security, availability and vendor lock-in are primary concerns for CIOs.
(iii) Sandbox/ trial implementations of non-critical applications emerge: Testing enterprise-readiness.

By 2015:

(i)   enterprises will expand the focus of cloud computing as a driver of business innovation.
(ii) Cloud computing will play a significant role in shaping client value propositions.
(iii) Enterprises will look at the cloud to drive innovation across the eco-system.
(iv) The Cloud will be increasingly used to drive collaboration and reduce business complexity.

The book looks at each trend and also highlights future research topics. For example, to take advantage of industries, practitioners and governments need to further develop Cloud applications and services and continue to invest in research and development. Metrics are needed to measure the impacts of these investments. How should organizations build trust to achieve collaborative applications and services? What are the legal implications of collaborative Cloud-based commerce, learning, and government? Note that the next generation Cloud provisioning models rely on advanced monitoring and automatic scaling decision capabilities to ensure quality of service (QoS), security, and economic sustainability.

Most papers in this book express concern over the customer's fears regarding the use of cloud applications. How to calm these fears in the future is also addressed. In this book, most authors focus on some issues in Cloud computing design, technologies, and implementation. For instance, although the security and availability guarantees from reputable organizations, such as Amazon and Microsoft, customers are still worried about their data over the Cloud. Do the keys of the encrypted data encrypt or only place it somewhere on the host machines? The owner of the data will be worried because his or her data and software are not under his or her control. In addition, the data owner may not recognize where the data is geographically located at any particular time. There is still a question mark over how data will be more secure if the owner does not control its data and software. To minimize this risk, customers must be given the opportunity to assess the trustworthiness of the Cloud. Another question is what happens if someone stops the company's servers for work or if the servers are faced with major problems that prevent them from working? What follows now are a few mentions of some reasons for why the customer's fears are increased.

Many phone customers might be shocked to find that phones may rely exclusively on the company's servers to handle and keep their data. The question is: what happens if I stop the company's servers for work or if the servers face major problems that prevent them from working? The truth is that, regardless of the capacity and capabilities of the company that manages these servers, the potential collapse of the system could take place everywhere and at any moment, and then this meltdown could happen. Thus, the second question, could cloud computing fail? To overcome this big question, another chapter discusses this as part of a future work.

Reputable companies attempted to mitigate customer fears by confirming that the cloud model is secure, the cloud services are protected, the data centers and hosted servers are encrypted, the communication channel between the customer and the cloud resources are secured, and that it is protected from any kind of attack. However, attackers claim that cloud resources are penetrated much more easily than non-cloud environments. For instance, Sony stated that the customers' credit card data was secure, but the attackers claimed that the customers' credit card data was sold online. Both parties make claims about the truth, but without information, all we can do is make choices about what to believe. If Sony is telling the truth about encrypting the data, it seems that the level of encryption was not sufficient (Bradley, 2011)

One of the effective solutions for security issues in Cloud computing is the way Amazon used the Cloud services for introducing a number of web services for customers. Amazon constructed the Amazon Web Services (AWS) platform to secure access for web services. The AWS platform introduces protection against traditional security issues in the Cloud network (Amazon Web Services, 2011).

Furthermore, Amazon only offers restricted Datacenter access and information to people who have a legal business need for these privileges. If the business need for these privileges is revoked, then the access is stopped, even if employees continue to be employed by Amazon or Amazon Web Services (Amazon Web Services, 2011).

In this publication, the authors proposed a number of perspectives to calm customer's fears against Cloud concerns.

First perspective: Preventing details about how the model-driven security policies should be enforced, this book recommends that governments keep their information assurance architectures secure and confidential. For instance, the UK Cabinet office published a number of Government Cloud documents, but did not publish the Information Assurance documents. The contributors of this book do not agree with the perspective of Ulrich Lang (Co-Founder & CEO at ObjectSecurity, USA) who stated Government Cloud documents should publish the Information Assurance documents the for the following reasons:

- There is no need for creating a public cloud if the documents are confidential, and creating public Government Cloud will not make sense.
- Building public or even private Government Cloud would cost billions and make a lot of undeserving time-servers (2011).

Second perspective: To date, financial organizations are not willing to adopt public cloud because it will be risky, as explained in the above sections. It is possible to use the private cloud in financial organizations. Some authors discuss the following questions:

- Are financial organizations willing to embrace cloud computing?
- What will be their preference - Private or Public Cloud?

A summary of the key points of this book:

- Importance of the transition from traditional to Cloud in the intended sense.
- Developing strategies and solutions to the problem of research by linking traditional relationships and concepts that facilitate access to information. For example:
    ○ Architectural Design for Cloud applications and services
    ○ How to implement the Cloud applications and services.
    ○ Cloud computing for large scale applications
    ○ Cloud technologies for P2P, services, agents, grids and middleware
    ○ Cloud technologies for software and systems engineering
    ○ Cloud for E-government
    ○ Databases, IR and AI technologies for Cloud
    ○ Social networks and processes on the Cloud
    ○ Representing and reasoning about trust, privacy, and security
    ○ Cloud computing techniques and approaches
    ○ Frameworks for developing Web applications
    ○ Security issues for Web applications
    ○ Scalability issues and techniques
    ○ Applications that illustrate interesting new features or implementation techniques
    ○ Performance measurements of Cloud applications
    ○ M-commerce applications, issues, and security

Shadi Aljawarneh
Isra University, Jordan

REFERENCES

Amazon Web Services. (2011, May). *Amazon Web Services: Overview of security processes.* White paper. Retrieved from http://awsmedia.s3.amazonaws.com/pdf/AWS_Security_Whitepaper.pdf

Bradley, T. (2011, April 30). Sony says data is protected, attackers say it's for sale. *CIO Magazine.* Retrieved from http://www.cio.com.au/article/384858/sony_says_data_protected_attackers_say_it_sale/

Economist Intelligence Unit. (2011, October). *EIU survey.* Retrieved from http://www.management-thinking.eiu.com/new-directions.html

Ellis, R. (2012). *Presentation and case study by Autonomy: Extending data protection to mobile devices.* Presented at 2012 Conference on IT Governance and Planning, London.

Forrester Research. (2011, April). *Sizing the Cloud: Understanding and quantifying the future of Cloud Computing.* Retrieved from http://www.forrester.com/Sizing+The+Cloud/fulltext/-/E-RES58161?objectid=RES58161

Ulrich Lang's Security Policy Automation & Model-Driven Security Blog. (2011, May 31). *Government clouds (G-Cloud) – Security through obscurity?* Retrieved from http://objectsecurity-mds.blogspot.com/2011/05/government-clouds-g-cloud-security.html

Varia, J. (2009). *The Cloud as a platform for platforms.* Retrieved from http://aws.typepad.com/aws/2009/07/the-cloud-as-a-platform-for-platforms.html

Acknowledgment

I would like to thank all those who have contributed towards the success of this book and 2011 volume contents of IJCAC, especially the international editorial board, members of the organizing committee, the IGI Global committee, and everyone else who made a contribution.

Shadi Aljawarneh
Isra University, Jordan

Chapter 1

An Abstract Model for Integrated Intrusion Detection and Severity Analysis for Clouds

Junaid Arshad
University of Leeds, UK

Paul Townend
University of Leeds, UK

Jie Xu
University of Leeds, UK

ABSTRACT

Cloud computing is an emerging computing paradigm which introduces novel opportunities to establish large scale, flexible computing infrastructures. However, security underpins extensive adoption of Cloud computing. This paper presents efforts to address one of the significant issues with respect to security of Clouds i.e. intrusion detection and severity analysis. An abstract model for integrated intrusion detection and severity analysis for Clouds is proposed to facilitate minimal intrusion response time while preserving the overall security of the Cloud infrastructures. In order to assess the effectiveness of the proposed model, detailed architectural evaluation using Architectural Trade-off Analysis Model (ATAM) is used. A set of recommendations which can be used as a set of best practice guidelines while implementing the proposed architecture is discussed.

1. INTRODUCTION

The advent of internet technologies has significantly changed the methods used in e-Science along with the emergence of new computing paradigms to facilitate e-Science research. Cloud computing is one of such emerging paradigms which makes use of the contemporary virtual machine technology. The collaboration between internet and virtual machine technologies enable Cloud computing to emerge as a paradigm with promising prospects to facilitate the development of large scale, flexible computing infrastructures, available on-demand to meet the computational

DOI: 10.4018/978-1-4666-1879-4.ch001

requirements of e-Science applications. Cloud computing has witnessed widespread acceptance mainly due to compelling characteristics such as; Live Migration, Isolation, Customization and Portability, thereby increasing the value attached with such infrastructures. The virtual machine technology has profound role in it. Amazon, Google and GoGrid (2010) represent some of commercial Cloud computing initiatives whereas Nimbus and OpenNebula represent academic efforts to establish a Cloud.

Cloud computing has been defined in different ways by different sources however, for the purpose of research described in this paper, we define Clouds as a high performance computing infrastructure based on system virtual machines to provide on-demand resource provision according to the service level agreements established between a consumer and a resource provider.

A Cloud computing system representing the above definition has been presented in Figure 1. A system virtual machine, as described in this definition, serves as the fundamental unit for the realization of a Cloud infrastructure and emulates a complete and independent operating environment. Within the scope of this paper, we define the cloud platforms focused at satisfying computation requirements of compute intensive workloads as *Compute Clouds* whereas those facilitating large scale data storage as *Storage or Data Clouds*. For the rest of this paper, we use terms *Cloud computing* and *Clouds* interchangeably to refer to our definition of compute clouds. As described in the above definition, Cloud computing involves on-demand provision of virtualized resources based on Service Level Agreements (SLA) thereby facilitating the user to acquire resources at runtime by defining the specifications of the resource required (Burchard, Hovestadt, Kao, Keller, & Linnert, 2004). The user and the resource provider are expected to negotiate the terms and conditions of the resource usage through SLAs so as to protect the quality of service being committed at resource acquisition stage.

As with any other technology, different models of Cloud computing have been proposed to

Figure 1. A Cloud computing system

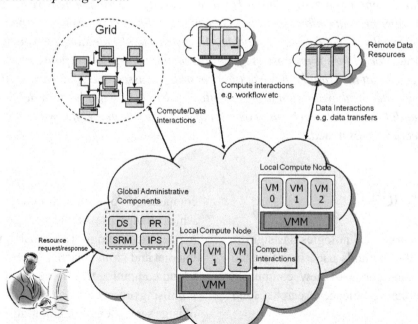

harvest its benefits. These are *Infrastructure as a Service (IaaS), Software as a Service (SaaS)* and *Platform as a Service (PaaS)*. Each of these models is focused at achieving specific objectives by introducing novel mechanisms at respective layers of the modern software architecture (Arshad, Townend, Xu, & Wei, 2010). With regards to these models, the Cloud computing system presented in Figure 1 resembles IaaS and therefore, inherits the characteristics of this model of Clouds. In the remaining sections of this paper, we use the term Cloud computing to refer to this model of Cloud computing.

However, as with any other emerging paradigm, security underpins extensive adoption of Cloud computing. Specifically, we highlight the importance of intrusion detection and severity analysis for Clouds in this paper. We also summarize our efforts to address this problem whilst taking into account unique characteristics of Clouds. Furthermore, specific requirements of Clouds for intrusion severity analysis have been summarized with a detailed description provided in Arshad, Townend, and Xu (2010b). In this paper, we focus on the requirement of minimizing overall response time for an intrusion by proposing an abstract model for integrated intrusion detection and severity analysis for Clouds. We also present the architectural evaluation for the proposed solution with the objective to evaluate the effectiveness of the proposed model. To the best of our knowledge, we believe that we are the first to conduct this research for Clouds.

The rest of paper has been organized as follows. The next section introduces the intrusion detection and severity analysis in general and introduces the challenges for intrusion detection and severity analysis for Clouds. In section 3, we present a summary of the requirements of Clouds for intrusion severity analysis followed by a description of our proposed method for automatic intrusion severity analysis for Clouds. A detailed description of this method has been presented in Arshad, Townend, and Xu (2010a). We describe

our proposed architecture in detail in section 5 whereas section 6 contains a detailed description of the evaluation performed for the proposed model.

2. INTRUSION DETECTION AND SEVERITY ANALYSIS

Intrusion detection is a well established research domain focused at improving the overall security of a system against malicious users. Historically, an intrusion detection system (IDS) strives to facilitate a system administrator by raising an alert whenever it detects an intrusion. Contemporary IDSs can be broadly classified as host or network based with respect to their location. As suggested by their names, a host based intrusion detection system is located on the host being monitored and therefore has the benefit of maximum visibility of the monitored system. However, it has the disadvantage of being prone to getting compromised in the event of a successful in intrusion taking control of the monitored system. A network based IDS on the other hand, is usually installed at the edge of a network and has the advantage of being isolated from the monitored system. However, it has the disadvantage of reduced visibility of the monitored system. Leveraging the isolation provided by virtual machines, Hypervisor based IDS have been proposed (Laureano, Maziero, & Jamhour, 2004; Litty, 2005) which combines the benefits of both host and network based IDS thereby improving the security of the monitored system.

Furthermore, clouds inherit unique characteristics such as diversity, mobility and flexibility from virtual machines, which present novel security challenges and, therefore, require dedicated efforts to address them (Garfinkel & Rosenblum, 2005). Among these characteristics, diversity provided by virtual machines (Figure 2) introduces challenges for intrusion detection and response systems.

As described in Figure 2, virtual machines provide the ability to host multiple different execution environments on a single physical ma-

Figure 2. A virtualized resource

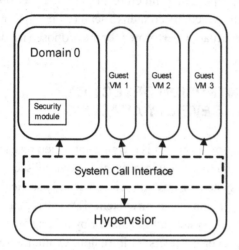

chine, which enables a cloud provider to be able to address diverse user requirements with same physical resources. However, it also poses a number of novel security challenges such as; evaluating the impact of an intrusion on the guest virtual machines (Arshad, 2009). From Figure 2, a security module residing in the domain 0 of a virtualized resource has to evaluate the impact of an intrusion on the guest virtual machines. This process becomes non-trivial given the potentially different security requirements of the guest virtual machines and the dynamic and flexible nature of Cloud infrastructures. We define the impact of an intrusion on a virtual machine as the *Level of Severity* (LoS) of the intrusion. Related to this, we define intrusion severity analysis to be *the process to evaluate the level of severity of an intrusion for a monitored virtual machine.*

Intrusion severity analysis, as defined above, has significant impact on the overall security of a system such as; selecting appropriate response mechanism and the overall intrusion response time. This will enable provision of an intelligent response selection mechanism, which facilitates triggering response mechanisms based on the severity of an intrusion for the victim application. Furthermore, a customized severity analysis also facilitates delivery of virtual machine spe-

cific quality of service with respect to security, which is vital for a user-oriented environment of clouds computing. This is envisioned to enable the cloud providers to devise Quality of Service (QoS) based pricing strategies, taking into account the quality of security service delivered in relation with the security requirements of virtual machines. Finally, the dynamic nature of Clouds demands such mechanism to minimize intrusion response time. Therefore, in order to preserve the flexible and dynamic nature of Clouds, the efficiency of such system is critical. To the best of our knowledge, we are the first to identify the intrusion severity analysis problem, highlight its importance and propose a solution to address it for Clouds (Arshad, 2009).

3. RESEARCH CONTEXT

With respect to intrusion severity analysis, Cloud computing has a number of distinct requirements. These requirements are primarily due to the unique characteristics of Clouds (Garfinkel & Rosenblum, 2005). In Arshad, Townend, and Xu (2010b), we provide a detailed description of these requirements along with their comparison with existing solutions, however, a summary of these has been presented below:

- *Comprehensive Severity Evaluation:* The ability to host multiple execution environments on a single physical machine governs diversity in the infrastructure, which demands a comprehensive severity evaluation.
- *Real-time Operation:* Clouds support flexible infrastructure where guest virtual machines can be created, migrated and deleted at runtime. This requires a security module to be considerate of the dynamic nature of the infrastructure.
- *Customization:* Due to the diversity inherited by Clouds, each guest virtual machine

can potentially have different security characteristics. Furthermore, we hold that security characteristics of an application dictate the severity of an intrusion on that application (Arshad & Townend, 2009). Therefore, it is necessary for a severity evaluation approach for clouds to enhance customization to a virtual machine level.

- *Automatic Approach:* With regards to flexibility and dynamic nature of clouds, the human intervention for intrusion severity analysis process needs to be minimized thereby eliciting the requirement for an automated approach is mandatory.
- *Minimized Response Time:* Due to the runtime behaviour of a Cloud computing infrastructure, intrusion response time becomes critical. Specifically, the intrusion response time needs to be minimized to facilitate dynamic and runtime nature of Clouds. It is, therefore, required for a severity evaluation method to be integrated with other security modules such as intrusion detection and response systems.

From the above described requirements, our focus in this paper is on *Automatic Approach* and *Minimized Response Time*. In order to fulfil these requirements, a solution to address intrusion severity analysis problem must minimize human intervention throughout the process without compromising the overall security of the infrastructure. However, most of the traditional intrusion detection and response systems are independent and isolated from each other (Stakhanova, Basu, & Wong, 2007). Typically, an intrusion detection system generates alerts for the attention of a human system administrator. The human administrator is then supposed to consult available resources and manually evaluate the severity of the intrusion of the victim application. This process consumes a substantial amount of time between intrusion detection and response and therefore, compromises the security of systems against intrusions with ex-

ponential frequency. This has been demonstrated by Denial of Service (DoS) and Distributed Denial of Service (DDoS) attacks.

Additionally, one of the objectives of Cloud computing is to provide on-demand resources without compromising their security. In order to achieve this objective, an integrated approach for intrusion detection and response systems is mandatory. This is because an integrated approach guarantees minimal response time by eliminating the human intervention. In the following sections of this paper, we present our proposed solution to address these requirements for intrusion detection and severity analysis for Clouds.

4. AN AUTOMATIC APPROACH FOR INTRUSION SEVERITY ANALYSIS FOR CLOUDS

The solution proposed to address intrusion severity analysis problem for Clouds has been described in detail in Arshad, Townend, and Xu (2010b). In this paper, we present a summary of the proposed solution. As described in Arshad, Townend, and Xu, (2010b), the proposed solution is based on the presumption that the severity of an intrusion for a particular virtual machine depends on a number of factors including; security requirements of the application hosted by the virtual machine, the state of any Service Level Agreement (SLA) negotiated beforehand and, the frequency of attack on a security requirement. There can be other parameters, however, these are regarded as the most important of the factors and therefore have been incorporated in the proposed solution. We hold that, the severity problem can be treated as a special case of traditional classification problem as it essentially involves segregating intrusion trails into different classes (Arshad, Townend, & Xu, 2010a). Finally, both supervised and unsupervised learning algorithms have been used to implement our proposed solution.

With respect to virtual machine specific security requirements, one option can be to render the security policy definition and management a responsibility of the virtual machine itself. This can be achieved by implementing a policy engine within each virtual machine, which will coordinate with detection and severity analysis modules in the privileged virtual machine. This approach is attractive due to the ease of implementation and simplicity of the resultant system. However, it breaks the isolation property as, in the event of a successful attack, an attacker can modify the security policies to facilitate its malicious objectives. Furthermore, a guest virtual machine needs to be trustworthy to be delegated such responsibility which is contradictory to our assumption that, all guest virtual machines are treated as compromised. Due to these limitations of this approach, an approach has been adopted that guarantees isolation while ensuring customization with respect to security policies. Using service level agreements to negotiate security requirements for a virtual machine has been proposed. Following this approach, a user is envisaged to specify the security requirements as part of the service level agreement at the resource acquisition phase. However, in order to accomplish this, quantification of security is required. The proposed security quantification is summarized in Table 1.

With respect to SLA state, the time remaining for completion of a job has been designated as the SLA state. This is because of the fact that severity of an intrusion is also affected by the time available for response. Ideally, the SLA state would be calculated by using different parameters such as; quality of service metrics and available resources etc. This requires establishment of complete monitoring infrastructure to monitor the status of these parameters and evaluate aggregate SLA state using some mathematical model. Due to these complexities, this has been rendered as out of the scope of this paper. However, it is assumed that SLA state is available as an aggregate metric that can be used for formal analysis such as the one described in this document. Finally, the frequency of attack attempts on a particular security requirement depicts either the value of the target or likelihood of success of attack attempt against the security requirement under attack. This therefore requires relatively immediate and more effective response mechanism to avoid recurrence of such attack attempts. For this reason, the frequency of attacks has been characterised as an important factor to dictate the severity of an intrusion.

As stated earlier, we propose to solve the severity problem by treating it as a classification problem. Related to this, a characteristic of supervised learning techniques is that they involve an initial training or leaning phase to serve as a basis for online classification. However, with the problem focused in this research, no previous knowledge of severity of intrusions for applications is maintained which makes it difficult to use supervised learning techniques. Furthermore, most of the unsupervised learning techniques are more suitable for offline analysis as the classifications tend to change over the length of analysis datasets. This characteristic makes them inappropriate for systems that require real-time classification such as the one under consideration in our research. Therefore, we decided to use both supervised and unsupervised classification techniques to achieve our objectives. An unsupervised classification technique i.e. K-means has been used to prepare the training datasets for further analysis and super-

Table 1. Proposed security requirements

Security Attributes	Requirements
Integrity	Workload State Integrity
	Guest OS Integrity
Availability	Zombie Protection
	Denial of Service Attacks
	Malicious Resource Exhaustion
	Platform Attacks
Confidentiality	Backdoor Protection

vised classification technique i.e. Decision Trees has been used for real-time severity analysis. The experimentation and evaluation of using these techniques for the proposed solution is presented in Arshad, Townend, Xu, and Jie (2010).

5. ABSTRACT MODEL FOR INTEGRATED INTRUSION DETECTION AND DIAGNOSIS FOR CLOUDS

The primary objective of this paper is to present the abstract model proposed to achieve integrated intrusion detection and severity analysis for Clouds. The abstract model is envisioned to fulfil the requirement of Clouds for minimized intrusion response time with respect to intrusion severity

analysis. In order to achieve this objective, the time between the point when an intrusion is detected and an appropriate response is activated becomes critical. In order to minimize this time interval, integration between intrusion detection and response systems is required. The integrated architecture presented in Figure 3 achieves this objective by defining different components involved in this process and outlining the interactions between these components. Additionally, the response selection mechanism is improved by incorporating rigorous intrusion severity evaluation; this ensures that the selected response is proportionate to the intrusion attempt and the security characteristics of the victim application.

Figure 3. An abstract model for integrated intrusion detection and severity analysis

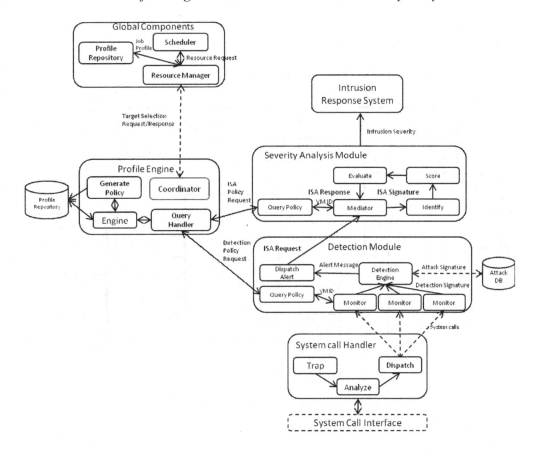

5.1. Assumptions

As described in Figure 2, domain 0 is considered to be the most privileged virtual machine within a virtualized resource. Furthermore, this status of domain 0 guarantees that it is entirely isolated from the monitored virtual machines i.e. guest virtual machines. Therefore, a guest virtual machine cannot intrude in the process executed within the domain 0. In order to harvest these benefits, the system represented by the abstract model in Figure 3 is envisioned to be incorporated with the domain 0 of a virtualized system. Additionally, the intended system can interact with the monitored virtual machines via system call interface which implies that the visibility of this system is limited to system calls executed by monitored virtual machines. The hypervisor is assumed to be trustworthy and because of maximum isolation from guest virtual machines, it is not possible for a malicious guest virtual machine to hijack the hypervisor.

It is also assumed that all the guest virtual machines are compromised and therefore, not trustworthy. This is because guest virtual machine can potentially have multiple different owners who, from a resource provider's perspective, cannot be trusted for the security of the infrastructure. Finally, it is assumed that service level agreements are negotiated and agreed upon during the resource acquisition phase.

5.2. Components of the Model

The detailed abstract model for integrated intrusion detection and severity analysis for Clouds has been presented in Figure 4. We present the description of different components of the abstract model below. However, evaluation of this model is described in a later section.

Figure 4. Quality attribute characterization for security

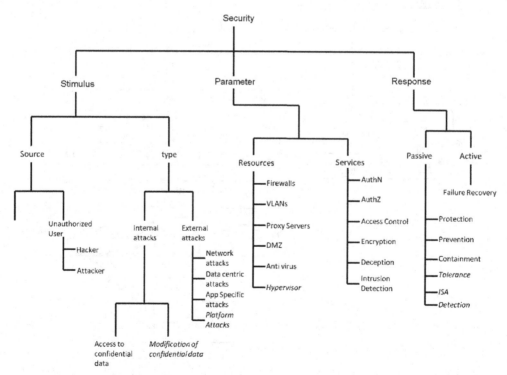

- *System Call Handler:* The system call handler is envisaged to be a module to intercept system calls executed by the guest virtual machine to facilitate further analysis on system calls. This module is significant because, as per our assumptions, both the intrusion detection and severity analysis modules are envisioned to use system call data to perform their respective functions. Therefore, this module acts as a pivotal component of the entire architecture. This module is envisaged to intercept all system calls executed by all monitored virtual machines, append virtual machine specific information such as virtual machine identifier (VMID) and, dispatch it to an intrusion detection system. The virtual machine specific information is used by different components of the model to achieve virtual machine specific behaviour.

- *Detection Module:* The detection module represents a system call based intrusion detection system. However, the type of detection engine i.e. anomaly or misuse based is implementation specific. In case of a misuse based intrusion detection system, the *Attack DB* represents a database containing signatures of known attacks whereas *Profile Engine* is envisaged to support anomaly based intrusion detection by providing profiles for normal and abnormal behaviours. Although the specific functionality of this module depends on the implementation, the general objective of this component is to detect if a particular system call event represents a malicious behaviour. The event, in this case, may contain a single system call or a sequence of system calls depending upon the type of IDS. In order to achieve flexibility, a *Monitor* is a dedicated component created at runtime for each virtual machine. A monitor is responsible for handling system

call events for a specific virtual machine. The detection module is envisioned to encapsulate information about malicious event as a *Diagnosis Request* which is sent to the severity analysis module for further processing.

- *Severity Analysis Module:* The intrusion severity analysis module represents a component responsible for evaluating the severity of a particular malicious behaviour for a victim virtual machine. This module works in collaboration with the detection module to achieve integrated operation. Upon receiving a *Diagnosis Request* from the detection module, this module is envisioned to perform rigorous severity evaluation. This involves translation of malicious event from system call to security properties of the victim and is supported by the *Profile Engine* to provide VM specific security characteristics. By doing this, a customized intrusion severity evaluation is achieved with minimum human intervention. One example implementation of this module has been provided by Arshad, Townend, and Xu (2010b).

- *Profile Engine:* Profile engine is envisaged to generate and manage virtual machine specific profiles. Among other attributes, these profiles are envisaged to contain the prioritized security characteristics for all the guest virtual machines. In order to achieve these objectives, a profile engine is envisioned to interact with global Cloud resource manager. As part of its functionality, profile engine communicates with both the intrusion detection and severity analysis modules to facilitate virtual machine specific operation.

- *Global Components:* with any Cloud infrastructure, there is a need for mechanism to manage the overall infrastructure. More recently, a number of efforts have been made to address this need. These include

Nimbus, OpenNebulla, and iVIC (Huai, Li, & Hu, 2007). *Global Components* represent such mechanisms focused at providing overall management of the Cloud infrastructure. As has been described in the Figure 3, these include components such as resource manager, scheduler and profile repositories. Furthermore, it can also include security components such as firewall to enhance overall security of the cloud infrastructure.

- *Intrusion Response System:* The objective of this research is to minimize the intrusion response time and improve the mechanism for response selection. In order to achieve this, the result of intrusion severity evaluation is fed into an intrusion response system. The intrusion response system is envisaged to use this knowledge to select appropriate response mechanism for a particular intrusion. Furthermore, the automated and integrated nature of the proposed architecture facilitates minimizing intrusion response time.

The architecture presented in this section provides an abstract model to facilitate virtual machine specific intrusion detection and diagnosis. This approach is envisaged to help fulfil the requirements of Clouds for intrusion severity analysis as described earlier in this paper. Furthermore, it is envisioned to improve the response selection mechanism used by intrusion response systems.

6. EVALUATION OF THE ABSTRACT MODEL

In order to assess the applicability and effectiveness of an architecture, evaluation is mandatory. Architectural evaluation is necessary in order to evaluate if the proposed architecture meets the expected or objected quality of service. Additionally, software architectures represent a key ingredient

in its life cycle and an architectural decision is often costly to rectify in the later stages of the SA life cycle. Therefore it is recommended that evaluation should be done at the design level so as to achieve a robust design which can facilitate achievement of overall objectives of a software life cycle.

A number of different methods have been proposed to evaluate software architectures. These methods differ from each other with respect to the stage of development of the software and the level of information available about the different components of the architecture. Mathematical models represent methods which using mathematical systems to evaluate an architecture. These methods require implementation specific data and therefore, are best suitable for software systems with known implementation data. Additionally, due their requirement for implementation specific data, mathematical models are more suitable for component based system. An alternative approach to mathematical models is scenario based approach for architectural evaluation. Scenario based evaluation methods do not require implementation specific data and rather use *scenarios*. A scenario is a brief description of some anticipated or desired use of a system. Due to this, these methods can be applied to any type of software architectures. As our proposed architectural is novel and does not have any components implemented to gather performance data, we chose scenario based evaluation methods to perform evaluation of the architecture proposed in this paper.

There are a number of methods proposed to perform scenario based evaluation for software architecture. A detailed comparative analysis of these methods has been presented in Roy and Graham (2008). However, these methods can be distinguished based on the quality attributes focused by the method and the number of quality attributes considered by the method. Quality attributes represent different parameters which can be used to assess a particular software system. These include security, flexibility, maintainability etc. With respect to scenario based methods, Software Architectural Analysis Method (SAAM) represents the pioneer method and is considered parent method for a number of subsequent architectural analysis methods such as Architectural Trade-off Analysis Method (ATAM) (Kazman, Klein, Barbacci, Longstaff, Lipson, & Carriere(1998), SAAM for Complex Scenarios (SAAMCS) (Lassing, Rijsenbrij, & Vliet, 1999) and Aspectual SAAM (ASAAM) (Tekinerdogan, 2004).

For the purpose of this research, we have chosen Architectural Trade-off Analysis Method (ATAM) to assess the effectiveness of the proposed model for integrated intrusion detection and severity analysis for Clouds. The evaluation is conducted with respect to four different quality attributes i.e. flexibility, diversity, security and performance. The choice of ATAM is motivated due to its ability to take into account multiple quality attributes for architectural evaluation whereas other scenario based evaluation methods are mostly focused at one particular quality attribute (Roy & Graham, 2008). For instance, Architectural Level Modifiability Analysis (ALMA) (Bengtsson, Lassing, Bosch, & Vliet, 2004) is envisaged to evaluate the modifiability of the software architecture. ATAM is primarily a risk identification mechanism which facilitates identification of scenarios where a desired quality attributes is affected by architectural decisions.

ATAM is envisaged to be conducted by a team of system and software engineers. Related to this, the initial steps of ATAM are focused at introducing the ATAM method to the team, outlining the objectives of the architecture, presenting the candidate architecture and identifying architectural approaches that can be vital to achieve quality attribute goals. As the research presented in this paper does not have a team, these steps have been performed individually. The objective of this research, as described earlier, is to evaluate the architecture for four quality attributes i.e. security, performance, diversity and flexibility. With

respect to architectural approaches and critical components, the key architectural components are; virtual machine hypervisor, global resource manager and system call interceptor. These have been considered critical due to their vital role to accomplish the objectives of the architecture. For instance, global resource manager is critical to establish and maintain virtual machine specific profiles which have a profound role to achieve the objectives outline by *diversity* quality attribute.

Furthermore, ATAM uses different tools such as *quality attribute characterization* and *utility tree* to accomplish its goals. And, the output generated as part of ATAM is in the form of *risks and non risks* and *sensitivity and trade-off points*. These outputs are then used to fine-tune the architecture for its overall objectives. Each of these tools and their specific outputs for the proposed abstract model has been described in the following subsections.

6.1. Quality Attribute Characterization

In order to evaluate an architecture against a given quality attribute requires understanding of different dimensions of that attribute. For instance, evaluating an architecture for security requires knowledge of how to measure security and understanding how can security be influenced by architectural decisions. This is termed as *quality attribute characterization* for ATAM and forms the basis, along with scenarios, of processes involved in ATAM. A quality attribute characterization for security has been presented in Figure 4. The characterization presented in this figure is an extended form of characterization proposed by Raza and Abbas (2008). The attributes in italic font represent our extensions to include virtual machine specific attributes. As described in this figure, attribute characterization consists of three categories; *stimulus*, *architectural parameters* and *responses*. The stimuli represent external events which require a response from the architecture. As part of security characterization, these include different types of attacks. The responses represent measures to achieve quality attribute requirements and have been represented by different methods to achieve security in Figure 4. Finally, the architectural parameters represent aspects of architecture which facilitate achieving the responses. Among other attributes, these include Hypervisor in Figure 4 which represents an architectural component to facilitate security responses such as intrusion detection and severity analysis described in this paper.

For the attribute characterizations for other quality attributes focused in this research i.e. performance, flexibility and diversity, the characterizations proposed by Kazman, Klein, Barbacci, Longstaff, Lipson, and Carriere (1998) have been used. A utility tree for the proposed architecture has been presented in Figure 5.

6.2. Utility Trees

As described earlier, ATAM uses scenarios as the basis of its activities. However, with multiple quality attributes, a number of potentially conflicting scenarios can be generated. For instance, in

order to security objectives of the proposed architecture, system call interception is envisaged to be used. However, this approach will result in a performance overhead for the monitored virtual machines. This presents a conflict between two desired quality attributes i.e. security and performance. In order to comprehend with such conflicts, ATAM introduces *utility trees*. The objective of utility trees is to identify, prioritize and refine the most important quality attribute goals. As shown in Figure 5, the high level nodes represent the most important quality goals which in this case are; performance, security, flexibility and diversity. Furthermore, the leaves of this tree represent scenarios to achieve respective quality attribute goals. The result of generating an utility tree is a prioritization and characterization of specific quality attributes as has been demonstrated by Figure 5.

6.3. Risks and Non-Risks

As part of the activities described till now, the architecture has been presented along with the quality attributes, scenarios for these attributes

Figure 5. Utility tree for the proposed architecture

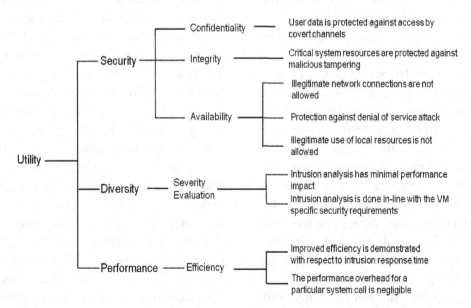

and the utility tree. The next step in ATAM is to analyze this architecture with respect to the quality attributes. A critical output of this process is the identification of *risks* and *non-risks* for the proposed architecture with respect to the quality attributes. A *risk* is an architectural decision with potentially problematic consequences whereas a *non-risk* is a good architectural decision which is based on assumptions implicit in the architecture. With respect to the proposed architecture, the risks and non-risks identified during the evaluation process are presented in Figure 6.

Each risk presented in Figure 6 is an outcome of analyzing the proposed architecture with respect to specific quality attribute objectives. For instance, consider the risk; *the decision to have a database supported intrusion detection engine is a risk if it is not capable of detecting unknown attacks.* This is a risk to the quality attribute security and multiple scenarios described as part of the utility tree in Figure 5. Specifically, this risk represents a threat to achieving the objectives of the proposed architecture with respect to security. This is because, efficient intrusion detection has a profound role to fulfil the purpose the proposed

architecture and requires an intrusion detection system capable of detecting all malicious events.

As for the non-risks, Figure 6 presents a list of non-risks for the proposed architecture. For instance, consider the non-risk; *the integrated operation of different components of the architecture is a non-risk for performance with respect to improved intrusion response time.* This presents an advantage to achieve the quality attribute performance with respect to the scenario described in the utility tree in Figure 5 i.e.; *improved efficiency is demonstrated with respect to intrusion response time.* This has been categorized as a non-risk because it is a property of the proposed architecture which has been inherited due to the assumption of integrated operation between different components of the architecture. Additionally, it helps achieve the objective of proposed architecture with respect to performance.

6.4. Sensitivity and Trade-Off Points

As described earlier, one of the reasons to select ATAM for our research was its ability to take into account multiple, potentially conflicting, quality

Figure 6. Risks and non-risks for the proposed architecture

Risks

• The decision to have system call handler as an interceptor for every system call is a risk to meet the performance objective of improved intrusion response time.

• The decision to have a universal intrusion detection engine for all the virtual machines is a risk to achieve the performance objective of improved intrusion response time.

• The decision to have a database supported intrusion detection engine is a risk if it is not capable of detecting unknown attacks.

• The evaluation component in severity analysis module is a risk to achieve objective of severity evaluation if the algorithm used does not take into account all the parameters. This is particular for cases when a particular identified attack type affects multiple security requirements.

• Initiating VM specific monitor is a risk if performance cost is excessive.

Non-risks

• The decision to have a separate light weight monitor for each virtual machine is a non-risk to achieve the performance objective of improved intrusion response time.

• The decision to have a detection engine supported with an attack database is a non-risk to achieve the objective of security given the coverage of knowledge database.

• The integrated operation of different components of the architecture is a non-risk for performance with respect to minimal human intervention.

• The integrated operation of different components of the architecture is a non-risk for performance with respect to improved intrusion response time.

• Separation of Concern based design to improve flexibility within the architecture is a non-risk.

• The severity analysis module supported by VM-specific policies is a non-risk to achieve the objective of severity evaluation.

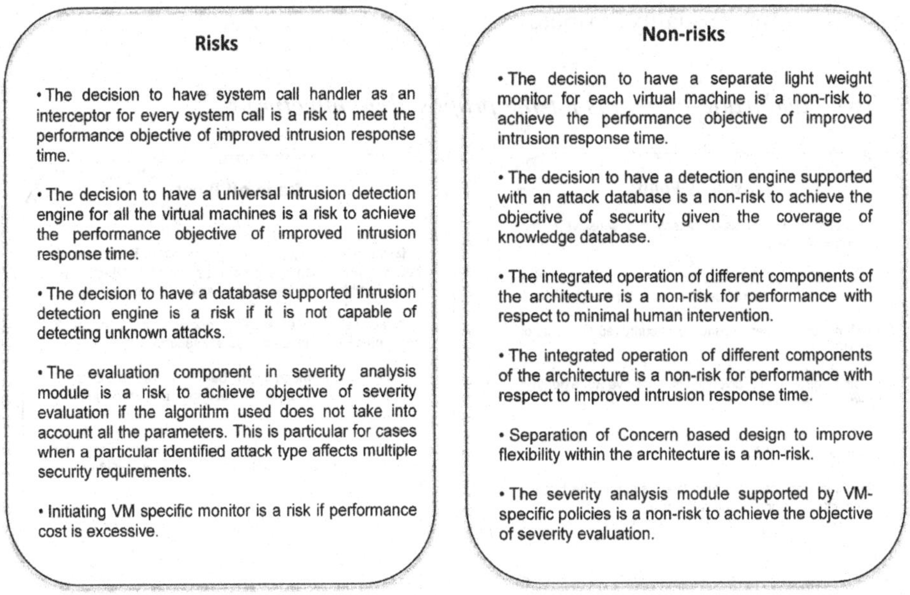

attributes for the evaluation of an architecture. Related to this, ATAM facilitates identifying *sensitivity* and *trade-off points* for the architecture. *Sensitivity points* represent properties of a component that is critical to the success of the overall system whereas *trade-off point* represents a property that affects more than one quality attribute or sensitivity point. The sensitivity and trade-off points for the proposed architecture have been presented by Figure 7.

With respect to sensitivity points, each point highlights the importance of a component of the architecture to achieve one or more quality attributes. For instance, consider the sensitivity point; *the performance overhead for a system call is sensitive to the latency due to interception by system call handler.* This highlights the importance of system call interceptor to achieve the performance quality attribute. Specifically, it emphasizes the importance of efficiency of system call interceptor to achieve the scenario for performance quality attribute i.e.; *the performance overhead for a particular system call is negligible.* Therefore, in order to minimize the performance overhead for system call execution, an efficient system call interceptor with minimum latency is critical.

As described earlier, trade-off points represent properties of components that can introduce conflicts between two or more quality attribute or quality attributes. For instance, consider the trade-off point; *latency due to system call interception is a trade-off between performance in terms of execution time and security in terms of detecting malicious system call execution.* This highlights the trade-off between two quality attributes i.e. security and performance. More specifically, it emphasizes the importance of an efficient system call interceptor for security and performance quality attributes. Therefore, there is need to prioritize scenarios with respect to these quality attributes.

7. RECOMMENDATIONS

As a result of the activities described above, a set of risks, non-risks, sensitivity points and trade-off points has been generated. In accordance with these results, a set of recommendations have been proposed to foster use of the proposed architecture to achieve objectives outlined earlier in this paper. These recommendations are focused at specific components of the architecture and can be used as best practice guidelines to implement the architecture proposed in this paper to achieve efficient integrated intrusion detection and severity analysis for Clouds.

Figure 7. Sensitivity and trade-off points for the proposed architecture

Sensitivity Points

• The efficiency of intrusion detection is sensitive to the coverage of Attack DB.

• The performance overhead for a system call is sensitive to the latency due to interception by system call handler.

• VM migration is sensitive to the security requirements of the VM.

• Comprehension of diversity is sensitive to algorithm used for evaluation

• Efficient VM migration is sensitive to the performance cost of initiating a VM-specific monitor.

• The intrusion response time is sensitive to the latency due to inter-component interactions.

Trade-off Points

• Having VM-specific light weight monitors is a trade-off between performance and rigorous intrusion detection by using complete VM-specific intrusion detection engines.

• Offline vs. Online severity analysis is a trade-off between performance and system security.

• Authentication mechanism between the participating nodes is a trade-off between security and performance.

• Latency due to system call interception is a trade-off between performance in terms of execution time and security in terms of detecting malicious system call execution.

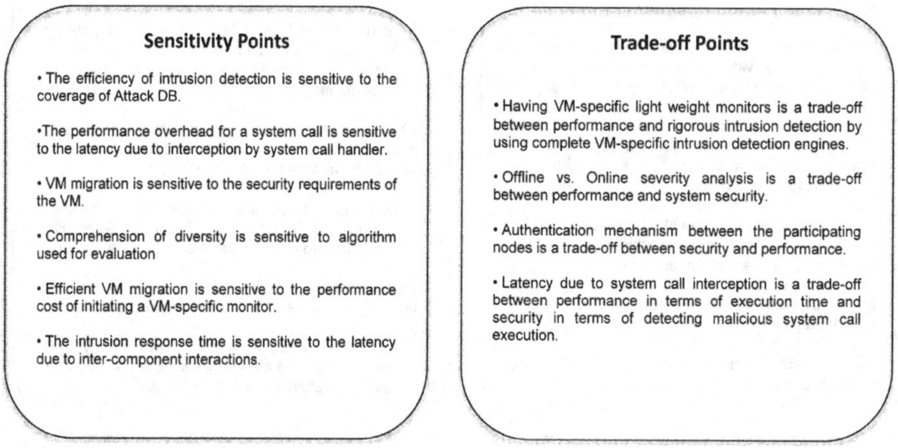

- The efficiency of the system call interceptor is critical to overall response time for an intrusion.

- In order to improve security of the Cloud infrastructure, the IDS should be capable of detecting maximum possible malicious events. Specifically, emphasis should be given to the malicious events represented by the threat model for a system.

- The trade-off between performance and security should be application specific. This highlights the need for appropriate mechanisms to optimize the balance between security and performance according to the security requirements of the application.

- The algorithm used for severity evaluation should take into account all the parameters as proposed in this paper. This is to ensure effective severity evaluation in the case when a malicious event affects multiple security requirements.

- The monitors as part of detection module should be light weight to minimize the performance overhead. Furthermore, the monitors should facilitate minimizing the overall performance overhead due to intrusion detection.

- The implementation of detection module is critical to avoiding it becoming a bottleneck for the overall detection module. Therefore, appropriate mechanisms such as *threads* should be used to foster multitasking within detection module.

- The trade-off between online and offline severity analysis should be application specific. This is because some applications such as mission critical computations are more vulnerable to performance overheads as compared to others. Therefore, quality of service requirements should be consulted whilst implementing the severity analysis module.

The set of recommendations listed above are primarily concerned with the implementation of the proposed architecture. These recommendations are focused at mitigating with the risks identified as part of the evaluation process described earlier in this paper. Furthermore, we believe, that the decisions to comprehend with the trade-off points should be application specific. This is because the impact of attributes involved in risks and trade-off points varies across different applications primarily depending upon the characteristics of the applications and their quality of service requirements. Therefore, we recommend that Cloud infrastructure providers should take into account these factors when implementing the architecture proposed in this paper.

8. CONCLUSION

Cloud computing represents an emerging computing paradigm which introduces novel opportunities to establish large scale, flexible computing infrastructures. As with any other emerging paradigm, security underpins the extensive adoption of Clouds computing. In this paper, we have presented our efforts to address one of the significant issues with respect to security of Clouds i.e. intrusion detection and severity analysis. Specifically, we propose an abstract model for integrated intrusion detection and 'severity analysis for Clouds. With respect to requirements of Clouds for intrusion severity analysis (Arshad, Townend, & Xu, 2010b), our proposed model is envisaged to fulfil requirements for minimal response time and human intervention. In order to assess the effectiveness of the proposed model, we have also presented architectural evaluation for this model using Architectural Trade-off Analysis Model (ATAM). The evaluation facilitates improved understanding of the architecture and helps identify risks for implementation of the proposed architecture. It also identifies properties of the components to achieve quality objectives along with conflicting properties which require prioritization. Finally, we also present a set of recommendations which can be used as a set of best practice guidelines whilst implementing our proposed architecture.

REFERENCES

Arshad, J. (2009). *Integrated intrusion detection and diagnosis for clouds.* Paper presented at the 39th Annual IEEE International Conference on Dependable Systems and Networks, Lisbon, Portugal.

Arshad, J., & Townend, P. (2009). Quantification of security for compute intensive workloads in clouds. In *Proceedings of the 15th International Conference on Parallel and Distributed Systems* (pp. 479-486). Washington, DC: IEEE Computer Society.

Arshad, J., Townend, P., & Xu, J. (2010a). An automatic approach to intrusion detection and diagnosis for clouds. *International Journal of Automation and Computing.*

Arshad, J., Townend, P., & Xu, J. (2010b). An intrusion diagnosis perspective to cloud computing. *International Journal of Automation and Computing.*

Arshad, J., Townend, P., Xu, J., & Jie, W. (2010). Cloud computing security: Opportunities and pitfalls. *International Journal of Cluster Computing.*

Bengtsson, P., Lassing, N., Bosch, J., & Vliet, H. V. (2004). Architecture-level modifiability analysis. *Journal of Systems and Software, 69*(1-2), 129–147. doi:10.1016/S0164-1212(03)00080-3

Burchard, L., Hovestadt, M., Kao, O., Keller, A., & Linnert, B. (2004). The virtual resource manager: An architecture for SLA-aware resource management. In *Proceedings of the IEEE International Symposium on Cluster Computing and the Grid* (pp. 126-133). Washington, DC: IEEE Computer Society.

Garfinkel, T., & Rosenblum, M. (2005). When virtual is harder than real: Security challenges in virtual machine based computing environments. In *Proceedings of the 10th Workshop on Hot Topics in Operating Systems* (p. 20).

GoGrid. (2010). *Scalable load-balanced windows and linux cloud-server hosting.* Retrieved from http://www.gogrid.com/

Huai, J., Li, Q., & Hu, C. (2007). CIVIC: A hypervisor based computing environment. In *Proceedings of the International Conference on Parallel Processing Workshops* (p. 51).

Kazman, R., Klein, M., Barbacci, M., Longstaff, T., Lipson, H., & Carriere, J. (1998). *The architecture tradeoff analysis method* (Tech. Rep. No. CMU/SEI-98-TR-008 ESC-TR-98-008). Pittsburgh, PA: Carnegie Mellon Software Engineering Institute.

Lassing, N., Rijsenbrij, D., & Vliet, H. V. (1999). On software architecture analysis of flexibility, complexity of changes: Size isn't everything. In *Proceedings of the 2nd Nordic Software Architecture Workshop* (pp. 1103-1581).

Laureano, M., Maziero, C., & Jamhour, E. (2004). Intrusion detection in virtual machine environments. In *Proceedings of the 30th IEEE EUROMICRO Conference* (pp. 520-525). Washington, DC: IEEE Computer Society.

Litty, L. (2005). *Hypervisor-based intrusion detection.* Unpublished doctoral dissertation, University of Toronto, ON, Canada.

Raza, A., & Abbas, H. (2008). *Security evaluation of software architectures using ATAM.* Paper presented at the IPID ICT4D PG Symposium, Joensuu, Finland.

Roy, B., & Graham, T. C. (2008). *Methods for evaluating software architecture: A survey* (Tech. Rep. No. 2008-545). Kingston, ON, Canada: Queen's University.

Stakhanova, N., Basu, S., & Wong, J. (2007). A taxonomy of intrusion response systems. *International Journal of Information Security*, *1*(1-2), 169–184.

Tekinerdogan, B. (2004). ASAAM: Aspectual software architecture analysis method. In *Proceedings of the Fourth Working IEEE/IFIP Conference on Software Architecture* (pp. 5-14). Washington, DC: IEEE Computer Society.

This work was previously published in the International Journal of Cloud Applications and Computing (IJCAC), Volume 1, Issue 1, edited by Shadi Aljawarneh & Hong Cai, pp. 1-16, copyright 2011 by IGI Publishing (an imprint of IGI Global).

Chapter 2
Energy-Efficiency in Cloud Computing Environments:
Towards Energy Savings without Performance Degradation

Ismael Solis Moreno
University of Leeds, UK

Jie Xu
University of Leeds, UK

ABSTRACT

Due to all the pollutants generated during its production and the steady increases in its rates, energy consumption is causing serious environmental and economic problems. In this context, the growing use and adoption of ICTs is being highlighted not only as one as the principal problem sources but also as one of the principal areas that could help in the problem's reduction. Cloud computing is an emerging model for distributed utility computing and is being considered as an attractive opportunity for saving energy through central management of computational resources. To be successful, the design of energy-efficient mechanisms must start playing a mayor role. This paper argues the importance of energy-efficient mechanisms within cloud data centers and remarks on the significance of the "energy-performance" relationship in boosting the adoption of these mechanisms in real scenarios. It provides an analysis of the current approaches and the outline of key opportunities that need to be addressed to improve the "energy-performance" relationship in this promising model.

INTRODUCTION

Many people today are devoted to a widespread adoption of Information and Communications Technologies (ICTs). However, due to the priorities of both providers and consumers, this has been focused principally on aspects such as processing speed, bandwidth, transfer rate, storage and memory capacity just to mention only a few, the environmental impact of their use has been relegated until recent years, when changing climate patterns and pollution problems have become high priority in the world's nations' agendas.

DOI: 10.4018/978-1-4666-1879-4.ch002

The increasing accumulation of greenhouse gases is changing the world's climate, creating serious problems such as droughts, floods and higher temperatures. In order to stop the accumulation of these gases in the atmosphere, it is necessary to stop the global growth of emissions, in which the generation of electricity plays a major role not only because of the carbon dioxide which results from the coal and oil used in this process, but also because it releases sulphurs and other pollutants into the atmosphere.

Additionally to the ICTs environmental repercussions, the worldwide economy is also being affected by the steady increases in electricity rates. The number of "*smart*" devices, peripherals, computers, data centers and the amount of communications are rapidly growing along with the electricity cost required to feed them. This problem is more perceptible within the industries and enterprises which have to support large amounts of computing infrastructure normally represented by enormous data centers provided with powerful cooling systems that also require great amounts of energy to work.

In this context, cloud computing an emerging model for distributed utility computing which is normally represented by large and power-consuming data centers designed to support the elasticity and scalability required by its customers. Cloud computing is becoming commercially attractive and its use is growing since it promises reducing the maintenance and management costs in comparison with traditional data centers. However, and despite that one of cloud computing commercial credentials is the reduction of energy consumption for customers, it still represents a serious problem for providers who have to deal with increasing demand and performance expectations. This creates the need for mechanisms to improve the energy-efficiency of cloud computing data centers while preserving desired levels of operation.

Green IT emerges as a new perspective for designing, developing and managing computing infrastructure aiming for more efficient processes and mechanisms to avoid waste of resources and considering the environmental implications of its use and disposal. Regarding with energy efficiency, a branch of Green IT named Energy-aware computing which is normally applied in embedded systems where strong energy constraints exist, has came forward to change the high-level computing systems point of view from "*performance-mainly*" to "*performance-energy*" balanced systems reducing the cost by an improved use of resources and the impact to the environment by diminishing the energy consumed while QoS is maintained.

Currently, some approaches have arisen to contribute to energy-efficiency improvement for data centers. Specifically, cloud computing approaches are exploiting the advantages of virtualization technology to maximize the use of underlying physical resources, dynamically resizing computing power in proportion to the customers' requirements. However these approaches do not consider some variables that in real cloud computing scenarios could lead to performance degradations or failures. These variables represent challenges that should be addressed to boost the adoption of these mechanisms in real scenarios where customer satisfaction has priority.

This paper argues about the importance of energy-efficient mechanisms within cloud data centers and remarks the "energy-performance" relationship significance. First, it describes how ICTs are negatively impacting the environment. Then green and cloud computing are introduced. Finally, the importance of energy-efficient mechanisms in cloud computing, the analysis of current approaches and the identified opportunities in this area are presented.

ICT'S ENVIRONMENTAL IMPACT

It is probably not a perceptible problem for most users, but ICTs affect the environment in different ways. According to Murugesan (2008), each of the stages of a computer's life, from its production, use and disposal produces environmental problems. Among these problems, the excessive electrical power consumption by hardware such as servers, networks, monitors and cooling systems appears to be the most critical since it results in increased greenhouse gas emissions. However, the pollution produced during the manufacturing of computing equipment and all the e-waste generated during its disposal should be taken in consideration in order to mitigate where possible the environmental impact of the ICTs helping to construct a more sustainable environment.

According to the results of the Smart 2020 report mentioned in Smarr (2010), it was estimated that the ICT Industry contributed about 2 percent of the total global greenhouse gas emissions generated in 2007 and also that these will grow at a rate of approximately 6 percent per year, even assuming successful efforts to lower the industry's carbon intensity over the next decade. This means that the total emissions will roughly triple between 2002 and 2020.

In Hickey (2008) Gartner Research Vice President Simon Mingay, mentions in accordance with the 2020 report that the global amount of carbon dioxide emissions needs to be reduced from 60 to 80 percent by 2050. But more immediately, a 25 percent reduction is necessary by 2020 in order to diminish the environmental effects. Reducing the footprint generated by the ICTs will play a major role in the achievements of these goals. As mentioned by the Climate Group in Webb (2008), ICTs could save 7.8GtCO2e or 15 percent of the global emissions in 2020.

Use of Hazardous Materials and E-Waste Generation

Electronic waste is becoming a serious fast-growing worldwide problem. Most unwanted computers and electronic equipment end up in landfills. Murugesan (2008) mentions that analysts predict that two-thirds of the estimated 870 million PCs manufactured in the world in the next five years will end up in landfills. The United Nations Environment Program has also estimated that 20 to 50 million tons of e-waste are generated worldwide each year and this number is increasing. Studies mentioned in Schneider (2010) and Chickowski (2007) show that this situation is more evident in some industrialized countries such as U.S. and some European nations where it has been demonstrated that the e-waste is rapidly growing in comparison with other kinds of municipal trash.

Beyond the amount of e-waste generated, the real problem is induced because some of the computer components contain toxic materials such as lead, chromium, cadmium and mercury as described in Murugesan (2008). If all the ICT residues in landfills are put into the ground, these toxic materials can filter dangerous chemicals into the waterways and the environment. Furthermore, if the material in the landfills is burned, toxic gases are released into the atmosphere polluting the air and contributing to the changes in climate patterns and global warming.

Chickowski (2007) mentions that the e-waste management practice of many developed countries is to send it to developing countries where very low-cost labor is used to split it apart and recover components, generally with low safety conditions for the workers and absolutely no regard for the environment in the local area. This irresponsible handling has generated many serious environmental problems for these countries and gradually for the entire world.

Energy Consumption by Computing and Cooling Systems

From the environmental perspective, the growing energy consumption becomes a serious problem not only because of the carbon dioxide that results from the coal and oil used in this process but also because it releases sulphurs and other pollutants into the atmosphere. Along with the International Energy Agency (IEA) and the World Coal Institute, coal is the first source of energy worldwide being used to generate about 41 percent of global electricity. It is set to continue, with coal feeding 44 percent of global electricity in 2030.

The energy consumption for computing could be divided according its use in two edges, the first regarding to the energy consumed by the clients conformed by PCs, peripherals and all types of mobile devices and the second refers to the energy consumed by servers, networks and cooling systems in data centers.

Regarding the *"client edge"*, energy consumption represents a serious problem because of the rising adoption of computers and mobile devices worldwide. In Smarr (2010), according to the Smart 2020 report results, it is mentioned that 57 percent of the total CO_2 emissions relating to ICTs will be produced by this sector. This is because in contrast with data centers, the "client edge" normally presents a lack of policies and rules to enforce the management and reduction of energy consumption relying only on end-user's responsiveness to activate and consciously use the energy-saving mechanisms installed in their computing infrastructure. The fact is that the most of the energy consumption in this edge occurs when PCs are idle, wasting on average 85 watts/hour even with the monitor off. This occurs mostly because of the need to be continuously available on the network or simply because the users do not take care about the importance of saving energy; they prefer computers to idle instead turning them completely off in order to avoid the time used for restoring their work environments.

On the other hand, due to the need to maintain the quality of service that customers expect and the continuous expansion of the industry, energy consumption in the *"data center edge"* is increasing along with their performance increase and the rising number of them in the world as can be seen in the information presented by Sun Microsystems (See, 2008). This increment will represent 18 percent of the total amount of CO_2 emissions generated by the ICTs for the next years. However, this number could be bigger if we consider the

issue that one of the most important directions for reducing energy consumption in the *"client edge"* is moving the workloads to controlled environments such as data centers for reducing the consumption in the latter but increasing it in the former one. Additionally, processing within this type of environments always comes along with other, but not less important factor represented by cooling systems. According to Patel, Bash, Belady, Stahl, and Sullivan (2001), these are necessary to maintain good levels of performance due to the large amounts of waste heat generated by the massive allocation of computing infrastructure. Moreover, Wang (2007) mentions that depending on the size, design and management of a data center the addition of cooling systems could double or triple the energy consumed, thus the environmental impact.

ENERGY CONSUMPTION ECONOMICAL IMPACT

Beyond the environmental issues, the growing energy consumption starts representing an economical concern in both client and data center edges because of the steady increases in electricity rates. In Nordman and Christensen (2010), it is mentioned that in developed nations such as U.S. the use of PCs is generating an electricity bill about $7 billion per year plus several billion dollars more for displays. Additionally, See (2008) presents an electricity cost forecast which takes as its baseline the $18,5 billions spent for supply data centers during 2005 and considers three different trends: the first is the servers growing rate at 14% a year along with U.S Energy Information Administration; the second is the increase per server consumption at 16% a year in accordance with Sun primary research; and the third is the increase in electricity cost at 12% per year provided by the U.S and Worldwide Server Installed Base 2006-2009 forecast, giving as a result that the energy costs for data centers could grow to $250 billion worldwide for 2012.

This rise on the energy prices combined with dynamic markets and high customer demand has led energy costs to be almost 30 percent of the total operation budget for some data centers. In accordance with Freeman (2009), this might result in the cost to power IT exceeding its acquisition cost in a matter of years limiting business' capacity to grow and change to support customer demands.

GREEN COMPUTING

With the aim of minimizing the negative environmental impact of ICTs, emerges a different perspective to perform and use computing infrastructure named "*Green Computing*" or "*Green IT*". Its principal objective is to find a balance between good QoS levels and a diminished impact to nature resulting in a sustainable eco-friendly computing environment.

In (Naditz, 2008), Simon Mingay defines Green IT as "*The optimal use of information and communication technology for managing the environmental sustainability of enterprise operations and the supply chain, as well as that of its products, services, and resources throughout their life cycles*". However, green IT can be explained since many different perspectives depending on the job position and the interests of who defines it. Regarding this, Molla (2008) presents a table which contains the definitions of green computing given by experts in different sectors of the ICTs. The authors classify these definitions in four different but interconnected perspectives which include "*sourcing perspective*" related to the environmentally preferable IT purchasing; "*operation perspective*" which includes improving energy-efficiency for computing and cooling systems; "*service perspective*", that refers to the role of IT in supporting a business' overall sustainability initiatives; and "*end of IT life*

management perspective", related to conscious e-waste disposal.

Beyond environmental benefits, the adoption of green computing practices and technologies result in economic and other rewards for individuals and enterprises. Murugesan (2008) mentions that some of these benefits could include savings in energy costs, improved systems performance, space savings in data centers, and the improvement of public image.

Energy-Aware Computing

Regarding energy-efficiency, a branch of Green IT named energy-aware computing has came forward to change the high-level computing systems point of view from "*performance-only*" to "*performance-energy*" balanced systems, reducing the cost through an improved use of resources and impact to the environment by diminishing the energy consumed while QoS is maintained. Energy-aware computing is a paradigm that intends to fill the gap between performance and energy waste by providing more management levels and application driven adaptability. As mentioned in Couch and Kumar (2008), "*The goal of energy-aware computing is not just to make algorithms run as fast as possible, but also to minimize energy requirements for computation, by treating it as a constrained resource like memory or disk*".

Although it is currently a trendy term, energy-aware computing is not new and it has been widely used in hardware design contexts. It emerged at circuit-level where the advent of portable and small-sized computer systems have created enormous energy constraints to increment the battery life duration. However, because growing adoption and use of ICTs have been indicated as one of the principal contributors in energy consumption (Webb, 2008), the term "*energy-aware computing*" is no longer exclusive to the circuit level and has risen to include computer and data center components. The designers left to be only concerned about the energy consumed by circuit blocks to extend batteries' life duration and started thinking about improving energy-demanding computer components such as CPUs, monitors, memory cards and networking equipment where the amount of energy consumed represents environmental and economical problems greater than just "*drained energy sources*".

As can be observed, since its beginning energy-aware computing has been strongly related to hardware improvements. Nevertheless, these are just tools that if not well utilized or managed could result in weak or zero energy efficiency enhancements increasing the total cost of ownership (TCO). This is because as mentioned in Carter and Rajamani (2010), generally speaking idle or underutilized resources consume considerable energy, often almost as much as they consume when they are active, resulting in a non-negligible amount of waste.

In order to take advantage of all these hardware improvements, the development of policies and software to administrate and maximize the use of those resources has begun to play a major role. Currently, some manual strategies such as turning on equipment when not in use, adjusting sleep mode and power settings in client computers, using energy monitoring software and others are being applied by managers in order to reduce their ICT infrastructure energy consumption (Murugesan, 2008; Schneider, 2010; Naditz, 2008). Additionally, some other specialized software-related approaches concentrated in three main categories: software development optimization, energy efficient network protocols, and virtualized data centers are emerging to enhance resources use while maintaining expected QoS levels.

Among all these current strategies and approaches, the implementation of virtualization in data centers is becoming one of the most important technologies for reducing the infrastructure cost including energy. In accordance with Evoy and Schulze (2008), this importance relies on the virtualization software's capability for handling and abstracting the details of sharing hardware resources with other instances maximizing their utilization, thus improving their efficiency. Additionally in IBM (2009), it is mentioned that there are substantial benefits for those companies or entities which implement virtualization, such as the reduction of the hardware cost and its operation; improvements in the use of resources; flexibility; responsiveness; security; and increase in the availability for disaster recovery. Because of this, the use of virtualized environments such as cloud computing data centers and virtual desktop platforms could be seen as good approaches for impulse mechanisms such as "*thin computing*" which aims to reduce the energy consumption in the "*Client Edge*" moving the workloads from an uncontrolled environment to a controlled one as data centers where rules and energy-aware mechanisms can be better applied.

As observed, energy-aware computing is not only related with the development of high-technology electronic devices, but is also an issue related to development and design of efficient software, networking protocols, and mechanisms to improve high-performance data center resource utilization.

CLOUD COMPUTING

Defined by The National Institute of Standards and Technology (NIST) (Amrhein, Anderson, & de Andrade, 2010), "*cloud computing is a model for enabling ubiquitous, convenient, on-demand network access to a shared pool of configurable computing resources (e.g., networks, servers, storage, applications, and services) that can be rapidly provisioned and released with minimal management effort or service provider interaction*". Additionally in Vaquero, Rodero-Merino, Caceres, and Linder (2009), cloud computing is defined as "*a type of parallel and distributed system consisting of a collection of interconnected and virtualized computers that are dynamically provisioned and presented as one or more unified computing resources based on service-level agreements established through negotiation between the service provider and consumers"*. However, beyond these technical definitions cloud computing is associated with a new business paradigm for distributed utility computing that change the infrastructure's location to the network reducing the hardware and software management expenses for customers applying a *"pay-per-use"* economic model.

On Need for Cloud Computing Energy-Aware Mechanisms

In its basic sense, cloud computing is a client-server architecture composed by large and power-consuming data centers designed to support the elasticity and scalability required by consumers. It is becoming attractive and its use is growing since it promises cost reductions for customers in comparison with permanent investments for traditional data centers. Gain (2010) references Gartner Research Firm, it is mentioned that *"cloud computing services revenue should total $56.3 billion for 2009, representing a 21.3% increase compared to 2008. The market is expected to explode to $150.1 billion in 2013"*. Additionally in accordance with Maggiani (2009), its use to support the increasing market of social networking and personal data storage is boosting its adoption among the general public. Both consumption scenarios are creating a huge infrastructure and energy demand that represents a challenging problem for coming years because of the environmental and economical implications discussed earlier.

Cloud computing is also considered a good platform to boost *"thin computing"* (Velte, Velte, & Elsenpeter, 2010) to reduce the energy consumption at *"client edge"* moving the loads from an uncontrolled environment to a controlled one embodied by data centers, where strong policies and energy-aware mechanisms can be applied. This could represent a non-negligible energy consumption increment for this edge. Because

the data center is one of the key areas for reducing the CO_2 footprint related to ICTs identified in the smart report 2020 along with the client edge and communications, the development of energy-aware mechanisms for cloud computing is closely related to the achievement of the 2020 target emissions reduction described in the same report. Moreover, energy savings in data centers is one of the most important concerns in the industry and enterprise sector becoming in sometimes a strong constraint for business expansion and economic growth.

Although one of the cloud computing commercial credentials is the reduction of energy consumption for clients as is mentioned in Velte et al. (2010), it still represents a serious problem for providers who have to deal with increasing demand and performance expectations. However, due to the business context that enclose cloud computing where customer satisfaction has priority, the design and development of energy-aware mechanisms becomes a non-trivial task. Saving energy without considering QoS implications could lead to a weak adoption or failure of this type of distributed service model. This is because in accordance with Weiss (2007) and Erdogmus (2009), the success of cloud computing demands a high degree of trust; terms such as privacy, ownership, availability and of course performance become very important to boost the user adoption and growing demand. All this creates the need for mechanisms to improve energy-efficiency at cloud computing data centers preserving at the same time the desired levels of operation.

CURRENT APPROACHES

The current energy-efficiency techniques for cloud computing data centers are close linked with distribution and scheduling algorithms and can be divided in two main categories: power and workload distribution.

Power distribution approaches look for an efficient fixed power allocation to maximize the data center performance ensuring the agreed response times. Basically, this kind of approaches varies the power delivered to the servers for incrementing or decrementing their number, maintaining the same power peak. This resource pool resizing could be dynamically leaded by load behavioral patterns or specific pre-established schedules in order to accomplish customer expected performance levels. However, according to Gandhi, Harchol-Balter, Das, & Lefurgy (2009), due to uncontrolled variables encountered in power distribution such as the outside arrival rate, weather, power to frequency relationship inherent in the technology, and the minimum power consumption of a server together with the advantages provided by virtualization for workload management, workload distribution approaches have led the trends in energy-efficiency for cloud computing data centers.

In essence, workload distribution approaches try to improve energy-efficiency by resizing the resource pool according to client requirements using workload consolidation and live migration. Based on the mechanism used to induce the energy waste reduction, these could by classified as *"dynamic processor scaling"* (DPS) and *"dynamic server's pool resizing"* (DSPR). This general classification is presented in Figure 1.

In *"dynamic processor scaling"*, energy savings are gained by adjusting the operating clock to scale down the supply voltages for the circuits. This is primarily achieved using slack reclamation with the support of dynamic voltage/frequency scaling incorporated into many recent commodity processors. However, this clearly depends on the hardware component settings, which are not available in all architectures and only represents a voltage reduction which means there still exists an energy drain. On the other hand, *"dynamic server's pool resizing"* promises the most power savings, as they ensure near-zero electricity consumption by being turned-off idle or low-utilized servers using technologies such as Wake-On-LAN

(WOL). However, Duy, Sato, and Inoguchi (2010) mentions that these approaches have had difficulties in assuring service-level agreements due to the lack of a reliable tools for predicting future demand and weak distribution policies to assist the turning off/on decision- making process.

Although there exist some other approaches related with the reduction of energy consumption for cloud computing, this paper only introduces those that include in their methodology not only energy savings but also performance preserving mechanisms and evaluations (Figure 1) (VMware, 2009; Ley, Bianchini, Martonosi, & Nguyeny, 2009; Nathuji, Isci, & Gorbatov, 2007; Li et al., 2009). While these approaches represent progress in achieving a balance between energy and performance, they still make assumptions that in real cloud scenarios could result in drawbacks. These assumptions include among others: virtual machine communication, hardware and workload heterogeneity, and implications of aggressive workload consolidation which should be addressed to achieve the adoption of energy-aware mechanisms in cloud environments where energy savings are required while maximizing the performance for the customers' satisfaction.

This represents a big challenge since the complexity of cloud computing as a single entity where many variables are involved (technological and commercial) and sometimes the adjustment or improvement of determined parameters could result in the disarrangement of others in addition with its dynamic evolution based on market demands. Flexibility and balance seem be the key especially for energy-efficiency in cloud environments. The first step is achieved by improving the use of resources through VM consolidation and live migration, minimizing the energy waste generated by idle servers. However in accordance with Cameron (2010), where it is said that *"The first generations of power-management hardware and software have saved energy often at the cost of performance..."*, for cloud computing it is necessary to start exploiting its fine-grained characteristics that in addition to current resource improvement mechanisms could achieve the balance and flexibility required for the success of energy-aware mechanisms characterized by environmental and economic benefits.

Power-Aware Scheduling of Virtual Machines in DVFS-Enabled Clusters

Von Laszewski, Wang, Younge, and He (2009) describe a scheduling mechanism which aims to reduce the power consumption in virtualized clustered environments by dynamically reducing processor speeds. The mechanism presented is composed of three algorithms that work together in order to allocate workloads in a virtualized cluster based on the required and available processor speed in the underlying physical nodes. The algorithms continuously monitor the VMs' status to adjust the processor speed on each node, reducing the power consumption. To achieve that, this approach uses profiles describing the available and maximum processor speed for each server in the cluster.

In aiming to maintain the performance levels, the Xen hypervisor performance governor is set to user space in order to enable the manual control of the frequencies according to the workload requirements. Additionally, the performance evaluation of varying the number of VMs and operating frequencies is presented. Here, nBench -a Linux CPU benchmark- is used to simulate intensive computing jobs and measure the CPU performance at the same time.

However, the performance results in this approach are never correlated with the energy reduction obtained. This makes it difficult to find out the level of the energy-performance balance achieved. Additionally, even though the authors mention homogeneous clusters, the explanation of how this characteristic affects the energy-performance improvements is weak. Moreover, they assume only one type of workload with fixed behavior. This is not necessarily true in a real

cloud scenario where different behavioral pattern applications can live together (Abdelsalam, Maly, Mukkamala, Zubair, & Kaminsky, 2009). Finally, no virtual machine live migration is used to reduce the number of servers, thus the energy savings rely only on the DVFS technology which is highly depending and varies on different hardware architectures (Duy et al., 2010).

Energy-Efficient Management of Data Centre Resources for Cloud Computing: A Vision, Architectural Elements, and Open Challenges

Buyya, Beloglazov, and Abawajy, (2010) present a conceptualization for the middleware layer between users and resources in a cloud environment with the aim of achieving energy savings while minimizing QoS degradation. To this end, the authors propose an architecture which is composed of 4 main components that include: the clients; a middleware layer named Green Service Allocator; the VMs; and the physical layer. They describe the Green Service Allocator as composed of a set of algorithms and monitors that work together to improve the workloads' distribution within cloud data centers for reducing power consumption while maintaining SLAs.

They propose the use of a bin packing algorithm named Best Fit Decreasing (BFD) with some modification allowing them to sort all VMs in decreasing order of current utilization and allocate each VM to a host based on a policy of least increase of power (LEPC) consumption. Finally, they handle the optimization of current allocation in two steps: first, VMs that need to be migrated are selected and second, the chosen VMs are placed on available hosts using the Modified BFD algorithm. All this process is supported by an energy monitor which turns on/off servers in proportion to the workload demands.

With the aim of reducing the performance degradation, a threshold value is assigned to each server and is monitored triggering workload migrations based on three polices: Minimization of Migrations (MM), Highest Potential Growth (HPG), and Random Choice. All these policies try to reduce the number of SLA violations, reducing the overhead caused by live migration and potential increase of utilization.

Although the authors mention the opportunity for saving energy in heterogeneous server environments and configure their experiment with different processor capacities, they do not explain how this heterogeneity is exploited by the proposed mechanism and how it impacts the final results. Furthermore workloads with different processor requirements are included in the experiment. However they cannot be considered fully representative of workload heterogeneity since aspects such as execution time, application architecture and networking along with their energy-performance implications are not considered.

Towards Energy-Aware Scheduling in Data Centers Using Machine Learning

Berral et al. (2010) presents a theoretical approach for handling energy-aware scheduling in data centers. Here, the authors propose a framework which provides an allocation methodology using techniques that include turning on/off machines, power-aware allocation algorithms and machine learning to deal with uncertain information while the expected QoS is maintained through the avoidance of SLA violations.

In order to save energy, the strategy proposed in this paper is simple; reduce the number of active nodes by turning off those that remain inactive using workload consolidation. To achieve this, they propose a scheduling algorithm named *"dynamic backfilling"* which allows the migration of workloads among servers in order to provide a greater consolidation and thus the reduction of active nodes. These workload movements are performed with regard to certain policies that include System Occupation (SO), Current Job

Performance (CJP) and Expected SLA Satisfaction (ESS) with the aim of improving the migration process and reducing SLA violations.

In order to reduce the performance degradation, machine learning techniques are introduced to predict the customer satisfaction level of each job before placing or moving them across the servers in the data center. Additionally working nodes thresholds are utilized to assist the turning on/off server frequency and adjust the overhead caused by these operations.

While in this approach the authors mention the inclusion of different workload types (grid and service) for the experiment, they do not describe how these different types are handled by the proposed mechanism. Apparently, they are treated as the same. However their results confirm that there exist significant differences in energy-performance among distinct types of workloads. Furthermore, this approach assumes heterogeneous data centers. Based on the idea presented in Nathuji et al. (2007), cloud environments could be composed by different server architectures with different performance and energy capabilities. Considering this could lead to improvements in benefit of the energy and performance balance.

Performance Evaluation of a Green Scheduling Algorithm for Energy Savings in Cloud Computing

Duy et al. (2010) presents an approach which aims to contribute in the saving energy problem by allocating VMs to the least number of turned on servers. The difference remarked in this work with respect to others that also try to reduce the energy consumption using *"dynamic servers' pool resizing"* is the introduction of an algorithm integrating a neural network predictor for optimizing server power consumption and reducing the performance impact in cloud computing environments. This neural network is used to anticipate the future load demand on servers by considering the historical demand with the aim of reducing the turning on/

off frequency, and the resulting overhead which could lead to serious performance degradation.

In this paper the authors describe the neural network, how it is composed, the training process, and how it works along with the "green" algorithm aiming to reduce the performance degradation. The simulations developed in which they used http workloads contained in the ClarkNet and NASA server logs are also presented. During these, performance was measured using the rate of drops, considering drop rate as the number of requests that exceed the capacity of each node to serve 1000 request/second for one single core.

In this approach, each time a user's request is submitted, a process that includes VM creation and scheduling is executed; this process is called "*step*". Here, the authors suppose that customer requests are processed within the same step being completed relatively quickly after their submission. Perhaps, this is because the type of workload utilized is represented by two types of HTTP requests (ClarkNet and NASA). In a real cloud scenario, different workload types could co-exist including those that persist for considerable periods of time (scientific, multi-tiered and multi-users applications). Considering this, in order to anticipate the number of turning on/off servers, not only predictions on the workloads' arrival rate are needed but also on the workloads' resource requirements and execution time. Additionally, a heterogeneous set of three different processors' capabilities is considered. However, it is not clearly explained how this heterogeneity is exploited by the proposed approach apart from the palpable performance implications due to the restriction of allocating only one VM per core to achieve SLAs accomplishment.

Optimal Power Management for Server Farm to Support Green Computing

Niyato, Chaisiri, and Sung (2009) present an approach which aims to contribute to the energy

saving problem for data centers. Here the authors argue that some efforts have been made as contributions to this, but all they are based on heuristic methods in which the optimal performance and power consumption cannot be guaranteed. Because of that, they introduce a mechanism which works in two different sections of distributed data centers. First, each data center works along with an optimal power management module to make decisions about server mode switching to minimize the power consumption (turning on/ off servers). Additionally, a module named job broker makes decisions on user's assignment to a specific data center with the aim of minimizing the total cost, which is composed of network and power consumption cost.

Optimal performance levels are pursued trough turning on servers in advance to reduce the workloads' waiting time. The decision on how many servers should be reactivated is obtained by formulating and solving the constrained Markov decision process (CMDP). Additionally, optimizations at the job broker look for the best workload placement within the different data centers trying to avoid job migrations among them that -as is mentioned by the authors- could lead to nonnegligible system performance degradation.

This approach considers the allocation of only one job per server. When the job is finished, the server sends a message to the scheduler indicating its status. Then the scheduler can assign a new job or deactivate it. This, in addition to the characteristic of awaking servers in advance, could be very beneficial to performance in scenarios where all the workloads had high computing demands. However, the energy savings in real cloud scenarios could be seriously affected because of the heterogeneity of workloads and the lack of mechanisms for handling heterogeneous hardware infrastructure. The allocation of jobs with low resource demands in complete servers could represent a serious resource waste problem.

GreenCloud: A New Architecture for Green Data Center

In Liu et al. (2009), *"GreenCloud"* is described as an approach which aims reduce the power consumption in data centers by reducing the number of turned on servers. In order to achieve that, the authors present an architecture composed of some components such as monitoring services, a migration manager, the managed environment, and the front end that provides information to users. Although the authors describe this architecture as their final proposal, in this paper they are mainly focused in describing the live migration algorithm which search optimal placement of virtual machines, minimizing the total cost; being the cost in this paper calculated considering physical machine cost, the virtual machine status and the virtual machine migration cost.

Maintenance of performance is pursued by a workload simulator which takes the resource requirements and collects real-time measurements from the data center in order to demonstrate the system performance to users, giving them the opportunity to adjust the parameters to obtain the desired performance. Because the test scenario is based on gaming applications, the performance measurements are presented in terms of Round Trip Time (RTT) which according to the authors is an essential concern in that specific type of applications.

This approach uses *"dynamic server's pool resizing"*. However, it is not explained how the overhead caused by the turning on/off process is handled. Moreover, in this paper the authors center their focus in only one type of workload -represented by gamming applications- where RTT is a high performance constraint. It could be interesting to analyze GreenCloud's behavior using different workload types. However, no mechanism is presented to handle workload heterogeneity. Finally, hardware heterogeneity is introduced during the calculation of migration where energy consumed by a physical machine is provided. Nevertheless, this is a static value that could be accompanied by inline monitoring to reveal the real server status in a specific time with a specific load.

Performance and Energy-Aware Cluster Level Scheduling of Compute-Intensive Jobs with Unknown Service Times

In Zikos and Karatza (2010) the authors evaluate three different job allocation policies which are based on shortest queue scheduling algorithm in a scenario represented by heterogeneous servers. The main idea is to analyze the energy and performance implications for scheduling intensive-computing jobs considering different processor profiles; some of them orientated to high performance computing and others to save energy mixed in the same cluster. The job allocation in this physical resource environment is evaluated considering first an energy-efficient approach represented by the policy named Shortest Queue with Energy-Efficient Priority (SQEE) in which processor with energy saving profiles are preferred and selected. Second, the allocation is evaluated considering a performance oriented policy named Shortest Queue with High Performance Priority (SQHP) in which processor with high computing oriented profiles are preferred and selected. And finally, another performance approach named Performance-Based Probabilistic–Shortest Queue (PBP–SQ) in which the selection probabilities are based on the computational capacity is evaluated.

Although this work does not deal directly with any other mechanisms for energy savings apart from energy-efficient processor use, since the focus is centered in the policies' evaluation, it could be complemented with both *"Dynamic Processor Scaling"* and *"Dynamic Server's Pool Resizing"*. The performance is evaluated in terms of response time and slowdowns of a job (Table 1). The relation with the specific policies and different levels of system loads are also presented.

Table 1. Energy-aware approaches with performance considerations

Approaches	Energy saving mechanism	Performance look up mechanism	Allocation -Migration policies	Energy / Performance Metrics
(Von Laszewski, et al., 2009)	Dynamic Processor Scaling	Monitoring to adjust servers processing capacity based on fixed requirements	--	Power consumed (Watts) / nBench Integer Index
(Buyya, et al., 2010)	Dynamic Server's Pool Resizing	Threshold monitoring to prevent SLA violations	LEPC, MM, HPG and RC	Energy consumed (Kwh) / SLA violation rate (%)
(Berral, et al., 2010)	Dynamic Server's Pool Resizing	Machine Learning to predict the resulting client satisfaction level	SO, CJP and ESS	Power consumed (Kw) / SLA accomplishment rate (%)
(Duy, et al., 2010)	Dynamic Server's Pool Resizing	Neural network predictor to reduce the turning on/off overhead	--	Energy consumed (Kwh) / Drop rate (%)
(Niyato, et al., 2009)	Dynamic Server's Pool Resizing	CMDP-based Algorithms to reduce the waiting time avoiding the job blocking	Network and power cost at job broker level	Power consumed (Watts) / Waiting time (seconds)
(Liu, et al., 2009)	Dynamic Server's Pool Resizing	Workload simulator to demonstrate the system performance	Migration Cost Calculation (MCC)	Energy consumed (Kwh) / Round Trip Time (ms)
(Zikos & Karatza, 2010)	--	Shortest Queue algorithm	SQEE, SQHP and PBP-SQ	Energy consumed (Energy units) / Response Time (seconds)

OPPORTUNITIES

Cloud computing is a commercial model which aims to provide computing infrastructure (software and hardware) as a service, reducing the customers' management costs. But for this to be successful, cloud providers need mechanisms not only for reducing the energy consumption to support the offered prices and demand but also for accomplishing with the required QoS to ensure customer satisfaction. Although some approaches to reduce the energy waste in cloud computing environments have been developed, these still neglect some opportunities that must be addressed to achieve a real balance between the energy consumed and the performance offered. For example:

Allocation and Migration Policies Considering Workload Types and Behaviors

Current approaches are focused on reducing the number of working nodes for powering down or turning off the inactive ones. In order to achieve this, virtualization is used to consolidate the greatest possible quantity of workloads in a single physical node. However, not one of these approaches considers the fact that different types of workloads (e.g. scientific, social network or enterprise applications) can be allocated into the same physical node. This unconscious aggregation can result in a negative influence among them, degrading the performance and incrementing the energy consumption at the same time. It might occur because each one of the workload types could have different resource requirements and life-time.

Furthermore, current approaches use live migration with the aim of improving workload distribution. Nevertheless, these migrations are performed considering neither the workload type nor the possible communication with others. For example, migrating streaming video applications could be more expensive than migrating other types of lighter applications such as a word processor (Srikantaiah, Kansal, & Zhao, 2008). Additionally, many of the applications hosted in cloud data centers are tiered or belong to complex workflows being in continuous communication among them. An unconscious migration and redistribution of this type of applications could result in a performance degradation caused by the increment of the network latency and a rise in the energy consumption by the use of networking equipment.

Finally, current approaches propose distribution mechanisms trying to consolidate the maximum number of workloads in a single server. This can be beneficial for reducing the number of active nodes; however, overloading servers could result in serious performance deficiencies, the increase of the waste heat and in consequence a rise in power consumption for cooling systems. As mentioned in Liu et al. (2009), *"... as higher utilization does equal increase power consumption and more waste heat"*.

The opportunity here is the design of resource-management policies and mechanisms considering what kind of applications can be mixed in a single host as well as the communications that exist among them to perform an optimized allocation. These mechanisms should consider the optimal usage point of each server to take advantage of them achieving the desired energy – performance balance. A study for characterizing the workload types and their behaviors is also required.

Dynamic Selection of Physical Nodes in Heterogeneous Environments

Current approaches consider homogeneity of physical resources giving to each node the same characteristics such as CPU speed, disk, memory,

etc. Thus all the servers present the same power consumption. However, in a real scenario, data center heterogeneity combined with workload distribution could provide better improvements in energy savings while preserving QoS (Liu et al., 2009; Srikantaiah et al., 2008; Ferreira, 2010). In other words, it is important to take advantage of the different hardware characteristics for distributing workloads giving preference on those servers with better energy-performance profiles, keeping those with high energy consumption turned off waiting for peak loads or critical applications. Additionally, most of these approaches consider only the processor as the unique resource in dispute, leaving out other important components such as memory and storage which due to the massive aggregation of workloads in the same node might represent bottle-necks which reduce the performance and increase the energy consumption as mentioned in Lee and Zomaya (2010).

The opportunity here is to design energy-performance profiles which integrate all the necessary information to help the scheduling algorithms for distributing the workloads on those servers that in a specific moment of time provide the best tradeoff between energy consumption and performance. Additionally, mechanisms for estimating the capacity of each server and their power profiles are necessary.

Prediction Mechanisms for a Smart Workload Distribution

Some approaches prefer not turn off/on the servers because they consider that this can cause a non-negligible overhead (Srikantaiah et al., 2008; Lee & Zomaya, 2010; Nathuji et al., 2007). Those that address turning on/off nodes have not considered the different workload types and their behavior. According to Buyya et al. (2010), turning resources off in a dynamic environment is risky from a QoS perspective. Due to the variability of the workload and massive aggregation, some VMs may not obtain required resources under peak load, thereby failing to meet the required QoS. Some other approaches such as Lefèvre and Orgerie (2010), have implemented prediction algorithms based on the average time of inter-submission time of previous jobs without considering either the resources needed by the job or its behavior relying on end user predictions for the time and resources required. This is risky since it can result in bad predictions and finally in greater power consumption or in a performance decrease.

The opportunity here is to develop mechanisms to predict the resources that a specific task will consume in order to allocate it correctly while keeping awake the proper number of servers, thereby avoiding the overhead generated during the shutting up/down process and achieving the QoS. Also Buyya et al. (2010), considers it essential to carry out a study of cloud services and their workloads in order to identify common behaviors, patterns, and explore load forecasting approaches that can potentially lead to more efficient resource provisioning and consequent energy efficiency.

Improved Resource Monitoring Considering Hardware's Performance

Resource monitoring is another important issue that should be addressed in order to improve energy efficiency in cloud computing environments. Some of the approaches presented here introduce resource monitoring to support an energy efficient workload distribution. However, all these monitors are just concerned about the energy consumed by the nodes without considering the underlying hardware's performance behavior. Monitoring mechanisms should contemplate optimal usage points for each node in cloud data centers to ensure a good balance between energy savings and performance.

The opportunity here is to design and develop energy–performance mechanisms for monitoring resources in cloud computing environments to support workload scheduling and distribution

algorithms considering additionally the relying infrastructure's performance and the workload heterogeneity that exists in cloud computing environments.

Live Migration's Overhead Reduction Mechanisms

There exist other problems directly related to live migration use. For example in some approaches, live migration seems not to be significant (Lee & Zomaya, 2010; Lefevre & Orgerie, 2010). This is because the migration process consumes a large amount of energy since it requires substantial attention from the hypervisor. According to Lefèvre and Orgerie (2010), hypervisor should copy the memory pages and send them to the new host. Additionally, if several migrations are required at the same time on the same node they are queued and processed one by one consuming a lot of energy. Finally, most approaches assume that the data centers are confined to a particular physical location. However, they recognize that a data center can span across multiple geographic locations (in a MAN or WAN for example). In this kind of scenario, migration becomes more complex due to the IP addressing problems impacting the virtual machines' performance adding a non-negligible overhead.

The opportunity here is to design and develop mechanisms to reduce the number of live migrations. Next generation of energy-aware mechanisms for cloud computing should avoid the continuous workload movements just to liberate resources. These migrations should be performed based on smart policies considering not only energy savings but also performance issues to accomplish the expected QoS. Additionally, mechanisms to reduce the overhead at hypervisor level and across different domains are also required.

CONCLUSION

Cloud computing is becoming one of the most important trends in service oriented and distributed systems. One of its strongest credentials is the "green" alternative offered to the customers. However, by itself cloud computing should not be considered as a green approach since in its core concept it is just a movement of loads and infrastructure from one place to another, including the energy consumed. What cloud computing represents is an attractive commercial opportunity to reduce the energy consumption at the "client edge" increasing it at data centers, where energy efficient mechanisms could be better applied. Because of this, some approaches have arisen trying to minimize energy consumption at cloud computing data centers. However, cloud providers are in need of mechanisms not only for reducing energy consumption to support the offered prices and demand but also for accomplishing with the required QoS to ensure the customer satisfaction.

In this paper the importance of energy savings without degrading the performance in cloud computing was discussed, since more than a technological advance it represents a business model where the satisfaction of customers has high priority. The state of art in energy-aware computing for cloud environments shows that the initials efforts for saving energy have started primarily focused in the reduction of energy waste generated by idle servers mainly supported by VM consolidation and live migration. These, in conjunction with scheduling algorithms have boosted up two main trends: "dynamic server's pool resizing" and "dynamic processor scaling".

Most of the approaches embraced in these trends have been designed aiming to achieve the highest possible energy savings with weak performance considerations. However, some others have been proposed with a strong concern for performance preservation introducing policies and evaluations in their methodologies. These approaches were described and analyzed in this paper.

From this analysis it is possible to conclude that there still exist some gaps that must be covered to achieve the energy-performance balance that is necessary in cloud computing environments. These represent a challenge related to the exploitation of cloud computing fine-grained characteristics that in addition with current approaches could lead to the success of energy-aware mechanisms, characterized by environmental and economical benefits. The introduction of some additional variables such as workload and hardware heterogeneity, workload networking and server's optimal utilization point that exist in real cloud scenarios as well as mechanisms to handle them still represent opportunities that should be addressed looking for the adoption of energy-aware mechanisms in cloud environments where energy savings are required maximizing the performance for the customer's satisfaction.

REFERENCES

Abdelsalam, H. S., Maly, K., Mukkamala, R., Zubair, M., & Kaminsky, D. (2009). Analysis of energy efficiency in clouds. In *Proceedings of the Computation World: Future Computing, Service Computation, Cognitive, Adaptive, Content, Patterns* (pp. 416-421).

Amrhein, D., Anderson, P., & de Andrade, A. (2010). *Cloud computing use cases white paper.* Retrieved from http://opencloudmanifesto.org/Cloud_Computing_Use_Cases_Whitepaper-4_0.pdf

Berral, J. L., Goiri, Í., Nou, R., Julià, F., Guitart, J., Gavaldà, R., et al. (2010, April 13). Towards energy-aware scheduling in data centers using machine learning. In *Proceedings of the 1st International Conference on Energy-Efficient Computing and Networking*, Passau, Germany (pp. 215-224).

Buyya, R., Beloglazov, A., & Abawajy, J. (2010, July 12-15). *Energy-efficient management of data center resources for cloud computing: A vision, architectural elements, and open challenges.* Paper presented at the International Conference on Parallel and Distributed Processing Techniques and Applications, Las Vegas, NV.

Cameron, K. W. (2010). The challenges of energy-proportional computing. *IEEE Computer, 43*, 82–83.

Carter, J., & Rajamani, K. (2010). Designing energy-efficient servers and data centers. *IEEE Computer, 43*, 76–78.

Chickowski, E. (2007). Safely eliminating e-waste. *IEEE Processor, 29*, 12.

Couch, A. L., & Kumar, K. (2008, December). *Workshop on power aware computing and systems.* Retrieved from http://www.usenix.org/publications/login/2009-04/openpdfs/hotpower08.pdf

Duy, T. V. T., Sato, Y., & Inoguchi, Y. (2010). *Performance evaluation of a Green Scheduling Algorithm for energy savings in cloud computing.* Paper presented at the IEEE International Symposium on Parallel and Distributed Processing.

Erdogmus, H. (2009). Cloud computing: Does nirvana hide behind the nebula? *IEEE Software, 26*(2), 4–6. doi:10.1109/MS.2009.31

Evoy, G. V. M., & Schulze, B. (2008). Using clouds to address grid limitations. In *Proceedings of the 6th International Workshop on Middleware for Grid Computing* (p. 11).

Freeman, L. (2009). *Reducing data center power consumption through efficient storage.* Retrieved from http://www.gtsi.com/eblast/corporate/cn/06_2010/PDFs/NetApp%20Reducing%20Datacenter%20Power%20Consumption.pdf

Gain, B. (2010). Cloud computing & SaaS in 2010. *IEEE Processor, 32*, 12.

Gandhi, A., Harchol-Balter, M., Das, R., & Lefurgy, C. (2009). Optimal power allocation in server farms. In *Proceedings of the Eleventh International Joint Conference on Measurement and Modeling of Computer Systems*, Seattle, WA (pp. 157-168).

Hickey, A. R. (2008). *Gartner: Green IT needs to be on midsize CIOs' radar screens.* Retrieved from http://www.crn.com/hardware/208700292

IBM. (2009). *Power systems: Introduction to virtualization* (Tech. Rep. No. 5733-SSI). Armonk, NY: IBM Corporation.

Lee, Y. C., & Zomaya, A. Y. (2010). Energy efficient utilization of resources in cloud computing systems. *The Journal of Supercomputing, 53*, 1–13. doi:10.1007/s11227-010-0435-x

Lefèvre, L., & Orgerie, A.-C. (2010). Designing and evaluating an energy efficient Cloud. *The Journal of Supercomputing, 51*(3), 352–373. doi:10.1007/s11227-010-0414-2

Ley, K., Bianchini, R., Martonosi, M., & Nguyeny, T. D. (2009). *Cost- and energy-aware load distribution across data centers.* Paper presented at the 22nd ACM Symposium on Operating Systems Principles.

Li, B., Li, J., Huai, J., Wo, T., Li, Q., & Zhong, L. (2009). EnaCloud: An energy-saving application live placement approach for cloud computing environments. In *Proceedings of the IEEE International Conference on Cloud Computing* (pp. 17-24)

Liu, L., Wang, H., Liu, X., Jin, X., He, W. B., Wang, Q. B., et al. (2009). GreenCloud: A new architecture for green data center. In *Proceedings of the Sixth International Conference on Autonomic Computing and Communications Industry Session*, Barcelona, Spain (pp. 29-38).

Maggiani, R. (2009). Cloud computing is changing how we communicate. In *Proceedings of the IEEE International Professional Communication Conference* (pp. 1-4).

Mello Ferreira, A. (2010). *An energy-aware approach for service performance evaluation.* Paper presented at the International Conference on Energy-Efficient Computing and Networking.

Molla, A. (2008, December 3-5). *GITAM: A model for the adoption of green IT.* Paper presented at the 19th Australian Conference on Information Systems, Christchurch, New Zealand.

Murugesan, S. (2008). Harnessing green IT: Principles and practices. *IT Professional, 10,* 24–33. doi:10.1109/MITP.2008.10

Naditz, A. (2008). Green IT 101: Technology helps businesses and colleges become enviro-friendly. *Sustainability: The Journal of Record, 1,* 315–318. doi:10.1089/SUS.2008.9931

Nathuji, R., Isci, C., & Gorbatov, E. (2007). Exploiting platform heterogeneity for power efficient data centers. In *Proceedings of the Fourth IEEE International Conference on Autonomic Computing* (p. 5).

Niyato, D., Chaisiri, S., & Sung, L. B. (2009). Optimal power management for server farm to support green computing. In *Proceedings of the 9th IEEE/ACM International Symposium on Cluster Computing and the Grid* (pp. 84-91).

Nordman, B., & Christensen, K. (2010). Proxying: Next step in reducing IT energy use. *IEEE Computer, 43*(1), 91–93.

Patel, C. D., Bash, C. E., Belady, C., Stahl, L., & Sullivan, D. (2001, July 8-13). *Computational fluid dynamics modeling of high compute density data centers to assure system inlet air specifications.* Paper presented at the Pacific Rim/ASME International Electronic Packaging Technical Conference and Exhibition, Kauai, HI.

Schneider, E. (2010). *Go green, save green: The benefits of eco-friendly computing.* Retrieved from http://www.apcmedia.com/salestools/SLAT-7DCQ5J_R0_EN.pdf

See, S. (2008). *Is there a pathway to a green grid??* Retrieved from http://www.ibergrid.eu/2008/presentations/Dia%2013/4.pdf

Smarr, L. (2010). Project GreenLight: Optimizing cyber-infraestructure for a carbon-constrained world. *IEEE Computer, 43,* 22–27.

Srikantaiah, S., Kansal, A., & Zhao, F. (2008). Energy aware consolidation for cloud computing. In *Proceedings of the HotPower Conference on Power Aware Computing and Systems* (p. 10).

The Climate Group. (2008). *SMART 2020: Enabling the low carbon economy in the information age.* Retrieved from http://www.smart2020.org/_assets/files/02_Smart2020Report.pdf

Vaquero, L. M., Rodero-Merino, L., Caceres, J., & Linder, M. (2009). A break in the clouds: Towards a cloud definition. *Computer Communication Review, 39*(1), 50–55. doi:10.1145/1496091.1496100

Velte, A. T., Velte, T. J., & Elsenpeter, R. (2010). Cloud computing basics. In *Cloud computing: A practical approach* (pp. 3–22). New York, NY: McGraw-Hill.

VMware. (2009). *VMware distributed power management concepts and use* (Tech. Rep. No. IN-073-PRD-01-01). Palo Alto, CA: VMware Inc.

Von Laszewski, G., Wang, L., Younge, A. J., & He, X. (2009). Power-aware scheduling of virtual machines in DVFS-enabled clusters. In *Proceedings of the IEEE International Conference on Cluster Computing,* New Orleans, LA (pp. 1-10).

Wang, D. (2007). Meeting green computing challenges. In *Proceedings of the International Symposium on High-Density Packaging and Microsystem Integration* (pp. 1-4).

Weiss, A. (2007). Computing in the clouds. *netWorker*, *11*(4), 16–25. doi:10.1145/1327512.1327513

Zikos, S., & Karatza, H. D. (2010). Performance and energy aware cluster-level scheduling of compute-intensive jobs with unknown service times. *Simulation Modelling Practice and Theory*, *19*(1), 239–250. doi:10.1016/j.simpat.2010.06.009

Chapter 3
The Challenge of Service Level Scalability for the Cloud

Luis M. Vaquero
Telefónica Investigación y Desarrollo, Spain

Juan Cáceres
Telefónica, Spain

Daniel Morán
Universidad Nacional de Educación a Distancia, Spain

ABSTRACT

This paper presents a brief overview of the available literature on distributed systems scalability that serves as a justification for presenting some of the most prominent challenges that current Cloud systems need to face in order to deliver their pledged easy-to-use scalability. Through illustrative comparisons and examples, this paper aims to make the reader's acquaintance with this long needed problem in distributed systems: user-oriented service-level scalability. Scalability issues are analyzed from the Infrastructure as a Service (IaaS) and the Platform as a Service (PaaS) point of view, as they deal with different functions and abstraction levels. Next generation Cloud provisioning models rely on advanced monitoring and automatic scaling decision capabilities to ensure quality of service (QoS), security and economic sustainability.

INTRODUCTION

Cloud computing is one of the most prominent terms employed in the latest years within the Information and Communication Technologies (ICT) field. This computing paradigm is commonly associated to new provisioning mechanisms of infrastructure resources. However, there is more to Cloud than mere infrastructure arrangement (Vaquero et al., 2009). The Cloud aims at offering every networked resource as a service to be consumed by a variety of actors far beyond any type of administrative domain or organization.

These general ideas may be familiar to those working in any type of distributed systems technology. The Cloud was born as the harvest of many different previous seeds; it is nourished

DOI: 10.4018/978-1-4666-1879-4.ch003

by the maturity and joint of technologies such as virtualization or the Grid.

According to the functionality and abstraction level, Cloud technologies can be classified into three categories (Vaquero et al., 2009): 1) Infrastructure as a Service (IaaS), where Cloud vendors provide interfaces for the provision and management of virtualized resources (virtual machines, storage or network); 2) Platform as a Service (PaaS) which offers development tools, application containers and specialized platform technologies, and hides the details of IaaS infrastructure management; and 3) Software as a Service (SaaS), where applications are offered on demand. Currently, there are many services in the market that offer these features, for example: IaaS Clouds such as Amazon Elastic Compute Cloud (EC2) and Simple Storage Service (S3), GoGrid or Flexiscale; incipient PaaS Clouds Google App Engine, Windows Azure or SalesForce; and SaaS such as Web mail, remote desktops, Application Stores or Web 2.0 social networks.

The Cloud is already delivering important advantages that place it in an optimal position to overcome hype and evolve towards maturity.

The illusion of a virtually infinite computing infrastructure, the employment of advanced billing mechanisms allowing for a pay-per-use mode on shared multitenant resources, the easiness to create and destroy new machines, the simplified programming mechanisms, etc. are already part of our common mindset and seem to be here to stay as a new ICT utility such as the electricity or water distribution systems. Figure 1 shows how Cloud types are layered by abstraction levels and management functions. In spite of the claimed advantages, the Cloud still needs some work with regards to a variety of different subjects.

Note that an upper layer may rely on intermediate layers or may skip some of them. For example, SaaS clouds can be directly constructed on physical hardware as it has been done traditionally, of course, but they could be also deployed using IaaS functions, or PaaS over IaaS. In the purest SaaS/PaaS/IaaS Cloud model: 1) software is offered as a service and it is composed by service components which are 2) hosted in platform containers (stack of OS, middleware, application containers and libraries); platform containers run in 3) virtual machines (VMs) which are hosted in

Figure 1. The cloud stack

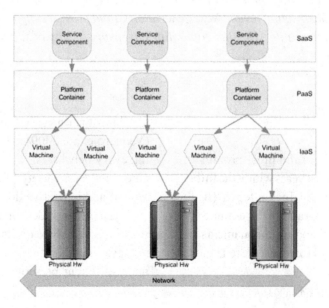

Figure 2. Typical lifecycle of a service in the cloud

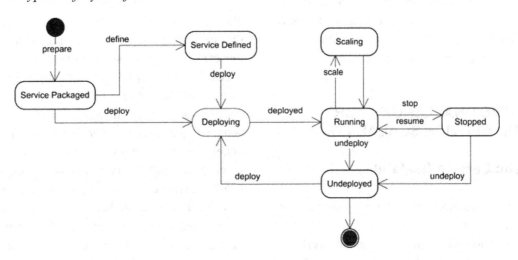

4) physical machines and are properly interconnected in the Cloud. In the IaaS model, service components have to be installed by the service provider in virtual images for a VM. The typical lifecycle of a service in the Cloud is shown in Figure 2.

It can be observed how one needs to develop and package appropriately a service for the Cloud, i.e., the transition from our in-house premises to the Cloud is no yet straightforward. After the service is packaged, the next step takes us to the definition of what we want our service to be. In IaaS Clouds, this second step can be skipped as they offer a virtual machine level API that constrains the specification of an application (understood as a set of services) to a set of detached "commands" for every VM. More advanced mechanisms have recently been proposed that try to increase the abstraction level of this definition phase (Galán et al., 2009; Rodero-Merino et al., 2010). Initiatives such as this one are the essential glue that sticks application components together and helps to specify their quality of service (QoS) and runtime behavior. The next step consists on deploying the service on top of the Cloud provider's infrastructure. Having the service deployed leads us to the running phase, the one in which the application will be laying most of the time

and that in which and appropriate control on the services' behavior is most desirable. When the service is running, two prominent processes are identified: 1) monitoring the services status in order to 2) react, by scaling up/down (Cáceres et al., 2010), and keep the promised quality of service and economic performance. In the sections below, we will see how these two elements are still linked to infrastructure or to platform elements and fall far below users' concerns making it very difficult to map business processes to application behavior regardless of the actual constituents of the service-oriented application (SOA). Some attempts tried to map Business Process Modeling (BPM) results to actual SOA components (Zdun et al., 2009; Milanović et al., 2009; Papazoglou & Heuvel, 2007) which usually lead to constraints and complex crossover relationships, but they do not fulfill business controllers or service architects. At a later stage, the service can be stopped (to be again resumed) or undeployed and destroyed.

The remaining of this brief survey is organized as follows. Next section presents an analysis of state of the art and "de facto" standard technologies providing advanced scaling mechanisms in IaaS Clouds to help enforce the reactivity required at runtime. We also offer a review on service-level monitoring tools that could be employed for feeding the scaling mechanisms with the appropriate set of data. Finally, we review current PaaS offers and analyze some examples on how aforementioned scaling mechanisms could be applied.

RUNTIME CHALLENGES FOR CLOUD SERVICE SCALABILITY

Holistic Service Scalability

Most of the available IaaS Cloud systems do not meet the requirement of an abstract interface that would help to reduce the management overhead associated to low level fine grained interfaces (RightScale, n.d.; Amazon Web Services, n.d.; Cáceres et al., 2010). These difficulties are not new to the Cloud itself, some previous technologies also found the same pitfalls.

Computational clusters aim at scaling in CPU time and/or availability by aggregating more off-the-shelf inexpensive machines within the same set of machines and missing the possibility of scaling at real time whenever an application demands resources that exceed the resources currently available in the Cluster. This scaling power is clearly not enough and, thus, new technologies and paradigms were developed such as the Grid. The Grid extended beyond the administrative borders by creating the concept of virtual organizations (VOs): a set of administrative domains that create the infrastructure and set up the procedures to cooperate in order to share resources often with scientific purposes (Foster, 2002). Although VOs certainly increased the scaling potential of some applications, they did not offer the chance of avoiding administrative procedures even when it is not correct (the Grid also showed its benefit for SOA), the Grid is often associated with computationally-intensive scientific applications. This way, applications could be parallel-programmed towards further scalability and the underlying computational resources optimized.

The available literature with regards to parallelization-derived scalability is extensive: see (Nascimento et al., 2008; Alexandre et al., 2009; Pan et al., 2009) for instances of very recent references. Additionally, the Grid infrastructure started to be itself scaled (Jin et al., 2008). However, parallelization is not always applicable and only some computationally intensive applications can benefit from this approach. The advent of Grid services (actually, service-oriented computing) resulted in newer forms of scalability: it was then possible to scale an application by deploying new service components on top of a given infrastructure belonging to a single owner or a federated set of infrastructures sharing idle resources (within VOs). However, applications on top of a Grid of services still had to be orchestrated on their own and specific means towards automated, user-level scalability (Lin et al., 2007; Lee et al., 2004; Alexandre et al., 2009; Dumitrescu et al., 2005; Saha et al., 2003). Different heuristics were employed to achieve the desired scalability e.g. best service selection (Lee et al., 2004) or the addition of new servers to face the increasing load (Deng & Wang, 2007).

There is no account so far of generic systems trying to solve application level scalability releasing users from the burden of creating ad hoc solutions for their particular application. The Cloud is the first computational paradigm concerned with this problem. Some very relevant illustrative examples are summarized in Table 1. The Cloud offers scaling middleware for their users to control the way their VMs or their set of VMs are scaled as the supported load grows or shrinks. Most of them enable automation procedures that free administrators from the boredom of continuously repeating the same actions to control the scale of the service.

As can be observed in Table 1, these (and other) Cloud systems are still too low level, too technologically focused and, thus, fall far from users' needs with regard to their interfaces and

Table 1. Examples of application-level scaling functions in the cloud

	Scaling Middleware	Automation	Abstraction
RightScale	Yes	Yes	VM/VM Array
GoGrid	Yes	No	VM
Amazon CloudWatch	Yes	Yes	VM

user experience (users still have to deal with virtual hardware and set VMs one by one).

More recent initiatives try to solve this problem by introducing a layer on top of current IaaS Clouds that provides users with the required abstraction level and allows them to control the service as a whole single entity, detaching business/performance logic from the underlying technological implementation (Rodero-Merino et al., 2010). These authors simplify the definition of services as whole entities so that business managers and architects do not have to dive into low level details, but can specify service runtime behavior by complying with rules that make sense to them. Rodero-Merino et al. (2010) provide a wider range of scalability mechanisms and a varied dictionary of actions that can be undertaken in order to keep the desired quality of service on top of several Cloud infrastructure providers. The code has been released and is available under the name of Claudia. Similar to the work of these authors, Lim et al. (2009, 2010) offer a rule based system to control the scalability of a collection of VMs. Their work provides Cloud scalability to a set of VMs by relying on conventional control theory techniques based on a series of infrastructure provided metrics. However, these control techniques are far from application providers' mindset, more so since there is a need to map from low level metrics to higher level ones. An important point raised by Lim et al. is that f proportional thresholding. This mechanism considers the fact that going from 1 to 2 machines can increase capacity by 100% but

going from 100 to 101 machines increases capacity by not more than 1%. The relative effect of adding a fixed-sized resource is not constant, so using static threshold values may not be appropriate. Higher level (closer to the application provider) guidance mechanisms are being considered to help solving the scenario above.

Also very recently, Marshall et al. have proposed a resource manager built on top of the Nimbus toolkit that also includes federation capabilities on top of local/private and external/public Clouds (Marshall et al., 2010). This framework dynamically adapted to a variety of job submission patterns increasing the processing power up to 10 times by federating on top of Amazon's (deploying new VMs adhered to the cluster on separate infrastructures). Unfortunately, this work is too "number crunching" oriented and does not offer a general framework for different application domains. The managing abstraction here is put beyond VMs and some focus is placed to the service itself, although to a lesser extent than that achieved by other approaches (Rodero-Merino et al., 2010). We feel that the work by Marshall et al. could easily be extended in order to make it more abstract.

In the same manner, another open source for federating data centers is Open Cirrus (Avetisyan et al., 2010). Open Cirrus allows access to hardware features and run daemons and aims to be offered to a wide variety of applications and open source stacks for the Cloud (like Claudia).

Some of the implemented scalability mechanisms aim to provide more expressive ways for defining service needs (Rodero-Merino et al., 2010), so the final scalability mechanisms can be tailored to better fit these needs. As Rodero-Merino et al. show, rule engines may prove to be useful, as they provide a way of expressing these needs in a human-friendly and computer-manageable way. The following source code snippet shows a set of rules for scaling service components based on business level measures (logic programming with constraints notation used for simplicity)

Box 1.

```
Full(X) :-
      MonitoringValue(X, Y), Limit(X, Z), Y>Z
CreateVM(BigDiskVM) :-
      Full(QueueLength), VMCount(BigDiskVM, X),
      Limit(BigDiskVM, Y), X<Y
CreateVM(HighBandwithVM) :-
      Full(ResponseTime), VMCount(HighBandwithVM, X),
      Limit(HighBandwithVM, Y), X<Y
Limit(QueueLength, 30)
Limit(ResponseTime, 210)
MonitoringValue(QueueLength, 64)
MonitoringValue(ResponseTime, 90)
```

along with a set of facts that would trigger some of them (see Box 1.):

In real time, MonitoringValue facts would be asserted by the VM monitoring probes, so when a measure raises the appropriate level, a new VM would be created reflecting the specific needs. As the snippet shows, a rule language working with an adequate ontology can provide a way to customize the service scaling mechanisms; in this case, the goal is achieved using different VM templates to do so, but different actions may be created in order to make particular resources of existing VMs grow (e.g. increase bandwidth or attach new disks). Representing QoS constraints with logic rules provide the additional advantage of making this knowledge easier to use in automated learning mechanisms, that, in the future, may provide automated ways of optimize the service performance.

Advanced Service-Level Monitoring

One of the limitations found when trying to offer whole service-level scalability is the scarcity of data sources feeding "scalability engines" with data at the appropriate abstraction level. Most distributed systems still rely on hardware metrics that are way too far from the needs of users to control their application as a whole from a business perspective. The typical service lifecycle shown above, reveals that users are often forced to constantly monitor their applications and they

still have to rely on low-level metrics (CPU, Disk, Memory, Network, etc.) or performing ad hoc mappings to "build" higher level ones (closer to their business end or mindset).

The literature regarding monitoring is very plentiful (Aiftimiei et al., 2007; Catalin et al., 2007; Nobrega et al., 2007; Imamamgic & Dovrenic, 2007). Log-based approaches such as Netlogger collect data and correlate them to obtain an overall system behavior but these systems present a large response time and huge overhead that makes impossible their employment for dynamic (near real-time) adaptation. Some other systems are mainly based on aggregating data form other monitoring systems (Sameer et al., 2006; Aiftimiei et al., 2007; Catalin et al., 2007; Imamamgic & Dovrenic, 2007) such as Ganglia, Nagios, etc.

Grids also brought up mechanisms for controlling the underlying infrastructure, relying on pure hardware metrics. These could be enough since memory and CPU offer a good correlation with the status of computationally intensive applications, but cannot be extrapolated to other Grid-service based (SOA) applications that scaled by expanding across the VO.

All these systems offer a system-level view of the resources but fail when it comes to offer a user view of the system itself they do not bring application/business levels metrics. In the same manner, mapping from this infrastructure metrics to high level ones that may make sense for users is very difficult and can hardly be accomplished in a general manner.

Although Clouds did not escape this trend, they pose inherent features that make them unique with regards to their monitoring needs towards application-level scalability. Cloud applications lay on top of a virtual infrastructure that falls beyond their control. Service providers need means to introduce non-intrusive custom probes to gather relevant metrics independently of the underlying infrastructure. Since an application components may dwell in different Clouds, the monitoring systems would need to offer federation capabili-

Table 2. Capacity of monitoring systems in the cloud

	Abstraction	Infrastructure-dependant	Dynamism	Federation
Log-Based	Hw.	Yes	No	No
GoGrid	Hw./VM	Yes	No	No
Amazon CloudWatch	Hw.	Yes	No	No
Globus GRAM	Custom	Yes	No	Yes (within the VO)
Nagios	Custom	Yes	No	No

ties, being totally independent of specifics from a given Cloud. The capability of monitoring systems to meet these requirements is shown in Table 2.

Some recent attempts have tried to build a "distribution framework" which also specified the interface of the data sources with the framework itself (Clayman et al., 2010). However, the specification of distribution interfaces does not solve any of the problems presented in Table 2. Indeed, the installation and configuration of such Cloud distribution frameworks is highly dependent on the underlying infrastructure and the pre-established communication channel.

From IaaS to PaaS Scalability

The works reviewed in previous sections are an illustrative set of current state of the art technologies, but they all are mostly aimed to provide IaaS solutions. Thus, as in Figure 1, they are still too tied to low level administration details: tasks as the reconfiguration of newly created machines or the connection between them have still to be accomplished by the service provider. Service providers have to build a VM Image from the scratch and have to install the entire software stack, including the operating system, to run their service components.

An increased abstraction level is granted to provide predefined components (e.g. preinstalled template images providing Web, Application, Content Management or Database servers already configured and prepared for scalability), bringing the service architecture closer to the Service definition (some Cloud providers, as GoGrid or Amazon EC2 are already providing customizable predefined images). Security, performance and scalability can be reinforced by supporting networking resources such as routers, firewalls or load balancers (Figure 3).

Platform as a Service (PaaS) Clouds, such as Google AppEngine, Windows Azure or Salesforce, deal with a higher level of abstraction in application hosting by managing infrastructure resources transparently: instead of giving access to the VMs to the service provider, the Platform manages automatically application containers (Java, Phyton, .net, etc.) that run the application components. Service Providers have only to focus on developing the application and packaging it to be executed in the application containers. Application containers may scale vertically by using more powerful VMs or horizontally by clustering VMs. Figure 3 shows an example of horizontal scalability schema based on replication of n-tier J2EE application servers: 1) Service Providers package software components to be executed in the platform, 2) the platform instantiate the proper runtime execution environments (OS, middleware, libraries, and application containers to run inside a virtual machine) and load balancing components, and 3) the platform deploys the application components by allocating the needed infrastructure resources. Therefore, the previous concepts on service level monitoring and scalability can be applied to runtime application containers by considering the underlying platform resource model: PaaS Clouds management functions decide

Figure 3. PaaS scalability model

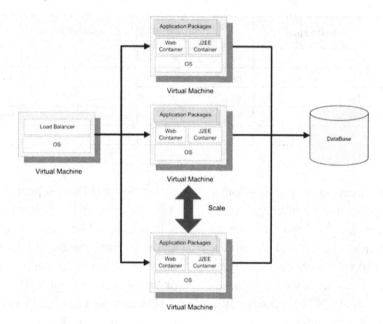

dynamically the amount of execution environments replicas which are needed to host the application components. To illustrate the feasibility of this schema, (Paz et al., 2010) evaluates and extends the replication protocols of the JOnAS J2EE Container for stateless Application Servers. The extension consists on taking the fault-tolerance replication protocol which follows a primary-backup schema (the primary server receives client operation request and propagates its state to the backup replicas, in case of failure client request are invoked on a backup replica) and allowing multiple primary servers in a cluster which share the same database (primary servers act as backup servers between each other). The profiling of the benchmarking results concludes the replication schema for scalability does not introduce too much overhead, and it identifies that the bottleneck in the benchmark application is due to the database that could be replicated for future benchmarks. Indeed, profiling techniques could be applied to identify the heuristics to govern the scalability of the application.

PaaS Clouds also provides development tools and APIs to access platform services (databases, messaging, caching, etc.). But, applications have to be implemented following a restricted programming model (i.e. HTTP-based web services) for taking advantage of the platform scalability. For example, Map Reduce (Dean & Ghemawat, 2008) is a programming model to parallelize datasets computation across large scale clusters applied to machine learning problems, report generation, web mining or news, image, graph and language processing. To support the Map Reduce programming model, a set of libraries, communication middleware, distributed data management and storage functions are needed. Obviously, not all the application problems can be solved using the Map Reduce model and most of the traditional transactional applications, widely used in enterprise domains, could be hardly adapted. In some cases, by relaxing the consistency or reducing the features of the databases, higher performance and scalability could be achieved.

CONCLUSION AND DISCUSSION

This work has briefly outlined the efforts made to provide service level scalability in the right terms for business controllers and systems architects to deal with the high variability and volatility of applications while retaining control and understanding on how the application is built and on how it evolves. While there has been a clear evolution from the early works on distributed systems, there seems to be a long way ahead towards the delivery of actual service level scalability at all phases of the SOA application lifecycle, from its conception until its maintenance during runtime.

Unfortunately, the latest efforts to bring scalability closer to business and user terms are focused on the runtime service lifecycle shown in Figure 2. More work is needed to bring easy scalability to other stages, such as developing or maintenance. Works in this way are granted and will help bridging the gap at all service lifecycle stages. This way service architects could include testing or developing stages while the service is running allowing them to control the way developers test their code and system engineers to perform application maintenance in the same environment that is being used for runtime. This implied the development, test, provision, run and maintenance of a service in the very same environment, reducing services time to market and costs.

Some recent initiatives have tried to deliver a more general and abstract framework, but neither of them covers the whole service lifecycle, from design to maintenance and iteration. Also, most of them imply installing a whole stack in order to federate data centers (even when they still allow for underlying technologies to be employed). Systems such as Open Cirrus and Elastic Sites were aimed at scaling the datacenter itself, rather than the applications running inside it. Only Claudia allowed a clean federation by installing a controller in our local premises or in the Cloud, thus it is no longer needed to control or install anything on the federated Clouds.

One of the major limitations of the works above (Galán et al., 2009; Rodero-Merino et al., 2010) is the static nature of the service definition. SOA applications are inherently dynamic entities in which new service components or their implementations can get in or out the running workflow at every stage, thus new VMs need to be added or removed to the service composition at runtime. Also, the rules governing the behavior of the different services composing the application may need to be changed on the fly.

Monitoring data that make sense to business controllers and system architects is a very complicated task. On the one hand, Cloud providers cannot make a one to suit every solution, since it is the specific business context/objective that makes a metric relevant or not and, thus, applicable to a set of rules governing a service. On the other hand, mapping from the offered infrastructure level metrics (e.g. CPU, memory, etc.) is also application specific and can often be done for a very restricted set of applications or business relevant performance indicators. Many important features are missing in this regard.

The literature available on monitoring up to date is very abundant, but neither of the available alternatives seems to meet all the requirements above. Distribution frameworks are needed and Cloud providers may incorporate them when they aim to face federation challenges, but the availability of including custom probes to monitor relevant metrics (that may also be very confidential, since monitoring the appropriate item may lead to success/failure and advantages over similar competitors). These probes/embedded agents need to be installed without relying on any help from the Cloud provider, as a consequence of virtualization, they need to be dynamically turned on/off and capable of delivering data from the sources to their processing points regardless of the actual location. In this regard, external probes (meaning external to the physical or virtual machines or networks to be monitored) are a preferred mechanism, since this potentially reduces the overhead they impose to the system.

A next generation of IaaS Clouds, which incorporates the scalability requirements described in this paper and some others (Buyya et al., 2009) related to security, convergence with other paradigms, stronger QoS support and portability, is expected in the medium term. We believe this next generation of IaaS Clouds could suppose the turning point towards its market adoption. On the other hand, the PaaS Cloud race has just started. Existing PaaS programming models are restricted to certain application domains. We envision new broad-spectrum programming models and frameworks (supporting SDKs, libraries and common services) will come out as enablers of next generation Internet applications, but this is a matter of current and future research that will determine the final adoption of the PaaS Cloud model.

REFERENCES

Aiftimiei, C., Andreozzi, S., Cuscela, G., Donvito, G., Dudhalkar, V., Fantinel, S., et al. (2007). Recent evolutions of gridice: A monitoring tool for grid systems. In *Proceedings of the Workshop on Grid Monitoring* (pp. 1-8). New York, NY: ACM Press.

Alexandre, R., Prata, P., & Gomes, A. (2009). A grid infrastructure for online games. In *Proceedings of the 2nd International Conference on Interaction Sciences* (pp. 670–673). New York, NY: ACM Press.

Amazon Web Services. (n. d.). *Amazon's Cloud-Watch monitoring system.* Retrieved from http://aws.amazon.com/cloudwatch/

Avetisyan, A., Campbell, R., Gupta, I., Heath, M., Ko, S., & Ganger, G. (2010). Open cirrus: A global cloud computing testbed. *IEEE Computer*, *43*(4), 42–50.

Buyya, R., Shin Yeo, C., Venugopal, S., Broberg, J., & Brandic, I. (2009). Cloud computing and emerging it platforms: Vision, hype, and reality for delivering computing as the 5th utility. *Future Generation Computer Systems*, *25*(6), 599–616. doi:10.1016/j.future.2008.12.001

Cáceres, J., Vaquero, L. M., Rodero-Merino, L., Polo, A., & Hierro, J. (2010). Service scalability over the cloud. In Furht, B., & Escalante, A. (Eds.), *Handbook of cloud computing*. New York, NY: Springer. doi:10.1007/978-1-4419-6524-0_15

Cirstoiu, C., Grigoras, C., Betev, L., Costan, A., & Legrand, I. C. (2007). Monitoring, accounting and automated decision support for the alice experiment based on the monalisa framework. In *Proceedings of the Workshop on Grid Monitoring* (pp. 39-44). New York, NY: ACM Press.

Clayman, S., Galis, A., Chapman, C., Toffetti, G., Rodero-Merino, L., & Vaquero, L. M. (2010). Monitoring service clouds in the future internet. In Tselentis, G., Galis, A., Gavras, A., Krco, S., Lotz, V., & Simperl, E., (Eds.), *Towards the future internet - emerging trends from European research*. Amsterdam, The Netherlands: IOSPress.

Dean, J., & Ghemawat, S. (2008). MapReduce: Simplified data processing on large clusters. *Communications of the ACM*, *51*(1), 107–113. doi:10.1145/1327452.1327492

Deng, Y., & Wang, F. A. (2007). Heterogeneous storage grid enabled by grid service. *Operating Systems Review*, *41*(1), 7–13. doi:10.1145/1228291.1228296

Dumitrescu, C., Raicu, I., & Foster, I. (2005). Di-gruber: A distributed approach to grid resource brokering. In *Proceedings of the ACM/IEEE Conference on Supercomputing* (p. 38). Washington, DC: IEEE Computer Society.

Foster, I. (2002). *What is the grid? A three point checklist*. Retrieved from http://dlib.cs.odu.edu/WhatIsTheGrid.pdf

Galán, F., Sampaio, A., Rodero-Merino, L., Loy, I., Gil, V., Vaquero, L. M., et al. (2009). Service specification in cloud environments based on extensions to open standards. In *Proceedings of the Fourth International Conference on Communication System Software and Middleware*, Dublin, Ireland (p. 19) New York, NY: ACM Press.

Imamagic, E., & Dobrenic, D. (2007). Grid infrastructure monitoring system based on nagios. In *Proceedings of the Workshop on Grid Monitoring* (pp. 23-28). New York, NY: ACM Press.

Jin, H., Tao, Y., Wu, S., & Shi, X. (2008). Scalable dht-based information service for large-scale grids. In *Proceedings of the 5th Conference on Computing Frontiers* (pp. 305-312). New York, NY: ACM Pres.

Lee, K.-W., Ko, B.-J., & Calo, S. (2004). Adaptive server selection for large scale interactive online games. In *Proceedings of the 14th International Workshop on Network and Operating Systems Support for Digital Audio and Video* (pp. 152-157). New York, NY: ACM Press.

Lim, H. C., Babu, S., & Chase, J. S. (2010). Automated control for elastic storage. In *Proceedings of the International Conference on Autonomic Computing* (pp. 19-24). New York, NY: ACM Press.

Lim, H. C., Babu, S., Chase, J. S., & Parekh, S. S. (2009). Automated control in cloud computing: challenges and opportunities. In *Proceedings of the 1st Workshop on Automated Control for Datacenters and Clouds* (pp. 13-18). New York, NY: ACM Press.

Lin, Q., Neo, H. K., Zhang, L., Huang, G., & Gay, R. (2007). Grid-based large-scale web3d collaborative virtual environment. In *Proceedings of the Twelfth International Conference on 3D Web Technology* (pp. 123-132). New York, NY: ACM Press.

Marshall, P., Keahey, K., & Freeman, T. (2010). Elastic site: Using clouds to elastically extend site resources. In *Proceedings of the 10th IEEE/ACM International Symposium on Cluster, Cloud and Grid Computing*, Melbourne, Australia (pp. 43-52). Washington, DC: IEEE Computer Society.

Milanović, M., Gašević, D., Wagner, G., & Devedžić, V. (2009, Novmeber 2-5). Modeling service orchestrations with a rule-enhanced business process language. In *Proceedings of the Conference of the Center for Advanced Studies on Collaborative Research*, Ontario, Canada (pp. 70-85). New York, NY: ACM Press.

Nascimento, A. P., Boeres, C., & Rebello, V. E. F. (2008). Dynamic self-scheduling for parallel applications with task dependencies. In *Proceedings of the 6th International Workshop on Middleware for Grid Computing* (pp. 1-6). New York, NY: ACM Press.

Nóbrega, A. D., Nyczyk, P., Retico, A., & Vicinanza, D. (2007). Global grid monitoring: The egee/wlcg case. In *Proceedings of the Workshop on Grid Monitoring* (pp. 9-16). New York, NY: ACM Press.

Pan, K., Turner, S. J., Cai, W., & Li, Z. (2009). Multi-user gaming on the grid using a service oriented HLA RTI. In *Proceedings of the 13th IEEE/ACM International Symposium on Distributed Simulation and Real Time Applications* (pp. 48-56). Washington, DC: IEEE Computer Society.

Papazoglou, M. P., & Heuvel, W. (2007). Service oriented architectures: Approaches, technologies and research issues. *The International Journal on Very Large Data Bases*, *16*(3), 389–415. doi:10.1007/s00778-007-0044-3

Paz, A., Perez-Sorosal, F., Patiño-Martínez, M., & Jiménez-Peris, R. (2010). Scalability evaluation of the replication support of JonAS, an industrial application server. In *Proceedings of the European Dependable Computing Conference* (pp. 55-60). Washington, DC: IEEE Computer Society.

RightScale. (n. d.). *Cloud management platform*. Retrieved from http://www.rightscale.com/

Rodero-Merino, L., Vaquero, L. M., Gil, V., Galán, F., Fontán, J., & Montero, R. S. (2010). From infrastructure delivery to service management in clouds. *Future Generation Computer Systems*, *26*(8), 1226–1240. doi:10.1016/j.future.2010.02.013

Saha, D., Sahu, S., & Shaikh, A. (2003). A service platform for on-line games. In *Proceedings of the 2nd Workshop on Network and System Support for Games* (pp. 180-184). New York, NY: ACM Press.

Vaquero, L. M., RoderoMerino, L., Cáceres, J., & Lindner, M. (2009). A break in the clouds: Towards a cloud definition. *ACM SIGCOMM Computer Communications Review, 39*(1), 50-55.

Zdun, U., Hentrich, C., & Dustdar, S. (2007). Modeling process-driven and service-oriented architectures using patterns and pattern primitives. *ACM Transactions on the Web, 1*(3), 14. doi:10.1145/1281480.1281484

This work was previously published in the International Journal of Cloud Applications and Computing (IJCAC), Volume 1, Issue 1, edited by Shadi Aljawarneh & Hong Cai, pp. 34-44, copyright 2011 by IGI Publishing (an imprint of IGI Global).

Chapter 4
Beyond Hadoop:
Recent Directions in Data Computing for Internet Services

Zhiwei Xu
Chinese Academy of Sciences, China

Bo Yan
Chinese Academy of Sciences, China

Yongqiang Zou
Tencent Research, China

ABSTRACT

As a main subfield of cloud computing applications, internet services require large-scale data computing. Their workloads can be divided into two classes: customer-facing query-processing interactive tasks that serve hundreds of millions of users within a short response time and backend data analysis batch tasks that involve petabytes of data. Hadoop, an open source software suite, is used by many Internet services as the main data computing platform. Hadoop is also used by academia as a research platform and an optimization target. This paper presents five research directions for optimizing Hadoop; improving performance, utilization, power efficiency, availability, and different consistency constraints. The survey covers both backend analysis and customer-facing workloads. A total of 15 innovative techniques and systems are analyzed and compared, focusing on main research issues, innovative techniques, and optimized results.

1. INTRODUCTION

Large scale data computing is a main technology driver for the many Internet services we see today. There data computing manifests as two classes of workloads. The first are usually called *customer-facing* applications that process many interactive requests from hundreds of millions of users within a short response time. The second are called *backend* applications that conduct data analysis jobs in batch mode, where each job may involve petabytes of data.

DOI: 10.4018/978-1-4666-1879-4.ch004

This paper reviews recent research directions in data computing for Internet services. We identify five inter-related research directions that are interesting to any cloud system with data computing workloads: (1) improving performance (response time, throughput, or job execution time); (2) improving system utilization (CPU, disk, and I/O bandwidth, etc.); (3) improving energy efficiency (saving power or energy); (4) improving system availability (availability and reliability); and (5) considering different consistency constraints. The last direction is important not only because consistency affects performance, but also due to the CAP theorem (Brewer, 2000): consistency constraints affect availability and partitioned fault tolerance.

Many Internet services now use Hadoop as their main data computing platform (Hadoop, n.d.). Hadoop is also extensively used by academia as a research platform and an optimization target. Supported by a vibrant community, with increasing contributions from companies and academia, Hadoop is developing rapidly. For instance, Hadoop was originally created to handle backend data analysis jobs, including only MapReduce, HDFS and a core. Components in the current Hadoop suite (Figure 1), except HBase, are still mainly for backend applications. However, much research work is ongoing to extend Hadoop for customer-facing applications. Since Hadoop is an open source software suite organized by Apache Software Foundation, research contributions are not hindered by proprietary barriers. This paper focuses on researches that have been or can be converted and integrated into Hadoop.

The rest of the paper is organized as follows: Section 2 discusses optimization techniques for backend data analysis workloads, to improve job execution time, system utilization and energy efficiency, through innovative techniques in scheduling, I/O organization and nodes disabling. Section 3 discusses optimization techniques for customer-facing interactive workloads. We survey five systems with different data models and review three optimization techniques for specific data model operations. Section 4 offers concluding remarks and points out future research problems. We only select techniques and systems that are representative. A total of 15 innovative techniques and systems are analyzed and compared, focusing on their main research issues, innovative techniques, and optimized results.

2. OPTIMIZATIONS ON BACKEND DATA ANALYSIS

This section analyzes seven techniques for optimizing data computing for backend data analysis workloads. The objectives are to improve job execution time, system utilization and energy efficiency, involving innovative scheduling, I/O organization, node disabling techniques. Section 2.1 introduces three Hadoop scheduling optimization techniques. Section 2.2 discusses three enhancements using storage techniques. Section 2.3 reviews a technique to improve energy efficiency. Section 2.4 summarizes these techniques.

Figure 1. Hadoop structure and subprojects (Hadoop, n.d.)

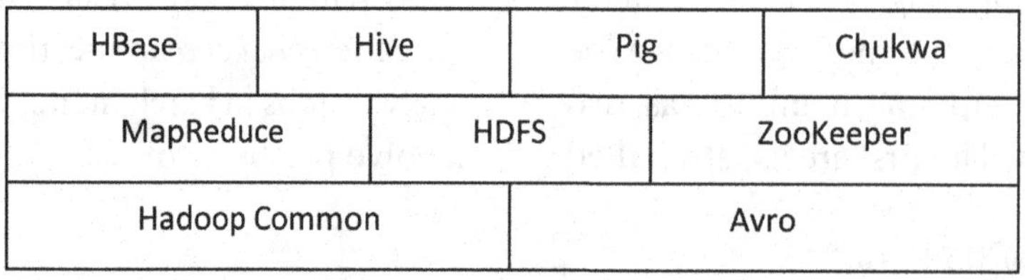

HBase	Hive	Pig	Chukwa
MapReduce		HDFS	ZooKeeper
Hadoop Common		Avro	

2.1. Scheduling Optimizations

The LATE Scheduler

The performance of Hadoop MapReduce framework depends heavily on its task scheduler. The task scheduler handles node crashes by re-running a failure task on another machine. If a node is available but its performance is poor, called a straggler, the scheduler uses a speculative execution mechanism to run a backup copy of its tasks in order to minimize jobs' response time.

The original Hadoop native task scheduler monitors task progress using a metric called progress score and a FIFO scheduling policy. It makes several implicit assumptions, e.g. cluster nodes are homogeneous and tasks progress linearly. These assumptions do not always hold, such as in a data center with various machines or a virtualized data center rented from Amazon's EC2. Heterogeneity seriously impacts the scheduler using a fixed threshold for selecting speculative tasks. Too many speculative tasks may be launched. The wrong tasks may be chosen for speculation first, resulting in resource contention and waste.

The LATE (Longest Approximate Time to End) scheduler (Zaharia et al., 2008) is based on an intuition: tasks with the largest estimated finish time should be backed up on fast nodes. Main innovative techniques in LATE include techniques to estimate a task's time to completion and to estimate which nodes are fast nodes; and a scheduling algorithm to rank running tasks based on their estimated time to completion, to allow fast nodes to request and get top ranked tasks to run. The LATE mechanism deals with both heterogeneity and node failures.

The original Hadoop native defines and maintains a task's *ProgressScore*. The LATE scheduler defines a *ProgressRate=ProgressScore/T* for each task, where T is the amount of time the task has been running. The LATE scheduler estimates a task's time to completion as *(1-ProgressScore)/ProgressRate*. The LATE scheduler uses three con-

figurable parameters, called *SlowNodeThreshold*, *SlowTaskThreshold* and *SpeculativeCap* to help improve scheduling results. *SlowNodeThreshold* is the percentile of speed below which a node is deemed too slow to receive speculative tasks. *SlowTaskThreshold* determines whether a task is slow enough to be speculated on. *SpeculativeCap* marks the number of speculative tasks that can be running at once, to prevent overloading. The LATE scheduling algorithm runs on the scheduler node and works as follows:

- When a worker node asks for a new task and the total number of running speculative tasks is less than the *SpeculativeCap*.
- Ignore the request if the total progress of the node is less than *SlowNodeThreshold*.
- Rank every running task that are not currently being speculated by estimated time to completion, i.e., *(1-ProgressScore)/ProgressRate*.
- Find out the highest-ranked task with progress rate under *SlowTaskThreshold* and launch a copy of this task.

The authors show through performance evaluation experiments that the LATE scheduler significantly improves job execution time over Hadoop's native scheduler. The Hadoop Sort benchmark job execution time could improve 27% to 162%, for different experiment configurations. In some cases, LATE scheduler may perform worse than no speculation, but it is still effective in most of Hadoop jobs.

Intuitively, there are several reasons why LATE improves performance. Only the slowest tasks and a small number of tasks will be re-executed. Node heterogeneity is considered when deciding where to run speculative tasks. By focusing on estimated time left rather than progress rate, LATE speculatively executes only those tasks which will improve job execution time, avoid unnecessary waste of resource.

Delay Scheduling and the Hadoop Fair Scheduler

It is becoming common to share a cluster between multiple users running both long jobs and small jobs over the same data set. Sharing leads to lower costs and avoid replication of data across clusters. Once a new job is submitted and there is no resource available, two approaches can be used to guarantee max-min fairness. One approach is to kill running tasks to release resource for the new job, which causes resource waste. Another approach is to let the new job wait for running tasks to finish, which negatively impacts fairness. Fair sharing among jobs may also conflict with preserving data locality. A job may be scheduled to a node according to fairness but without the required data on that node.

(Zaharia et al., 2010) first investigate a naïve fair sharing algorithm which uses the waiting approach. It strictly follows a queue fairness order and ranks jobs in increasing order of number of running tasks. It schedules jobs with required data on a node to run before the jobs without data locality. Tests identify two locality problems in this algorithm: (1) the head-of-line scheduling problem occurs in scheduling small jobs, as it is unlikely to have a small job's required data on the node; (2) the sticky slots problem is the situation that jobs never leave their original slots.

The authors present a simple delay scheduling algorithm which introduces a skip count in each job. A job without data locality can be delayed and increments its skip count by 1. A job without data locality can be delayed up to a maximum skip count D times. Analysis shows that the fraction of jobs without data locality decreases exponentially with the maximum skip count D. To achieve a certain level of locality, the amount of skipping is a fraction of average task length and decreases linearly with the number of slots per node.

The authors discuss the design issues of Hadoop Fair Scheduler (HFS). HFS uses a two-level scheduling hierarchy. At the top level, HFS allo-cates task slots across pools using weighted fair sharing policy where pools can be configured with different weights. A parameter called *minimum share* is used in each pool to guarantee a pool's minimum number of slots. At the second level, each pool allocates its slots among jobs using either FIFO with priorities or fair sharing. The delay scheduling algorithm is improved from the prior simple one with two changes in practice. One is introducing the maximum wait time W to replace the maximum skip count D in order to make sure a job can be scheduled in a predictable time. The other is using two maximum wait times W_1 and W_2 to explore the node locality and rack locality, respectively.

They test HFS performance using three different kinds of workload, an IO-heavy one, a CPU-heavy one and a mixed workload. The evaluation shows that delay scheduling achieves nearly optimal data locality in all kinds of workloads and can increase throughput by up to 2× while preserving fairness. They also note that delay scheduling can be ineffective when a large part of tasks is much longer than average, or the number of slots per node is small. The authors believe that using small tasks more frequently and hardware such as multicore and faster Ethernet can make improvements.

The Triple-Queue Scheduler for Heterogeneous Workloads

(Tian et al., 2009) investigate the characteristic of MapReduce jobs in a practical data center containing thousands of nodes. I/O-bound jobs and CPU-bound jobs can run simultaneously in the same cluster. These heterogeneous jobs are actually complementary since they demand different part of resources. Mixing these jobs on the same cluster can increase system utilization.

The authors test MapReduce workloads and identify three categories. They measure the current speed of a node running n map tasks by calculating the amount of the data needed to be processed in map-shuffle phase divided by the Map Task Completed Time (MTCT) and compare the result with the Disk I/O Rate (DIOR). If DIOR is bigger, this kind of task is classified as CPU-bound. If DIOR is smaller, this kind of task is called Class Sway, which means the map part needs more CPU and the shuffle part needs more disk I/O. Then they calculate the amount of data to be processed only in map phase divided by MTCT which is also compared with DIOR. If DIOR is smaller, then this kind of task is classified as I/O-bound.

The authors present a new triple-queue scheduler which consists of a workload prediction mechanism MR-Predict and three different queues: a CPU-bound queue, an I/O-bound queue and a wait queue. The MR-Predict mechanism automatically predicts the class of a new coming job based on the above classification. Jobs in the CPU-bound queue or I/O-bound queue are assigned to separate queues. If a job in the CPU-bound queue is classified as I/O-bound while being monitored, it will be moved to the I/O-bound queue to be scheduled.

The authors compile statistic of jobs to verify the workload classification. They examine three different benchmarks: TeraSort, Grep-Count and WordCount, and find out that they have different characteristics. The average CPU utilization rate of Grep-Count is almost above 90% in the test, so Grep-Count is CPU-bound. TeraSort having average CPU utilization rate almost under 40% is IO-bound. In test of WordCount, the CPU utilization decreases dramatically after the reduce task begins. They mix these three jobs to run on the same cluster using their triple queue scheduler. The experiments show that the triple queue scheduler improves job execution throughput by 30% in the map-shuffle phase, and reduces the whole execution time by 20%.

2.2. Optimizations on Storage

SplitCache

(Kim et al., 2010) observe that the same computation is performed multiple times on the same data. This significant redundancy in computation across jobs in execution results in waste of computing resource and also increases network traffic. The scheduling policy in the MapReduce framework takes the location information of the input data into account due to the limit of network bandwidth. However, this locality-aware scheduling policy is not effective if the input data is located in remote storage nodes that cannot provide a computation service.

The idea of SplitCache mechanism is to cache input data when these data first appear among applications and reuses them for future demand. This mechanism consists of two components: a cache server running on each worker node and the cache-aware task scheduler taking cache information into account. The cache server manages the remotely located input splits in its local storage, registers the information of caches to the cache-aware task scheduler and gives this information to map workers when needed. The cache-aware task scheduler considers not only the locations of input data but also the locations of caches in computing nodes when dispatching tasks to map workers. In contrast to the traditional cache system in which computation itself cannot be moved toward data, the cache-aware task scheduler moves computation toward cached data to maximize data locality and to reduce network traffic overhead.

The authors use TPC-H benchmark to test the performance of their system. The system with SplitCache achieves 65.5% faster execution and 87% reduction of network traffic, compared to a system without SplitCache. Since SplitCache is implemented exploiting components of Hadoop, it is not difficult to deploy the hierarchical management structure for scalability. Many applications that process the same data with varying conditions or constraints can benefit from the SplitCache mechanism.

SuperDataNodes

Current Hadoop design adopts a coupled approach, i.e., each node has both predefined computing and storage capabilities. (Poter, 2010) points out two limitations of this approach. First, the ratio of computation to storage might change over time, or might not be known in advance. Second, when the workload changes, it might be desirable to power down or re-purpose some of the Hadoop nodes for other applications. It is desirable to have a decoupled approach that can change the ratio of computing/storage capabilities while maintaining some data locality.

The main idea of SuperDataNodes is gathering disks to form a pool providing storage capability in a decoupled fashion. A SuperDataNode has an order of magnitude more disks than a traditional Hadoop DataNode. A number of virtual machines each running unmodified Hadoop DataNode processes use these SuperDataNodes as their storage. The author believes using SuperDataNodes will efficiently utilize the vast amount of storage pool and take advantage of high bandwidth network interfaces.

Adopting SuperDataNodes approach provides several benefits. First, SuperDataNodes decouple the amount of storage in HDFS from the number of nodes, which makes dynamic provisioning more flexible. Second, they can support archival data. Furthermore, they can increase uniformity for job scheduling and data block placement. Finally, they ease the management of clusters hosting both Hadoop and non-Hadoop applications.

A prototype of SuperDataNodes is developed in a Hadoop cluster, and tests find that Super-DataNodes not only are capable of supporting workloads with high storage-to-processing ratio, but in some cases actually outperform traditional Hadoop deployments through better management of a large centralized pool of disks. Using a SuperDataNode reduces the total job execution time of a Sort workload by 17% and that of a Grep workload by 54%, compared to a traditional Hadoop cluster.

SuperDataNodes do have some limitations. They need high-bandwidth connectivity within a datacenter rack. They may cause a possible reduction in fault tolerance due to consolidation of storage into a single node. SuperDataNodes cost significantly more than traditional nodes because of their larger memory and disk footprint.

DiskReduce

In HDFS, each data block has two other copies (by default) to provide high reliability. Though simple, this triplication policy causes a disk redundancy overhead up to 200%. Motivated by previous research on RAID, (Fan et al., 2009) proposes the idea of DiskReduce, to reduce dramatically disk redundancy overhead while retaining double node failure tolerance and the performance advantage of using multiple copies of the data. Their work utilizes the fact that files are immutable after written to HDFS and all blocks of a file are replicated in background. In the original HDFS, the background process looks for blocks with insufficient copies, while in DiskReduce, it examines the blocks with high overhead such as those triplicates which can be turned into lower overhead ones. This process is inherently asynchronous and redundant blocks remain in place until data reliability is ensured.

DiskReduce is an application of RAID in HDFS, designed to minimize the change to original HDFS implementation. Its prototype can accommodate different double failure tolerant encoding schemes, including a simple RAID 5 and mirror encoding and a RAID 6 encoding. It can also be extended to support other failure tolerant codes using different encoding/decoding module. It also supports delay encoding to get better trade-off between locality and disk usage.

The authors evaluate the DiskReduce prototype for the encoding function. They set up a 32-node partition, write a file of 16 GB on each data node and spread over the same 32 nodes using RAID groups of 8 data blocks. In total, 512 GB data are written into HDFS modified with DiskReduce. Without DiskReduce, disk overhead would be 512 × 2 = 1024 GB. With DiskReduce, experiment results show that the encoding process recovers 400 GB for RAID 5 and mirror and 900 GB for RAID 6, bringing disk overhead down from 200% to 113% and 25%, respectively.

The authors also try to bound the performance degradation from decreasing copies with encoding using three factors: backup tasks, disk bandwidth and load balance. By examining a trace of block creation and use times on the Yahoo! M45 Hadoop cluster, they find that 99% of accesses are made to blocks younger than one hour old, and that less than 6~12% of disk capacity is needed to delay encoding for an hour. Thus delaying encoding by about one hour is a good rule of thumb for balancing capacity overhead and performance benefits of multiple copies.

2.3. Optimizations on Energy Efficiency

Nodes Disabling

There is still little research work on improving energy-efficiency in data computing for Internet services, especially on Hadoop related platforms. In Leverich and Kozyrakis (2009), the authors made an early attempt focusing on Hadoop, especially HDFS. They point out that the energy efficiency of a cluster can be improved in two ways. One is turning active nodes into low-power standby modes according to the current workload needs. The other is tuning the status of each node to match the workload and avoid energy waste on oversized components. In practice, the characteristics of Hadoop make these methods not so effectively. To guarantee performance, a Hadoop cluster normally maintains idle components that remain active to ensure data reliability, which cause

extra power consumption. Mechanisms which deal with the hardware and software failures may also negatively impact energy efficiency.

A particular concern is the impact of data layout in HDFS. Since the HDFS distributes data across the whole disks of a cluster, it is hard to determine when to turn a node or component off safely. In addition, Hadoop uses a block replication factor (3 by default) to indicate the number of copies of any data-block and maintains two replication invariants in a Hadoop cluster: (1) no two replicas are stored on one node and (2) replicas must be stored in at least two racks. The authors introduce another invariant: at least one replica of a data-block must be found in a covering subset. The idea of a covering subset is to ensure the reliability of data in a sufficient set of nodes so that other nodes can be turned off or in a low-power mode.

The authors implement their changes based on Hadoop 0.20.0 and establish a 36-node cluster for evaluation. They follow an energy model based linearly on CPU utilization. The workloads are webdata_sort and webdata_scan from the gridmix batch-throughput benchmark of Hadoop. Experiments show that by disabling nodes, they achieve energy savings from 9% for webdata_sort to 51% for webdata_scan, at the expenses of decreasing the performance accordingly. Another experiment shows the Hadoop cluster uses the same amount of time to process 1 TB of data, both with 36 nodes active all the time and with 18 nodes disabled, resulting in 44% less energy consumed.

Some issues should be considered in future research on the energy efficiency of Hadoop. There is a trade-off between data replication strategy and power consumed by the node with replications. Scheduling policy also can take energy into account. MapReduce jobs and online-storage applications may verify in power consumption model and optimization should be investigated in both kinds of workloads.

Table 1. Summary of optimizations on backend data analysis

Approach	Job execution time	Job throughput	System utilization	Power consumption
LATE scheduler	2× faster on average	Not mentioned	Speculation may increase overhead	Not mentioned
Hadoop Fair Scheduler	Not mentioned	Increase by up to 2× on average	Better utilized under multi-user sharing	Not mentioned
Triple-Queue scheduler	Reduce 20% on whole bag of jobs	Improve by 30% on average	Better utilized in mixed workloads	Not mentioned
SplitCache	65.5% faster on average	Not mentioned	Add file split caches on local storage	Not mentioned
SuperDataNodes	Vary from 54% less to 92% more on different workloads	Not mentioned	May improve disk utilization under virtualization	May decrease by decreasing TaskTracker nodes
DiskReduce	Almost no change	Almost no change	Decrease HDFS replication overhead from 200% to 25%	Not mentioned
Nodes disabling	May increase due to fewer active nodes	May decrease due to fewer active nodes	CPU almost full utilized	Increase nodes energy efficiency on average

2.4. Summary

The above optimizations on backend data analysis target four objectives: job execution time, job throughput, system utilization and power consumption, as compared in Table 1.

LATE scheduling can significantly speed up Map-Reduce jobs by speculatively executing slow jobs on fast nodes. The Hadoop Fair Scheduler can double the job throughput by utilizing data locality via a delay scheduling scheme. The Triple-Queue Scheduler improves both execution time and throughput by mixing different types of workloads to better utilizing hardware.

SplitCache improves job execution time by adding file split caches on local storage. Super-DataNodes has varying effects on performance. Its main objective is to allow dynamically adjustment of the ratio of computation and storage capabilities in a Hadoop cluster. DiskReduce can reduce the HDFS replication space overhead from 200% to 25%, utilizing a RAID-like idea.

Power consumption has not been investigated extensively in the Hadoop setting. The node disabling technique shows that this is a promising area. Making nodes active on-demand can not only achieve better hardware utilization, but also increase energy efficiency.

3. OPTIMIZATIONS ON CUSTOMER-FACING APPLICATIONS

Customer-facing applications require short response time (0.1~2 seconds), high throughput (serving hundreds or more simultaneous requests), high availability, but usually relax consistency to scale to large scale data on top of thousands of commodity servers. Several systems have been proposed to tackle such problems, a main feature of which is new data models. In this section, we review five such systems and three optimization techniques on specific data model operations.

3.1. Data Models

Several distributed structured systems have been proposed support large-scale customer-facing applications, each with a slightly different data model, with various relaxed consistency.

BigTable

BigTable (Chang et al., 2006) starts a new era of distributed structured data store to pursue high performance, high availability, and acceptable consistency, to scale to petabytes data and thousands of commodity servers.

BigTable's design goals include wide applicability, scalability, high performance, and high availability. Wide applicability means that BigTable should serve for both customer-facing requests and batch-processing jobs with a wide range of tables sizes, cell sizes, compression options, in-memory or not, and table schema complexity.

The main innovations of BigTable are: (1) a new data model, (2) a centralized hierarchy metadata management scheme, and (3) a composition of interesting storage data structures.

BigTable introduces a data model simpler than relational data model. BigTable organizes data as a sparse table ordered by primary key. Each table has several column families, which are groups of sparse columns and serves as units of locality groups and access control policies. Each data cell has several versions of data identified by timestamps. BigTable supports locating data by row key, column name, and cell timestamp. BigTable supports scanning a continuous range of data. BigTable supports client programming via a script to transform and filter data. When updating data, BigTable supports one-row transaction.

BigTable employs a centralized hierarchy metadata management scheme to help data placement and load balancing. BigTable employs a centralized master server to split logic tables as continuous ranges called tablets, dispatch tablets to thousands of tablet servers, and manages metadata representing the dispatching information. BigTable uses a hierarchy tablets metadata style. Each tablet has about 1KB metadata containing the tablet identifiers, end key, and location. METADATA tables are introduced to store the tablets metadata. A ROOT table is introduced to store the tablets dispatching information of all METADATA tables. The ROOT table is stored in a well-known place and never split.

BigTable uses a composition of several interesting storage data structures to implement the data model, which form several building blocks: SSTable, GFS, and Chubby. SSTable is introduced to store column families with the help of corresponding memtables and logs. For a column family, the data is stored in an SSTable sorted by column name and timestamp, so that column families result in a column-oriented database style. SSTables have data blocks with typically 64KB size and a block index. Write requests are finished after writing the memtable and the logs. When the memtable is full, a minor compaction is done and the data is written to the disk as an SSTable file. Major compactions are done to merge a few SSTables into one. Logs can be replayed to recover the data. Read operations are performed by reading a whole data block. GFS is introduced as the underlying distributed file system to provide scalable and reliable distributed storage. SSTable files and logs are stored in GFS and availability comes from the GFS file block replica. BigTable uses the distributed lock service Chubby to manage the membership of live tablet servers. Tablet servers also use Chubby to elect the master and store the position of the ROOT table.

BigTable achieves good performance. On a single PC-class tablet server, BigTable can in a second serve 1212 random reads of 1000-byte rows, 4425 sequential reads, and 15385 continuous scans. The random reads are slower than sequential reads because each read needs to read a 64KB data block and for random reads only one row in the block are the right result. BigTable can perform 8850 random writes per second because the writes only involve the memtable and logs.

BigTable has been deployed on many clusters in Google and used by many applications. Google has 388 non-test BigTable clusters with 24,500 tablet servers in August 2006 for more than sixty Google products. One group of 14 clusters with

8069 tablet servers achieves more than 1.2 million requests per second. Web Indexing, Google Docs, Orkut, and Google Finance also use BigTable.

BigTable can be used as the input source and output sink of MapReduce jobs. Google Analytics uses a table containing statistics of visiting log for web sites and a summary table generated by MapReduce jobs to serve for web masters. Google Earth has a 70TB table to store raw imagery for preprocessing by MapReduce jobs at batch mode. A 500GB table in memory for data index is created to serving user request within a small latency. Personalized Search stores user queries and clicks in a Table, generates user profiles by MapReduce jobs, and personalizes search results.

HBase

HBase is an open source implementation of BigTable (HBase, n.d.). HBase is a subproject of Hadoop built on top of HDFS and ZooKeeper to provide scalable and reliable structured data storage (ZooKeeper, n.d.). HBase inherits the BigTable's data model and hierarchy metadata management policy. HBase calls the continuous range of a table as region, and employs HMaster and HRegionServer to construct the system.

HBase creates the HFile format since version v0.21 to implement Google SSTable. An HFile represents a column family in a table region. HFile splits data as 64KB blocks, appends data block index at the end of HFile, and stores key-value pairs in data blocks. The key in data blocks is composed of the rowid, column name, and timestamp, so as to form the columnar storage. HFile employs Bloom Filters to accelerate the look up of a row. When an HFile is loaded, the index is loaded to memory to check the data block location in the HFile without multiple disk accesses. When performing a single row reading, HBase reads a data block once a time to serve as pre-fetched data. HBase uses HDFS as the reliable file storage, and stores HFile and WAL (write ahead log) to HDFS.

HBase uses ZooKeeper to elect the master and manage membership.

There are more than 25 companies powered by HBase in June 2010 (HBase, n.d.). Adobe has 30 nodes running HDFS, MapReduce, and HBase for social services to structured data storing and processing for internal use. Another Adobe production cluster has been running since 2008. PowerSet has a 110-node Hadoop cluster running HDFS, MapReduce, and HBase. Twitter runs HBase across its entire Hadoop cluster, and runs a number of applications including people search rely on HBase internally for data generation. Twitter's operation team uses HBase as a time series database for cluster-wide monitoring/performance data. Yahoo! uses HBase to store document fingerprint for detecting near-duplications for real-time query traffic.

There are more than 20 projects adding new features to HBase. Some projects add ORM features to HBase, others add indexes for HBase, others build storages on HBase, and some projects add new RESTful interface or improve HBase usability for other programming languages.

Dynamo

Dynamo (DeCandia et al., 2007) is Amazon's highly available key-value storage system to provide an "always-on" experience. Dynamo aims to applications that only need primary-key load/store but have critical demand on performance, availability, efficiency, and scalability. In Amazon's platform, the target applications include best seller lists, shopping carts, customer preferences, session management, sales rank, and product catalog. Relational database provides too much useless complex functionality while is not easy for replica, scale, and partition for load balance.

Dynamo only provides two operations, including a GET to retrieval objects list with context for a given key, and a PUT to insert or update objects. Dynamo uses technologies introduced by the DHT system Chord (Stoica et al., 2001), including consistent hashing and ring topology to distribute objects, and virtual nodes for load balance. Dynamo uses a full membership model.

Two interesting features stand out with Dynamo. It provides for varying weaker consistency models. The system by default supports per-object (roughly similar to per-record) transaction semantics. However, each object can have multiple versions. Version conflicts detection and resolving are left to the application programmer. Also, Dynamo has N replica located in the ring, allows updates to be propagated to all replicas asynchronously to achieve eventual consistency, and uses vector clocks to find and resolve versions. Put and get are performed by a coordinator to write and read W and R quorums. Changing the value N, R, and W has impact on performance, availability and durability. For example, small W such as 1 provides high performance, high availability, low consistency, and low durability. Small R, such as 1, with W=N results in a high performance read storage. In Amazon, typical NRW is (3,2,2).

Dynamo is used for 2 years at 2007, and the successful responses ratio without timing out is 99.9995%. A Dynamo system can achieve response time within 300ms for a percentile of 99.9% of requests, for a peak load of 500 simultaneous requests per second without downtime.

PNUTS

PNUTS (Cooper et al., 2006) provides data storage with per-record atomicity consistency on commodity servers across data centers. PNUTS provides both ordered tables like BigTable and hash tables like Dynamo, but all with flexible schemas, because tables have records with attributes which can be added at runtime with low overhead. PNUTS relaxes consistency to achieve low latency of huge number of concurrent requests and high availability.

PNUTS is designed for customer-facing requests that read and write single records or small groups of records, such as no more than hundreds of records. PNUTS supports table scans in a small range with predicates. PNUTS does not support referential integrity or ad-hoc queries such as join and group-by.

PNUTS provides per-record timeline consistency model, which means for a given record, all replicas apply all updates to the record in the same order and always update forward to newer versions. Each record has a master replica, which is chosen according to the majority of writes, and all updates must be forwarded to the master. Each record has a sequence number consisting of a generation representing the newly inserted times and a version representing updates times on the existing record. PUNTS choose this model because web applications typically manipulate one record at a time. PNUTS provides APIs to support this model, such as read-any to get a version of data, read-critical to get a record equal or newer than a given version, and read-latest. PNUTS supports scan for range query. PNUTS provides write operations to create or update records and test-and-set-write to update a record only when the record is exactly the required version.

PNUTS partitions tables as tablets by continuous ranges or hashing, places tablets to servers called storage units, and moves tablets for load balance and data recovery. Storage units can use different physical storage layers, currently including MySQL-based ordered table and UNIX filesystem-based hash table.

PNUTS implements the above model with asynchronous record-level geographic replication by reliable message delivery services, the Yahoo! Message Broker (YMB). When a record is written, the master publishes the updates to the message broker to commit this record, which propagates the updates to all non-master replicas. With the broker, PNUTS provides record update notification.

PNUTS has some applications in Yahoo!. For instance, PNUTS stores the user databases containing billions of user IDs with user profiles, usage statistics, and application-specific data, supports large amounts of concurrent requests with low overhead, and ensures data availability. Other applications include Social Applications for friend relationship, Content Meta-Data management, Listings Management for Shopping aggregate listings of items, and Session data. PNUTS with hash tables has entered production in Yahoo!.

Cassandra

Cassandra (Lakshman & Malik, 2010) is a distributed structured data storage system built on top of large amount of commodity servers to provide scalability, availability, high write throughput, and good read efficiency. Cassandra provides a simple data model extended from BigTable, employs the fully-distributed design of Dynamo, and provides some features such as the ability of crossing geographically distributed data centers likes PNUTS.

Cassandra is designed for Facebook Inbox Search, which needs to support billions of writes per day with low latency for users in the world by replicating data across data centers.

Cassandra extends the BigTable data model. Cassandra provides simple column families like BigTable, and adds super column families to provide column families in the simple column families. Data is referred in the form of (column family:super column:column). Another extension is specifying the sort order of a column by time or name. In BigTable, rows for a column are sorted by the row key. In Cassandra, rows can be sorted by time to support query result shown in time sorted order. Cassandra provides get, put, and delete operations. Cassandra provides atomicity among column families per key per replica.

Cassandra employs a ring topology, partitions keys by an order preserving hash function to support range queries, and places data in several hosts for availability and load balance. Cassandra has a coordinator in charge of managing replica with replication policies such as "Rack Unaware", "Rack Aware" within a datacenter, and "Datacenter

Aware". Cassandra uses Zookeeper to elect the coordinator. Cassandra uses Scuttlebutt, an anti-entropy gossip-based mechanism with bootstrap seeds nodes to detect failure nodes. When nodes enter or leave the ring, they split or merge the key ranges. Cassandra employs local file system to store a file format with block, block index, and Bloom Filter for fast retrieval with persistency, commit log for durability and recoverability, and an in-memory data structure for efficiency.

For writes, Cassandra sends the requests to all replicas and waits for a quorum of replicas to finish. For reads, based on the required consistency guarantees, Cassandra either sends the requests to the closest replica or to all replicas and waits for a quorum of responses.

Cassandra is used in Facebook Inbox Search for 250 million users in 2010, and stores more than 50TB of data on a 150-node cluster in two data centers. MapReduce jobs build inverted index and store the resulted index in Cassandra. Inbox Search has two kinds of queries: (a) term search returns messages containing a given word, (b) interaction search returns messages between two given users. For query (a) the user id is the key, the separated words are the super columns, and message identifiers of the messages containing the word are the columns. For query (b) the user id is the key, the recipients identifiers are the super columns, the individual message identifiers are the columns. The median response times of the two queries are 18.27ms and 15.69ms, respectively. For more than 50 GB data, the Write and Read operations of MySQL need about 300ms and 350ms, while in Cassandra they are about 0.12ms and 15ms (Lakshman & Malik, 2009).

Cassandra is the backend storage system for multiple Facebook services in 2010. Cassandra is contributed by Facebook and is now an open source project hosted in Apache. Cassandra is in use at Digg, Facebook, Twitter, Reddit, Rackspace, Cloudkick, Cisco, etc. (Cassandra, n.d.). The largest production cluster has over 100 TB of data in more than 150 machines.

Summary of Data Models

These data models achieve different level of high performance, high availability, and consistency, as compared in Table 2. For performance, BigTable, HBase, and Cassandra's write performance is an order of magnitude better than random read due to the write policy based on in-memory data structures and logs. PNUTS has better read performance than write in ordered tables due to the underlying MySQL storage. Dynamo is the only system which does not support range queries due to consistent hashing for data partition. For availability, all systems achieve high availability through replica, while Dynamo achieves the best availability. For consistency, all systems support only one-row or one-object transaction. Dynamo is the only system that manually resolves conflicts caused by concurrent writes to multiple replicas. For data models, Dynamo is the only system treating the object as a byte stream without columns.

The above systems are suitable for customer-facing applications needing only one-row or one-object transaction with relaxed consistency, but with high demand on performance and availability on commodity servers. The workloads needing multiple-row transactions and complex queries are beyond the ability of these systems and should be served with relational databases.

BigTable and HBase are suitable for both back-end data analysis and customer-facing workloads needing read, write, and scan. Dynamo is suitable for applications that have high demand on availability, durability, and transaction resolving, but only need simple GET/PUT operations, such as shopping carts. PNUTS supports both ordered tables and hash tables, and supports replicas across data centers to benefit a wide range of customer-facing applications. Cassandra uses order preserving hashing on a ring topology across data centers to support range queries, and introduces column families and super column families to simplify application logic needing multiple indexes.

3.2. Operation Optimization

The main differing feature of the above five systems is their data models. Research work is also

Table 2. Comparison of five systems

Name	Performance	Availability	Consistency	Data Models
BigTable	Good write, Good in-memory read, poor random read, very Good scan	High availability comes from GFS data block replica	Per-record transaction	Sparse ordered table with column families
HBase	Good write, Good in-memory read, poor random read, very Good scan	High availability comes from HDFS data block replica	Per-record transaction	Sparse ordered table with column families
Dynamo	Read or write performance could be adjusted by different quorum configurations	High availability of 99.9995%, Can trade among performance, availability, and durability	Per-object transaction, manually resolving version conflicts	Hashed key with byte stream value
PNUTS	Good scan for ordered table, better read than write for ordered table, good read and write for hash table	High availability comes from replica across data centers	Per-record timeline consistency with operation on versioned records	Sparse ordered table and hash table
Cassandra	With more than 50GB data, write latency is 2500 times of MySQL, read latency is 23 times of MySQL	High availability due to replica placement with multiple policies	Per-record transaction within a replica	Sparse ordered table with column families and super column families

done on common techniques that can be used in one or more such data models, but focusing instead on optimizing specific data model operations, including bulk insertion, range query on primary key, and multi-dimensional range queries on distributed ordered tables.

Planned Bulk Insertion

The planned bulk insertion approach (Silberstein et al., 2008) identifies the problem of optimizing bulk-inserting records into Distributed Ordered Tables (DOTs). DOTs are first introduced in this paper to refer to distributed structured storage systems that horizontally partition data over a large cluster of shared-nothing machines, such as those used in BigTable and PNUTS.

Bulk insertion in a table may have very poor aggregate throughput due to the fact that the insertion data is hot in a small range, resulting in a small number of partitions, such that only a few of servers are involved in the insertion.

The planned bulk insertion approach introduces a planning phase before the actual insertions. The planning phase creates new partitions and distributes partitions among machines to balance the insertion load. The planning phase minimizes the sum of partition movement time and insertion time. This paper shows that the problem is a variation of NP-hard bin-packing problem, and reduces it to a problem of packing vectors.

The planned bulk insertion approach is evaluated on a prototype system deployed on a cluster of 50 machines. Performance experiments show that the planned bulk insertion method performs bulk inserts up to 5-6 times faster than the naïve method of inserting records at random order, and 10 or more times faster than the naïve method of inserting sorted records. The planned bulk method scales well as the number of storage servers and the data size increase, while the naïve method for sorted records does not scale.

The planned bulk insertion approach could be extended to deal with bulk updates and deletes.

Currently the approach optimizes a single bulk insertion at a time, while optimizing multiple bulk operations is the future work.

Adaptively Parallelizing

The adaptively parallelizing approach (Vigfusson et al., 2009) aims to optimize the range queries of primary key over Distributed Ordered Tables by choosing the best parallelism at runtime.

Range queries are naturally supported by Big-Table and PNUTS for lots of queries, such as query for time-ordered data, secondary indexes, and hierarchical clustering. For DOTs like BigTable and PNUTS, using maximum parallelism is not the best strategy. For a single client, a small set of servers saturate the client by returning result fast than the client, so maximum parallelism is a waste. For large amount of concurrent queries, maximum parallelism of each query causes disk contention in servers. This paper tries to find out the ideal parallelism level for each query to achieve high and consistent throughput for all queries.

The adaptively parallelizing approach has two aspects. First, an adaptive server allocation technique is introduced to adaptively determine the minimum number of parallel scanning servers for a single query to satisfy the client. The number is determined by query selectivity, client load, client-server bandwidth, and so on. The best number is determined by adjusting the initial number of servers upward or downward to match the server side aggregate sending rate and the client consuming rate. Second, a multi-query scheduling technique is used to adaptively schedule servers for different queries to minimize disk contention on servers. The scheduler has multiple policies according to query execution time and priority, and chooses policies to favor certain queries over others. The scheduler uses greedy scheduling algorithms to bound the makespan of a query, and optimizes Priority Biased Round Robin (PBRR) metric. A feature of this work is that it does not need to maintain data statistics or system statistics and is suitable for DOTs without statistics.

This approach is implemented in the PNUTS system. Performance experiments use synthetic data and a real data set from Flickr's photo metadata in a cluster with 10 storage servers and four client machines. The synthetic data has 20 million 5KB records, and the real data has 10 million Flickr photos ordered by date with a 214-bytes average record size. These experiments show that the approach can get the best parallelism, and benefits queries more when they are longer, more elective, or from a faster client. The approach improves clients' throughput by a factor of 2 to 3. The PBRR scheduler gives the ability to prioritize queries.

CCIndex

CCIndex (Zou et al., 2010) is a scheme to improve the performance while keeping acceptable space overhead and availability of multi-dimensional range queries on DOT systems.

Multi-dimensional range queries are common requirements. For example, finding out hottest pictures in this week in a photo-sharing application requires a multi-dimensional query like "timestamp > 1267660008 and rank > 1000". Multi-dimensional range queries are supported in relational databases, but are not naturally supported by DOT systems, which only support range queries on primary key. An intuition way to support multi-dimensional queries is to scan primary key while filtering other search columns by predicates. However, scanning is ineffective for low selectivity queries. Building secondary index for search columns has the problem of reading rows randomly in the table after a range scan over the index. Clustering index eliminates random reads, but introduces several times of space overhead. If we disable the underlying data replica, the availability is a big problem. So, none of the existing schemes can solve the problem of multi-dimensional range queries on DOT while

simultaneously considering high performance, low space overhead, and high availability.

CCIndex, short for Complemental Clustering Index, is introduced to tackle this problem. CCIndex creates several Complemental Clustering Index Tables, each for a search column with the complete row data, which eliminates the random reads and makes range query over this column a range scan. CCIndex leverages the region-to-server mapping information and estimates result size by the region number covered by the range so as to choose an optimized execution plan for a given multi-dimensional range queries. CCIndex disables the underlying data replica mechanisms, e.g., the replica in distributed file system in Big-Table, to avoid too much space overhead, and creates a replicated Complemental Check Table for each search column to form an incremental row-level data recovery mechanism.

The CCIndex builds a prototype on Apache HBase. Theoretical analysis and micro-benchmarks show that CCIndex consumes 5.3%~29.3% more space, has the same availability, and achieves 10.4 times higher range queries throughput compared with the built-in secondary index scheme in HBase. Synthetic application benchmark based on the Nagios monitoring applications shows that CCIndex achieves query throughput 1.9 ~ 2.1 times that of the same application built on MySQL Cluster.

Summary

The achieved operation optimization objectives of the above approaches are compared in Table 3. These approaches achieve significantly improvements in performance with factors from 2 to 11.4. CCIndex also considers space overhead and availability and achieves acceptable results.

Currently there are few of work to optimize operations for these data models. The above work only consider bulk insert, range queries over primary key, and multi-dimensional range queries. Future work could focus on more operations in

Table 3. Comparison of operation optimization objectives

Approach	Performance	Space overhead	Availability
Planned bulk insertion	5~6 times faster than naïve method for random insertion, 10 times faster for sorted insertion		
Adaptively Parallelizing	It beats other methods and allows all clients throughput by a factor of 2 to 3. The PBRR scheduler gives the ability to prioritize queries.		
CCIndex	11.4 times throughput for single range queries; 1.9~2.1 times for multi-dimensional range queries	Typically 5.3% ~ 29.3% more space	The same availability as secondary index

different situations. Current research mainly considers the performance objective, and future work could consider more objectives.

4. CONCLUSION

This paper surveys recent research work in data computing for Internet services, for both backend data analysis and customer-facing workloads. We identify five research directions, including improving performance, utilization, power efficiency, availability, under different consistency constraints.

For backend workloads, current optimization works include developing new scheduling policies, introducing storage techniques and making system more energy efficient. Scheduling policies focus on improving job execution time, job throughput, or system utilization by using improved speculative execution, exploring data locality, or mixing workloads having heterogeneous charismatic. Scheduler with cache enhancement makes jobs' execution faster and reduces network traffic. Storage pool and virtualization are introduced to decouple storage and computation. RAID and delay encoding techniques are used to decrease replication overhead while maintain performance. Nodes' average energy efficiency can be improved by disabling some nodes.

For customer-facing workloads, several systems are built with different data models to achieve good performance, availability, and just-enough consistency. These systems are suitable for applications needing only one-row or one-object transactional consistency for structured data storage. These systems achieve good performance on commodity servers by splitting data via hashing or by primary key ranges, balancing load, and using interesting persistent and in-memory data structures. These systems achieve availability through data replica placed with multiple policies. These systems also introduce several interesting consistency models. Some researches focus on improving specific operations of these systems, such as bulk insert, range query on primary key, and multi-dimensional range queries.

From this survey, we can conclude that future research on data computing for Internet services, especially research over Hadoop related platforms, could take the following six directions:

- Extending functionality. In particular, most of Hadoop is still backend oriented. Many customer-facing functionalities can be added to Hadoop. The challenges are to identify reusable common components.
- Improve performance with even larger scalability. There are still much room to improve response time, throughput and job execution time, especially when the

Hadoop system scales from the current hundreds to thousands of nodes to close to a quarter of million nodes in a few years.

• Enhance system utilization, from CPU, memory, disk to network utilization. This is key to maintain the low cost advantage of the Hadoop platform.

• Improve availability. Availability need to keep pace when the system scales by two orders of magnitudes, without too much negative impact on performance and utilization.

• Improve energy efficiency. This is a relatively new area and is wide open for research and innovation.

• Understanding consistency. Besides a few naïve consistency models such as strong consistency and eventual consistency, we understand little on the consistency models needed in Internet-scale data computing, as well as their implications for performance, utilization, and availability.

These six directions are not mutually exclusive. They are often inter-related. A potentially fruitful way of research is to identify one issue as the main direction, with the others as constraints. In any event, data computing in general, and Hadoop-related research in particular, will promise to be an exciting research area in the coming years.

REFERENCES

Brewer, E. A. (2000). Towards robust distributed systems. In *Proceedings of the 19th ACM Symposium on Principles of Distributed Computing*, Portland, OR (p. 7).

Cassandra (n. d.). *The apache cassandra project.* Retrieved from http://cassandra.apache.org/

Chang, F., Dean, J., Ghemawat, S., Hsieh, W. C., Wallach, D. A., Burrows, M., et al. (2006). Bigtable: A distributed storage system for structured data. In *Proceedings of the 7th USENIX Symposium on Operating Systems Design and Implementation.* (Vol. 7, pp. 205-218).

Cooper, B. F., Ramakrishnan, R., Srivastava, U., Silberstein, A., Bohannon, P., & Jacobsen, H.-A. (2006). PNUTS: Yahoo!'s hosted data serving platform. *Very Large Data Base Endowment, 1*(2), 1277–1288.

DeCandia, G., Hastorun, D., Jampani, M., Kakulapati, G., Lakshman, A., & Pilchin, A. (2007). Dynamo: Amazon's highly available key-value store. In *Proceedings of the 21st ACM SIGOPS Symposium on Operating Systems Principles* (Vol. 21, pp. 205-220).

Fan, B., Tantisiriroj, W., Xiao, L., & Gibson, G. (2009). DiskReduce: RAID for data-intensive scalable computing. In *Proceedings of the 4th Annual Workshop on Petascale Data Storage*, Portland, OR (pp. 6-10).

Hadoop (n. d.). *The apache hadoop project.* Retrieved from http://hadoop.apache.org/

Hadoop ZooKeeper. (n. d.). *The apache hadoop zookeeper project.* Retrieved from http://hadoop.apache.org/zookeeper/

HBase. (n. d.). *The apache hbase project.* Retrieved from http://hbase.apache.org/

Kim, S., Han, H., Jung, H., Eom, H., & Yeom, H. Y. (2010). Harnessing input redundancy in a MapReduce framework. In *Proceedings of the ACM Symposium on Applied Computing*, Sierre, Switzerland (pp. 362-366).

Lakshman, A., & Malik, P. (2009). Cassandra: Structured storage system over a P2P network. In *Proceedings of the 28th ACM Symposium on Principles of Distributed Computing* (p. 5).

Lakshman, A., & Malik, P. (2010). Cassandra: A decentralized structured storage system. *ACM SIGOPS Operating Systems Review, 44*(2), 35–40. doi:10.1145/1773912.1773922

Leverich, J., & Kozyrakis, C. (2009). On the energy (in)efficiency of Hadoop clusters. In *Proceedings of the Workshop on Power Aware Computing and Systems*, Big Sky, MT (pp.61-65).

Porter, G. (2010). Decoupling storage and computation in Hadoop with SuperDataNodes. *ACM SIGOPS Operating Systems Review, 44*(2), 41–46. doi:10.1145/1773912.1773923

Silberstein, A., Cooper, B. F., Srivastava, U., Vee, E., Yerneni, R., & Ramakrishnan, R. (2008). Efficient bulk insertion into a distributed ordered table. In *Proceedings of the International Conference on Management of Data* (pp. 765-778).

Stoica, I., Morris, R., Karger, D., Kaashoek, M. F., & Balakrishnan, H. (2001). Chord: A scalable peer-to-peer lookup service for internet applications. In. *Proceedings of the Conference on Applications, Technologies, Architectures, and Protocols for Computer Communications, 29*, 149–160.

Tian, C., Zhou, H., He, Y., & Zha, L. (2009). A dynamic MapReduce scheduler for heterogeneous workloads. In *Proceedings of the 8th International Conference on Grid and Cooperative Computing*, Lanzhou, China (pp. 218-224).

Vigfusson, Y., Silberstein, A., Cooper, B. F., & Fonseca, R. (2009). Adaptively parallelizing distributed range queries. *Very Large Data Base Endowment, 2*(1), 682–693.

Zaharia, M., Borthakur, D., Sarma, J. S., Elmeleegy, K., Shenker, S., & Stoica, I. (2010). Delay scheduling: A simple technique for achieving locality and fairness in cluster scheduling. In *Proceedings of the 5th European Conference on Computer Systems*, Paris, France (pp. 265-278).

Zaharia, M., Konwinski, A., Joseph, A. D., Katz, R. H., & Stoica, I. (2008). Improving MapReduce performance in heterogeneous environments. In *Proceedings of the 8th USENIX Symposium on Operating Systems Design and Implementation* (pp. 29-42).

Zou, Y. Q., Liu, J., Wang, S. C., Zha, L., & Xu, Z. W. (2010). CCIndex: A complemental clustering index on distributed ordered tables for multi-dimensional range queries. In *Proceedings of the 7th IFIP International Conference on Network and Parallel Computing*, Zhengzhou, China (pp. 247-261).

This work was previously published in the International Journal of Cloud Applications and Computing (IJCAC), Volume 1, Issue 1, edited by Shadi Aljawarneh & Hong Cai, pp. 45-61, copyright 2011 by IGI Publishing (an imprint of IGI Global).

Chapter 5
SaaS Multi-Tenancy:
Framework, Technology, and Case Study

Hong Cai
IBM China Software Development Lab, China

Berthold Reinwald
IBM Almaden Research Center, USA

Ning Wang
IBM China Software Development Lab, China

Chang Jie Guo
IBM China Research Lab, China

ABSTRACT

SaaS (Software as a Service) provides new business opportunities for application providers to serve more customers in a scalable and cost-effective way. SaaS also raises new challenges and one of them is multi-tenancy. Multi-tenancy is the requirement of deploying only one shared application to serve multiple customers (i.e. tenant) instead of deploying one dedicated application for each customer. This paper describes the authors' practice of developing and deploying multi-tenant technologies. This paper targets a technology that could quickly enable existing Java EE (Enterprise Edition) applications to be multi-tenancy enabled thus having the benefit of quick time to market. This paper describes the overall framework of multi-tenant SaaS platform, how to migrate an existing Java EE application, how to provision the multi-tenant application, and how to onboard the tenants. The paper also shows experiments which compare the economics of multi-tenant SaaS deployment versus traditional application deployment (one application for one tenant) with precise data.

1. INTRODUCTION

In recent years, we have witnessed the evolution of research on Utility Computing, Grid Computing, virtualization technology, SaaS (Wikipedia, n.d.), next generation Web technologies, and Services Computing that all together have yield a new research area named Cloud Computing (Wikipedia, n.d.). Cloud Computing promises to provide easy to use service oriented user experience to allow massive end users to use IT resources and

DOI: 10.4018/978-1-4666-1879-4.ch005

IT applications as services without on-premise installing the IT infrastructures.

A technical definition of Cloud Computing is "a computing capability that provides an abstraction between the computing resource and its underlying technical architecture (e.g., servers, storage, networks), enabling convenient, on-demand network access to a shared pool of configurable computing resources that can be rapidly provisioned and released with minimal management effort or service provider interaction." This definition states that Clouds have five essential characteristics: on-demand self-service, broad network access, resource pooling, rapid elasticity, and measured service.

This definition by default assumes the Cloud services as Infrastructure as a Service (IaaS). IaaS could be very successful in development and testing environment, because those public Cloud offering cheap and instant infrastructure services could ease the reach to IT resources. Amazon EC2 (Amazon Web Services, n.d.) is a well-known and typical public Cloud offering. On the other side, through two years practices in the Cloud market, we have observed that customers in different geographies typically have different market requirements of using Cloud services. For example, in the emerging market there are strong needs of e-commerce, accounting and ERP applications for small and medium business (SMB), so these SMBs can save investment on IT infrastructure and spend a bigger budget on their direct business operations such as marketing and businesses development. With this intent, SaaS is often their first resort. Today, some business applications and software have been developed to a stage that most of the functions are good enough to most of the customers. The ultimate goal of SaaS is to provide software to those segmentations of customers with flexible deployment (without on-premise installation) and pricing model (usually by paying monthly or annually subscription fee). However, SaaS has different maturity levels with varying capabilities and cost structures (Chong & Carraro, 2006). In its basic levels (Level 1/2), lower layer IT infrastructure resources are shared. In Level 3/4, multi-tenancy is the core feature with single instance of application serving multiple customers (tenants) to achieve the minimum operational cost, and maximize the revenue of the SaaS operator.

A key concept in SaaS multi-tenancy is "isolation point". Essentially, isolation points are artifacts or resources that need to be isolated for different tenants in shared Web application instances. An example of isolation point is a visitor counter implemented with static field. Requests to different tenants need to trigger the counters to increase their counts separately. Details of isolation points will be introduced in Section III. To serve multiple tenants, a SaaS application needs to have all its isolation points identified and isolated at runtime. Multi-tenancy is the core technology of SaaS, and we need to understand there are different options to implement multi-tenancy to enable Web applications which have Web UI, business logic, and database. One option of implementing multi-tenancy is taken by Salesforce.com who provides Web based SaaS application development based on a template of the application (such as CRM application). The problem of this option is that the application developers have to be trained to learn the new programming model, and pay more efforts to transforming those legacy applications with the new programming and runtime platform. The limitation is that there's little fine granularity customization exposed such as adding, updating, or deleting a class, a method, or a field. The second option of realizing multi-tenancy is through designing multi-tenancy entry points at design time (Osipov, Goldszmidt, Taylor, & Poddar, 2009) so that different tenants may have different service components through different entry points. This approach will allow the developers to follow the original program model they are familiar. The limit of this option is that the designer and developer of the application need to design all the artifacts that need to be isolated before implementation. The third option targets for existing Web application vendors (Guo, Sun, Huang, Wang, & Gao, 2007; Sun, Zhang, Guo, Sun, & Su, 2008; Cai, Zhang, Zhou, Cai, & Mao, 2009). In this approach, an existing Web application will be transformed to be multi-tenancy enabled. At the same time, application server runtime and database server runtime will be instrumented to route the requests for different tenants. This option could relieve the burden of application vendors who have existing Web applications with massive customers using them. These application vendors could use the SaaS multi-tenancy enabling tool and runtime to try their application in SaaS market quickly. In this way, they expect to serve more customers while still achieving SaaS operation efficiency. Our paper focuses on the third option of making multi-tenant applications. The ultimate goal of this option is to provide a transparent approach for Web application vendors.

This paper is organized as follows. In Section 2, we define the major problems we will face if we want to achieve the transparent SaaS migration approach. In Section 3, we define the conceptual model of SaaS multi-tenancy enablement. In Section 4, we walk through the end to end method, tooling, migration activities, and roles of SaaS multi-tenancy transparent enablement system. In Section 5, we give the related work.

2. PROBLEM STATEMENT

Today, Cloud and SaaS are often the business of a few very large enterprises offering well-established services, such as Google App Engine from Google, CRM application from Salesforce, and Amazon services from Amazon. There are different market categories, including massive markets, niche markets, and long tail markets. From a SaaS service provider perspective, most concerns relate to the cost structure, especially to the operation

cost of the SaaS environment. This operation cost further consists of the following parts: (1) The hardware resources required by each customer (tenant), such as server (CPU, memory), storage, and network; (2) The labor costs of SaaS operators for their daily work, especially the time spent on solving customer issues; (3) The opportunity cost of a non-available SaaS operation system, and the resulting revenue lost; and (4) Additional costs related to providing certain services related to customer-specific requirements. In this paper, we elaborate on points (1) and (2). In order to enable the sustainable growth of a SaaS market, design criteria need to direct all of SaaS service providers, SaaS service consumers, and 3rd party service providers. The goals are then to isolate the logic and data of customers' applications, then provide a certain level of customization, and finally allow an efficient operation with a significant cost reduction for SaaS operators. From a SaaS service consumer point of view, the design must allow different levels of SaaS offerings with different pricing structures. For example, very small and medium businesses can accept fine grain applications and data sharing, so long as their specific operation data is protected and isolated. But larger business entities may require individual databases or application deployment schemes.

The first problem we want to solve in this paper is the foundation of SaaS multi-tenancy which is isolation and customization for Web applications. And we assume that these applications are deployed in distributed middleware components in a product system. In Sun, Zhang, Guo, Sun, and Su (2008) we have seen methods of making application server to support multi-tenancy. But there are some limits there and the methods are extended by this paper. We know that in large-scale Web application production environment, we usually use clusters of middleware such as a cluster of Web server, a cluster of application server, and database server (hard to make itself a cluster). We not only need to make application server to support multi-tenancy, but also need to propagate tenant context information to the distributed middleware components in the Cloud.

And the second problem is how to provide separation of concerns when realizing SaaS multi-tenancy. In the whole life cycle of software engineering, there are multiple roles involved. The introduction of multi-tenancy may trigger additional issues. We need to separate the concerns of different roles so that they could focus on their own tasks without being bothered by the introduction of multi-tenancy aspect. A developer does not need to learn multi-tenancy but should focus on the application itself. The middleware provider would like to provide an add-on to the existing middleware (such as application server, database server, etc.) without maintaining two editions (one for traditional Web applications, one for multi-tenant Web applications). The SaaS/Cloud platform operator would like to minimize the effort of deploying and provisioning (multi-tenant) applications without additional deployment and administration efforts. The customers (tenants) of the multi-tenant applications may wish to do some customization through self-service approach without interference of the SaaS/Cloud operator.

3. SAAS MULTI-TENANCY ENABLING FRAMEWORK

3.1. Topology of Modern Web Applications

The production level topology of modern Web applications normally has the similar configuration as shown in Figure 1. It usually contains a cluster of Web servers, a cluster of Web application servers, a database server (could be in master/slave mode), an LDAP server, and sometimes a cluster of distributed caches.

The topology of a Web application deployment consists of Web application servers and database servers. On a Web application server, we usually have a Servlet container, as well as an EJB container. A Web application is deployed on a Web application server. A database server is usually installed on another machine devoted to database processing, and serving the role of a remote resource. Beyond application server and database server, there could exist other remote servers such as messaging server, distributed cache server, etc. For example, a remote LDAP server may provide directory access service; a remote message queue

Figure 1. Topology of a production stage Web application deployment

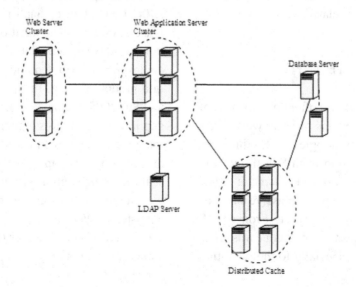

server may provide messaging service. From multi-tenancy perspective, a critical need of multi-tenancy enablement is intercepting tenant context information (such as tenant identifier) at the front end of a Web request. At the application server tier, the tenant content information will be carried by a thread and be used to route to tenant specific application server classes throughout the life cycle of the thread. The tenant context needs to be propagated to other remote resources (servers) for a Web application to complete its Web application processing.

3.2. Isolation and Customization in SaaS Multi-Tenancy

Isolation and customization are two major needs for SaaS applications to be multi-tenancy enabled. Within those two factors, isolation is the foundation of all enablement technologies, and customization provides tenant unique faces or behaviors.

1) Concept of Isolation and isolation points: An isolation point means a specific Web application artifact (class, method, or field) that has its tenant specific behavior or value. So it needs to be isolated for a specific tenant. The isolation points in standard Java Web applications could be categorized in Figure 2. The isolation points could be identified at the application transformation phase, and be stored in a metadata repository. Only based on isolation can SaaS applications provide customization capability in a consistent approach through application provisioning and configuration without being hard coded into the application.

But we want to point out in this paper that the low level isolation points (such as global variables and literals) will be handled by our JVM level SaaS multi-tenancy enabling technique automatically. And only high level isolation points (such as file

name and some UI artifacts) will be exposed and allow tenant administrator to make some simple configuration as illustrated in Section 4.

2) Customization: For example, a visitor counter on the home page of a Web application could be implemented using a tenant specific label plus the counted number using static field. Customization is handled during tenant onboarding phase. When a new tenant subscribes to this application, the tenant administrator could customize this counter's label through setting a tenant specific string.

3.3. The Architecture of SaaS Multi-Tenancy

As described in 3.1, modern Web applications are often deployed in a distributed and clustered environment. A Web application topology often consists of Web application server, database server. Sometime, a topology may also consist of message queue server, LDAP server, etc. And sometimes the application server may also connect to other remote application servers.

On the Web application server side, there are four key components to handle multi-tenancy at runtime. The first component is multi-tenancy interceptor which will be used to intercept a Web request and get the tenantID (to which tenant this Web request is sent for) from the Web request. The second component is a TenantContext object. This object is used to store the tenant context information in a thread object that will be shared during this Web request along the servlet chain (in a Java EE programming model). The third component is multi-tenant resource isolation broker (Figure 3). This component is responsible for handling isolation on the application server so that the same instance of application could be used to serve multiple tenants while at the same time their states are isolated. The last component is remote tenant context stub. This component is used to propagate

Figure 2. A list of typical (Java) Web application isolation points

	Type of isolation points	Where it is used
Type 1: Application Logic level isolation	App Scope Objects	Global variables (including static field, global objects, cache, etc.)
		Literals (constant)
		Application cache
		Servlet Context
		Servlet Context Application configuration files
	EJB	SessionBean
		EntityBean
	Authentication & Authorization	JAAS w/ global unique login name (e.g. email)
		JAAS wo/ global unique login name
		Other DB/LDAP Based security
	Remote Service Call	HTTP/REST/SOAP
		JMS
		Socket
		RMI
	File	Standard java IO interface
		Logging
	Type of isolation points	**Where it is used**
Type 2: Resource level isolation	Database	Pure JDBC
		Common Data Access Framework (e.g. JPA)
		EntityBean
	LDAP	Basic LDAP
	Message	JMS
		MDB

the tenant context object to remote servers (such as a database server) so that remote servers could also be multi-tenancy enabled.

3.4. The Conceptual Model of Multi-Tenancy in Web Programming Model

As shown in Figure 1, in production environment a Web application is often deployed in a distributed system consisting of many middleware stacks running on many physical machines or VMs in the Cloud environment.

In order to make those existing Web applications become multi-tenancy applications, we not only need to get the tenant context information, but also need to propagate the tenant context information to those remote distributed systems offering remote resources that the Web applications will use. Since there are many different commercial or open source implementation of the Web application programming model, we need a conceptual

Figure 3. Architecture of multi-tenancy framework in a distributed Web application topology

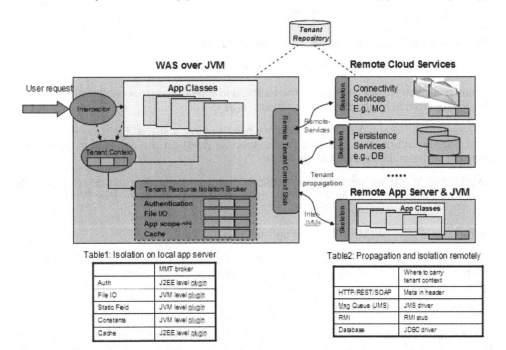

Table1: Isolation on local app server

	MMT broker
Auth	J2EE level plugin
File IO	JVM level plugin
Static Field	JVM level plugin
Constants	JVM level plugin
Cache	J2EE level plugin

Table2: Propagation and isolation remotely

	Where to carry tenant context
HTTP/REST/SOAP	Meta in header
Msg Queue (JMS)	JMS driver
RMI	RMI stub
Database	JDBC driver

model to capture the most critical and common artifacts for enabling multi-tenancy. We propose a conceptual model of multi-tenancy in Figure 4.

We could see in Figure 4 that the conceptual model of SaaS multi-tenancy includes multi-tenant interceptor, tenant context, (in memory) shared map to store tenant information, original application logic (using threads), JVM or Web container APIs, and remote servers. Multi-tenancy interceptor could be implemented with a Web filter and capture tenant information by knowing which tenant this Web request is for. Tenant context is an object to store tenant information. Tenant context could be bound with a thread and used by application business logic codes. The invocation of JVM and Web container APIs will be intercepted and added with some multi-tenancy specific pre or post operation. Tenant context information can also be propagated by a client (e.g., JDBC client) of remote service to the server (database server) of that service.

4. THE LIFECYCLE AND TECHNOLOGIES OF ENABLING SAAS MULTI-TENANCY

4.1. SaaS Application Level Multi-Tenancy

4.1.1. Application Multi-Tenancy Transformation Tool

As pointed out, the foundation of SaaS multi-tenancy is isolation. As shown in Figure 2, there are many types of isolation points. It's very hard for a developer to browse all the source code and derive that list. So we build a multi-tenancy transformation tool to facilitate this effort for SaaS application developers.

We build the SaaS multi-tenancy tooling as a simple Eclipse based plug-in that only adds a new "*isolation point*" view to the traditional Eclipse IDE (Integrated Development Environment). A snapshot of the SaaS multi-tenancy tooling is shown in Figure 5.

Figure 4. The conceptual model of SaaS multi-tenancy covering interceptor, tenant context, multi-tenant partitions, routing mechanism, and tenant propagation

We define a Web request as q, and the multi-tenant interceptor as I.

We define the tenant context for the Mth tenant as $t(M)$.

From each Web request, we derive the tenant, and bind the tenant context information to a thread T generated by the application server : $I(q) \rightarrow \bar{T}$.

For each multi-tenant Web application, we need to have a shared global map P to store tenant specific information (such as tenant specific field) on the application server. We define the partition for tenant M in the map as P(M).

We denote W(M) as some Web application business logic. We use $\Rightarrow$ to denote the transformation (routing) of an invocation from an original JVM level or Web container level API R_{jo} to a new multi-tenancy aware API R_j,

W(M) call R_{jo} $\Rightarrow$ W(M) call R_j and update P(M).

Sometime, W(M) needs to access a remote resource. It starts so by calling the local resource client C_{jo} which has been replaced with a multi-tenant version C_j on the application server,

W(M) call C_{jo} $\Rightarrow$ W(M) call C_j and update P(M).

At this time, the tenant context information should be propagated from the local resource client to the remote resource server. We denote the remote service server of type i with S_i. In the multi-tenancy scenario, $S_i(M)$ is used to denote the Mth partition for tenant M in the remote service server (e.g., a DB server).

Part 1 (marked ①) of the tooling canvas is the traditional Java Web project explorer. Part 2 and 3 (marked ② and ③ respectively) make up the "*isolation point*" view. Part 2 lists all the candidate *isolation point*s for the application developer. Part 3 consists of some useful buttons (such as "confirm" *isolation point*s, "search" *isolation point*s within the results by key words, and "export" all confirmed *isolation point*s to a metadata storage. This provides a heuristic approach to narrow down the isolation points.

4.1.2. Instrument Web Middleware to Enable Multi-Tenancy

JVM Instrumentation approach for isolating constants and fields in the applications:

The Java instrument method provides services that allow Java programming language agents to instrument programs running on the JVM. Those isolation points are defined and used by applications. So in a normal setting, the developers should be responsible for identifying these isolation points. As mentioned earlier, they could also add annotations to it so that later other roles (deployers) could use them later.

The basic idea of JVM instrumentation is to start an agent listener when the JVM initiates and then intercept these values of constants (or fields) in the application and change them according to the tenants' configuration. It follows the below steps:

Step 1: Code instrumentation

- A developer identifies isolation points in the code (literal, field, method, class) via inline annotation or external configuration
- A developer uses an agent to instrument the code based on the previous annotation, by compiling from source, rebuilding from binary, or injecting at runtime.

Figure 5. The snapshot of SaaS multi-tenancy migration tool as an Eclipse plug-in

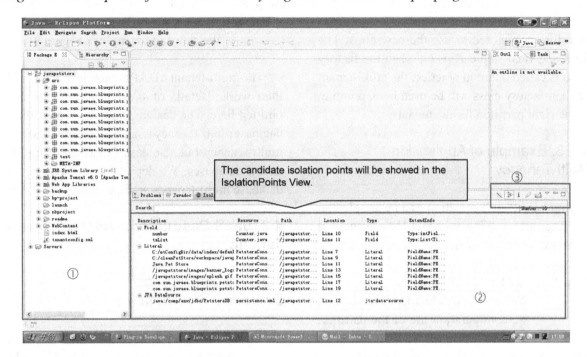

Step 2: Tenant initialization
- ○ A deployer sets up a tenant by
  - Configure literal value
  - Specify customized method/ class implementation.

Step 3: Tenant service
- ○ For each user request, resolve it to a certain tenant context
- ○ Process this request in the resolved tenant context

o Web container instrumentation (routing for container management application artifacts):

As described before, with a muli-tenant interceptor, we can get the tenant context information, and we can bind that information to a thread object. After that, any resource provider used by the application server can be wrapped as a multi-tenant resource provider which could get the tenant context information and execute some operation before or after the original resource provider. Because a Web container can be configured to use a new resource provider so now the

behavior of the Web container could be modified to "route" the Web container instructions to different tenants' space.

Some commonly used items that need to be isolated and customized in the Web applications include,

- LDAP security partition,
- Data source partition or persistence partition,
- Partition of certain indices within the application (such as user submitted images), and
- Partition of UI resources such as images under different folders.

The common method of isolation of application server-managed items consists of three steps. The first step is to set up partitions for those items and store a piece of content that belongs to a specific tenant into their partition. Here, the information is often stored in XML files. In the second phase, the listener or object factory that are the family of using those isolation items will be updated to the multi-tenant ones. The application server is therefore configured to use the new class. The third and final step carries on when the factory class is called upon; in practice, the multi-tenant version factory class will be used here, pointing to the right partition for the tenant.

4.1.3. Example of Application Multi-Tenancy

Giving LDAP authentication as an example, the multi-tenant enabling approach is executed in this way:

Step 1: Set up the LDAP partitions.
- We use the Apache LDAP server, and the configuration file of the partition is shown in Figure 6.

Step 2: Set up the runtime tenant context.
- We show an example of how to create tenant context in Figure 7.

Step 3: Change the LDAP factory class in the configuration file of Tomcat (server.xml).
- We show an example in Figure 8.

The multi-tenant LDAP authentication should then work. Details of the implementation are omitted here. The design principle of providing database multi-tenancy is not only enable database multi-tenancy but also ease the usage of multi-tenant databases. We depict the architecture of database multi-tenancy in Figure 9.

4.2. SaaS Resource Level Multi-Tenancy

4.2.1. Architecture of Database Multi-Tenancy

As shown in Figure 9, database resource pools could be built over the IaaS layer. A Web UI could be provided to manage the tenantID – Data Node mapping. At runtime, whenever there's a database

Figure 6. Example configuration file for LDAP partitions

```
<defaultDirectoryService ...>
  <systemPartition>
    <jdbmPartition id="system" cacheSize="100" suffix="ou=system" optimizerEnabled="true"
syncOnWrite="true">
      ...
    </jdbmPartition>
  </systemPartition>
  <partitions>
      <jdbmPartition id="host1" cacheSize="100" suffix="o=host1" optimizerEnabled="true"
        syncOnWrite="true">
    <indexedAttributes>
    </indexedAttributes>
  </jdbmPartition>

      <jdbmPartition id="host2" cacheSize="100" suffix="o=host2" optimizerEnabled="true"
        syncOnWrite="true">
    <indexedAttributes>
      ......
    </indexedAttributes>
  </jdbmPartition>
  ......

  <ldapService id="ldapService"
        ipPort="10389"
        allowAnonymousAccess="false"
        saslHost="ldap.example.com"
        maxSizeLimit="1000">
  ......
  </spring:beans>
```

Figure 7. Snippet of setting up runtime tenant context

```
public abstract class TenantContext{
        private static ThreadLocal<TenantContext> context = new ThreadLocal<TenantContext>();
        private static Map<String, TenantContext> hostTenantMap = null;

        public synchronized static void initialize(){
                if(hostTenantMap == null){
                        TenantConfigParser parser = new TenantConfigParser();
                        hostTenantMap = parser.parse(System.getProperties().get("catalina.base") +
File.separator + "tenantConfig");
                }
......
}
```

Figure 8. Snippet of configuration in server.xml, class "com.ibm.multitenancy.ldap.AuthenticationLda-pUtil" is used to handle LDAP process that support multi-tenancy

```
<Resource name="jndi/MultiTenency"
          auth="Container"
          type="com.ibm.multitenancy.ldap.AuthenticationLdapUtil"
          factory="com.ibm.multitenancy.ldap.LdapBeanFactory"
          connectionName="uid=admin,ou=system"
          connectionPassword="secret"
          connectionURL="ldap://localhost:10389"/>
```

Figure 9. Architecture of multi-tenant database servers

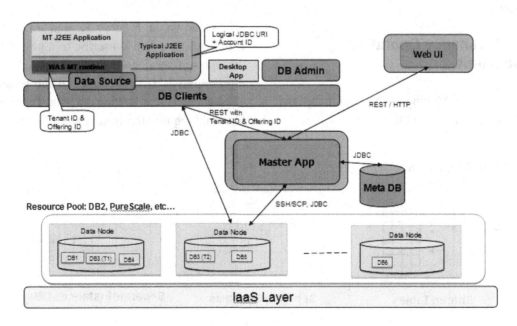

request, it will call the master application of the multi-tenant database first and make the mapping.

4.2.2. Different SLAs for Multi-Tenant Database System

Different options of database sharing relate to different cost of database deployment. The customers should have the flexibility to select and pay for the most proper SLA (Figure 10) of multi-tenant database, and should be able to switch to other options very easily. In the field of multi-tenant database, there are three options.

- **Shared tables:** typically chosen by SaaS applications it provides highest degree of multi-tenancy, but support only row-level isolation and limited customization
- **Separate schemas**: provide schema-level isolation, tenant-specific customization, backup/restore. It still share same physical database resources, has slightly higher cost for footprint (catalog) and maintenance
- **Separate instances/DBs:** Provide highest isolation, less sharing and highest cost

Database multi-tenancy should support all the options.

4.3. Multi-Tenancy Operation Management

1.   (Subscriber) Making the New Multi-tenant Application into an Offering:

An offering is different with an application in that an offering is based on an application (functional components) and contain other artifacts such as SLA and matching price (non-functional components). The 3-tuple {application, SLA, price} makes up a SaaS billing policy. A snapshot of massive multi-tenancy administration console is shown in Figure 11. The multi-tenancy administration console may cover many functions such as creating a new offering, updating an offering, etc. When creating a new offering, some compulsory information must be filled or selected such as the name of the offering, types of SLA this offering supports, whether this offering needs to be activated immediately, entry point of provisioning service (to be used to make the tenants onboarding). In the example in Figure 11, we can select how many different types of SLAs this offering can support. The types of SLA may vary from the most cheap and low resource consuming "economic" type (shared table approach), to "intermediate" type (separate schema), to "advanced" type (separate database), till the most delegated "deluxe" type (separate database instance).

2.   (Tenant Administrator) Subscribing to a Multi-tenant Offering:

As shown in Figures 12 and 13, a tenant administrator could subscribe to a new offering through self-service approach or through a client representative. If the account of the tenant has already been created, then the tenant administrator could log into the system and select the name

Figure 10. SLAs of database servers

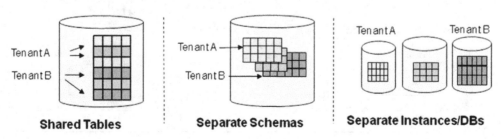

Figure 11. Making the new SaaS multi-tenant application a SaaS offering

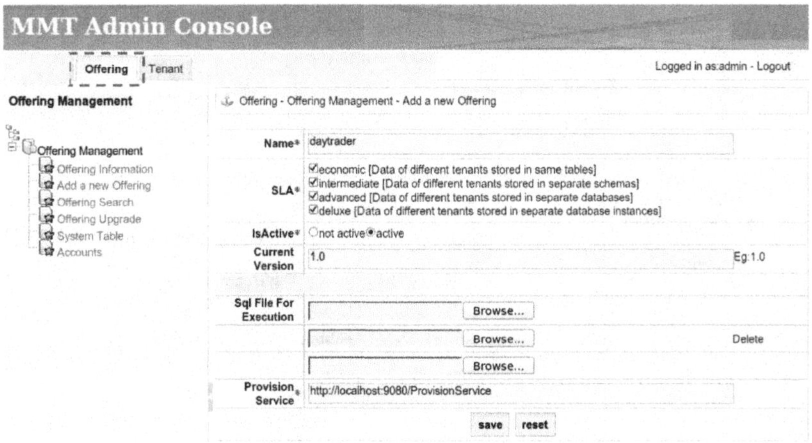

of the new offering, then select the type of SLA the tenant wants to use. At the backend, the SaaS multi-tenancy administration console could automatically provisioning necessary resources and create partitions for the new tenants of the offering.

3.   (Tenant Administrator) Customizing the Multi-tenant Application for the Tenant:

As said, the difference between a multi-tenant application with a traditional Web application comes from the introduction of the concept of "tenant". Before a tenant administrator makes tenant specific customization, a tenant cannot

Figure 12. Select a SLA for the new tenant

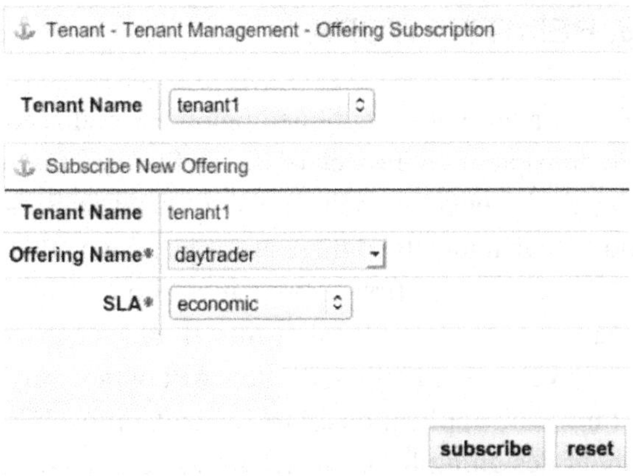

use the application. Figure 13 shows the snapshot of the UI for a tenant administrator to make customization.

The contents shown in the UI in Figure 13 vary from applications to applications. The contents will be automatically generated based on the metadata of the multi-tenant application which is derived from the application transformation phase (described in subsection 4.1.1). Not all isolation points have customization points. Only high level isolation points that have tenant specific initial or permanent values need to be exposed on this page for tenant administrator to make customization. Examples of the customization points include logo (image or literal) of a tenant, labels (literals), etc. A tenant administrator will not be bothered by low level isolation points such as static fields, literals, etc.

5. ECONOMICS OF SaaS MULTI-TENANCY

5.1. The Experimental Environment and Application

We adopt a well know Java Web application RUBiS as the testing application. RUBiS is a Java Web

Figure 13. The UI for a tenant administrator to make configuration

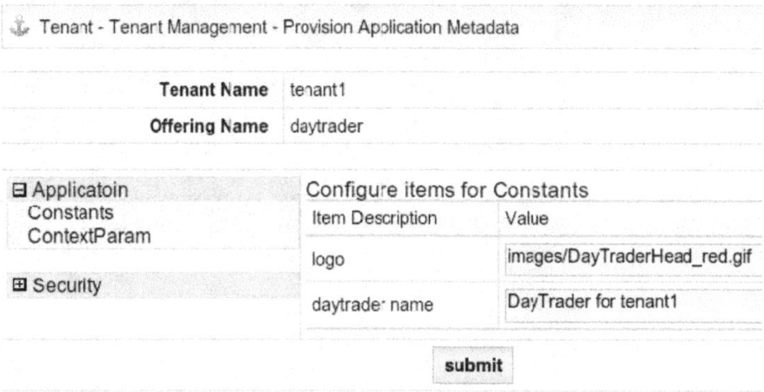

application that simulates the user behavior on eBay Web site such as browsing and ordering. The experiment environment is set below.

The RUBiS application has an application footprint of 4Mbytes, plus 90 Mbytes Hibernate cache. So it has a total of 94M application scope data. On the database side, we have a system data of about 90 MBytes (600k category records, 1.8 M region records). The tenant-specific data is about 29 Mbytes including user data, items data, bids data, transaction data, comments, etc.

We set the endurable response time to 2.5 seconds. If the response time goes beyond that, we would assume that current machine cannot support that many tenants. We use a workload of bidding mix meaning 15% read-write interactions. We assume each tenant has 100 registered users with 5% active user ratio.

We have designed three scenarios for testing:

S1: Multi-Tenant: this is the scenario of our fine granularity multi-tenancy technology. This scenario provides the maximum resource sharing.

S2: Multi-AppInstance: this is the scenario of deploying multiple copies of the same application on a same application server. This approach provides medium isolation.

S3: Multi-Stack: this is the scenario of deploying multiple middleware, and then deploying one instance of application over each middleware. This approach provides the maximum isolation.

We deploy a Web application server and a DB2 server each on a virtual machine with 1*4 cores CPU and 8G memories.

5.2. The Experimental Results

Through the experiments we have gained useful insights into the number of maximum supported tenants under each scenario. The comparison is shown in Table 1.

From Table 1, it could be seen that using RUBiS as the benchmark application, we could achieve 10 times gain of active tenants (and 100 times total tenants) on the same box compared with the deployment of multiple application instances.

6. RELATED WORK

The related work mentioned below is related to the transparent approach of SaaS (fine granularity) multi-tenancy enablement. As said in Section 1, we understand there are at least two other approaches. The first approach is about totally changing the programming model and requires the developers to follow the new API or new way of building applications. The second approach is about designing for multi-tenancy at application

Table 1. Comparison of different resource sharing approaches

System Configuration: 4 cores and 8 GB memory limit for WAS and DB2			
Tenant Characteristics: 100 registered users per tenants 5% active user ratio (5 users)			
Active tenant ratio: 10%			
	Multi-Stack	**Multi-AppInstance**	**Multi-Tenant**
# concurrent tenants (**footprint/Tenant**)	9 ◊ due to memory footprint	32 ◊ due to memory footprint	300 ◊ due to performance bottle-necks
# registered tenants	9 ◊ inactive tenants consume runtime resources	32 ◊ inactive tenants consume runtime resources	3,000 ◊ inactive tenants don't consume runtime resources
Scaling Nbr of Tenants	1x	3x	**300**
MMT Performance Over-head	./.	1x	App level: <2% DB level: <4%
Cost for 3,000 Tenants (**$/Tenant**)	333 x systems + 3,000 x middleware + 3,000 x apps + maintenance & operations	93 x systems + 93 x middleware + 3,000 x apps + maintenance & operations	1 system + 1 middleware + 1 MT app + maintenance & operations

design time in which new applications cannot be reversed to normal application deployment. So these two approaches were not comparable to the work in this paper since the method we described here refers to a transparent approach that could leverage the existing code base and at the same time allow transformation to multi-tenant application with minimum effort using the multi-tenancy transformation tool.

Database multi-tenancy maturity models and implementation methods were introduced in Chong and Carraro (2006). But Chong and Carraro (2006) only covered database multi-tenancy and didn't cover the application side multi-tenancy and didn't cover the end to end life cycle management of a Web application, specifically the enabling of multi-tenancy on the application server.

The study of comprehensive isolation and customization of SaaS multi-tenancy were introduced in Guo, Sun, Huang, Wang, and Gao (2007) and Sun, Zhang, Guo, Sun, and Su (2008). But the approach has some limits. E.g., change of source code is needed when transforming static fields in a Web application.

A complete end to end transparent SaaS multi-tenancy migration method was introduced in Cai, Zhang, Zhou, Cai, and Mao (2009). But it didn't cover the life cycle management of SaaS multi-tenant application. It mainly discussed multi-tenancy runtime environment.

This paper includes SaaS multi-tenancy enablement work in the whole life cycle of Cloud platform management which takes the latest progress of Cloud virtualization technologies.

7. CONCLUSION

The contributions of this paper are multi-folds.

First, this paper describes the key concepts in application multi-tenancy which are isolation and customization. Besides the introduction of isolation points which are foundation of application multi-tenancy we introduced the core SaaS multi-tenancy models consisting of tenant interceptor, tenant context, tenant map, tenant propagation, remote resources (such as database server, LDAP server, message queue server) in a distributed multi-tier Web system.

Second, this paper introduces the end to end process of making an existing Web application to be multi-tenancy enabled, and separating concerns of different roles involved. So in the end to end process, the application developer could still focus on the application itself, and could use the SaaS multi-tenancy enablement tool to transform the application and derive the multi-tenant metadata from the application. The SaaS operator could then provision the application and make it an offering on the SaaS platform. A tenant administrator could use the SaaS multi-tenancy life cycle management application to make simple application configurations to fit the needs of that tenant. The end users of the tenant use the application and have the feeling that they are using an application specifically deployed for the tenant.

Third, this paper puts the SaaS multi-tenancy operation in the complete life cycle of Cloud platform services. It describes the method of creating multi-tenancy enabled virtual images, creating the application model, transforming the application model into topology model, instantiating the virtual images, then activating the image and making additional installation and configuration including the multi-tenancy related actions.

REFERENCES

Amazon Web Services. (n. d.). *Amazon elastic compute cloud (Amazon EC2)*. Retrieved from http://aws.amazon.com/ec2/

Cai, H., Zhang, K., Zhou, M. J., Cai, J. J., & Mao, X. S. (2009). An end-to-end methodology and toolkit for fine granularity SaaS-ization. In *Proceedings of the IEEE International Conference on Cloud Computing* (p. 101).

Chong, F., & Carraro, G. (2006). *Architecture strategies for catching the long tail*. Retrieved from http://msdn.microsoft.com/en-us/library/aa479069.aspx

Guo, C. J., Sun, W., Huang, Y., Wang, Z. H., & Gao, B. (2007). A framework for native multi-tenancy application development and management. In *Proceedings of the 9th IEEE International Conference on E-Commerce Technology and the 4th IEEE International Conference on Enterprise Computing, E-Commerce, and E-Services* (pp. 551-558).

Osipov, C., Goldszmidt, G., Taylor, M., & Poddar, I. (2009). *Develop and deploy multi-tenant web-delivered solutions using IBM middleware: Part 2: Approaches for enabling multi-tenancy*. Retrieved from http://www.ibm.com/developerworks/webservices/library/ws-multitenantpart2/index.html

Sun, W., Zhang, X., Guo, C. J., Sun, P., & Su, H. (2008). Software as a service: Configuration and customization perspectives. In *Proceedings of the IEEE Congress on Services Part II* (pp. 18-25).

Wikipedia. (n. d.). *Cloud computing*. Retrieved from http://en.wikipedia.org/wiki/Cloud_computing

Wikipedia. (n. d.). *Software as a service*. Retrieved from http://en.wikipedia.org/wiki/Software_as_a_service

This work was previously published in the International Journal of Cloud Applications and Computing (IJCAC), Volume 1, Issue 1, edited by Shadi Aljawarneh & Hong Cai, pp. 62-77, copyright 2011 by IGI Publishing (an imprint of IGI Global).

Chapter 6

Cloud Computing in Higher Education:
Opportunities and Issues

P. Sasikala
Makhanlal Chaturvedi National University of Journalism and Communication, India

ABSTRACT

Cloud Computing promises novel and valuable capabilities for computer users and is explored in all possible areas of information technology dependant fields. However, the literature suffers from hype and divergent definitions and viewpoints. Cloud powered higher education can gain significant flexibility and agility. Higher education policy makers must assume activist roles in the shift towards cloud computing. Classroom experiences show it is a better tool for teaching and collaboration. As it is an emerging service technology, there is a need for standardization of services and customized implementation. Its evolution can change the facets of rural education. It is important as a possible means of driving down the capital and total costs of IT. This paper examines and discusses the concept of Cloud Computing from the perspectives of diverse technologists, cloud standards, services available today, the status of cloud particularly in higher education, and future implications.

INTRODUCTION

The birth of the web and e-commerce has led to the networking of every human move and thus the personal lives started moving online. Today Internet has become a platform to mobilize the entire human society. Enormous data are required to be processed every day and therefore it requires too many hard wares and soft wares at every individual level. This leads towards high cost and increase in pollution. To reduce cost and inculcate green environment concept, attention is required to hold the pool of data accessed and process the same. Hence, reshaping the data center and evolving into new paradigms to perform large

DOI: 10.4018/978-1-4666-1879-4.ch006

scale distributed computing is the need of the hour (Magoules, Pan, Tan, & Kumar, 2009). An infrastructure for storage and computing on massive data and to pay for what you want, advance into a realistic solution, to centralize the data and carry out computation on the super computer with unprecedented storage and computing capability. Gartner, Inc., defines the solution as Cloud Computing, a style of computing where massively scaleable IT-enabled capabilities are delivered 'as a service' to external customers using Internet technologies (Gartner, 2010). Cloud computing is an important development on par with the shift from mainframe to client-server based computing.

McKinsey suggests that using clouds for computing tasks promises a revolution in IT similar to the internet and World Wide Web. Burton Group concludes that IT is finally catching up with the Internet by extending the enterprise outside of the traditional data center walls. Writers like Nicholas Carr argue that a so-called big switch is ahead, wherein a great many infrastructure, application, and support tasks now operated by enterprises will-in the future-be handled by very-large-scale, highly standardized counterpart activities delivered over the Internet. Cloud computing is also potentially a much more environmentally sustainable model for computing. The ability to locate cloud resources anywhere frees up providers to move operations closer to sources of cheap and renewable energy.

As higher education faces budget restrictions and sustainability challenges, one approach to relieve these pressures is cloud computing. With cloud computing, the operation of services moves "above the campus," and an institution saves the upfront costs of building technology systems and instead pays only for the services that are used. As capacity needs rise and fall, and as new applications and services become available, institutions can meet the needs of their constituents quickly and cost-effectively. In some cases, a large university might beam a provider of cloud services. More often, individual campuses will obtain services from the cloud. The trend toward greater use of

mobile devices also supports cloud computing because it provides access to applications, storage, and other resources to users from nearly any device. While cost savings and flexibility are benefits to the use of cloud computing, the downside of such service adoption could include possible risks to privacy and security. But ultimately cloud computing could provide a means to stretch limited resources and make them more useful, to more people, more of the time. The growing breadth of institutional sourcing options requires IT leaders to evaluate more options and providers. As technologies like virtualization and cloud computing assume important places within the IT landscape, higher education leaders will need to consider which institutional services they wish to leave to consumer choice, which ones they wish to source and administer "somewhere else," and which services they should operate centrally or locally on campus. One important option is the development of collaborative service offerings among colleges and universities. Yet, substantial challenges raise at least some near-term concerns including risk, security, and governance issues; uncertainty about return on investment and service provider certification; and questions regarding which business and academic activities are best suited for the cloud.

The common perception of infrastructure that must be bought, housed, and managed has changed drastically. Institutions are now seriously considering alternatives that treat the infrastructure as a service rather than an asset and, are not bothered about where the infrastructure is located and who manages it. A key differentiating element of a successful information technology is its ability to become a true, valuable, and economical contributor to cyber infrastructure (Foster & Kesselman, 2004). Cloud computing embraces cyber infrastructure, and builds upon decades of research in virtualization, distributed computing, grid computing, utility computing, and, more recently, networking, web and software services (Cloud Portal, 2010). Cloud computing is a next natural step of integration of current diverse technologies and applications. The literature asserts that cloud computing in higher education is different and it is important.

Information technologists are skeptical about hype. Most of the people in this segment have heard of, tried, used service bureaus, application hosts, grids, and other sourcing techniques in higher education. But what is different about the cloud? The first key difference is its technical aspects. The maturity of standards throughout the stack, the widespread availability of high-performance network capacity, virtualization technologies are combining to enrich the sourcing options in higher education (Geelan, 2009). As Service Providers and Users are quite different, the generation raised on broadband connections, Google search, and Facebook community are likely to grip the idea of cloud-based services in higher education. Such users, who are raising the sales of netbooks, are likely to move towards lower cost lightweight computing, web delivered services, open-source operating systems and applications. According to Gartner, "The consumerization of IT is an ongoing process that further defines the reality that users are making consumer-oriented decisions before they make IT department-oriented decisions" (Gartner, 2010). The consumerization of IT along with the emergence of SaaS and other web based service options will force way in higher education. At the same time, the focus on managing IT costs and return on investment are driving commercial enterprises to move swiftly. The top two trends identified in higher education software survey were SaaS and web services/SOA (INTEROP, 2008).

Finally, recognizing these technical, generational consumer, and enterprise economic trends, developer communities and system integrators in higher education are shifting away from established software vendors, and the established vendors are working to "cloud-enable" their products (INTEROP, 2008). McKinsey & Company (2010) suggests that using clouds for computing tasks promises a revolution in IT similar to web and e-commerce (Uptîme Institute, 2009). Burton Group (2010) concludes that, IT is finally catching up with the Internet by extending the enterprise outside of the traditional data center walls. According to Nicholas Carr, the "Big Switch" is ahead, wherein a great many infrastructure, application, and support tasks now operated by enterprises will be handled by very-large-scale, highly standardized counterpart activities delivered over the Internet (Carr, 2008). As the topic on cloud computing in higher education has become the central focus point among researchers we felt a dire need towards a review of the topic (Rewatkar & Lanjewar, 2010). In this paper a thorough review of the literature on the topic of Cloud Computing has been attempted by us. A framing of the roles that Higher Education might play in this emerging area of activity has also been thoroughly analysed. It also explores what shape a higher education cloud might take and identifies opportunities and models.

CLOUD HIGHER EDUCATION

Cloud Computing: The Concept and Definition

In recent times the most discussed topic and the next emerging revolutionary application anticipated is all about cloud computing and it's utility mainly in the higher education sector. However a thorough search over the web portals about cloud computing in general and more specifically in higher education leaves us highly excited as well as equally confused (Google, 2010). These are common phenomenon observed with things that are new, things that promise to transform, and things with ambiguous names. McKinsey & Company (2010) uncovered 22 distinct definitions of cloud computing from well known experts. But one of the biggest problems we have in IT is the vagueness and lack of precision in all of our work around these complex topics. A better accepted definition of Cloud computing is of Gartner's (2010) that defines it as a style of computing where scalable and elastic IT capabilities are provided as a service to multiple customers using Internet technologies. This characterizes a model in which providers deliver a variety of IT-enabled capabilities to consumers. Cloud-based services can be exploited in a variety of ways to develop an application or a solution. Using cloud resources one can rearrange and reduce the cost of IT solutions. Enterprises will act as cloud providers and deliver application, information or business process services to customers and business partners.

According to NIST cloud computing is a model for enabling convenient, on-demand network access to a shared pool of configurable computing resources (e.g., networks, servers, storage, applications, and services) that can be rapidly provisioned and released with minimal management effort or service provider interaction (National Institute of Standards and Technology, 2010). This looks to be a clear definition as it is Internet-based computing, whereby shared resources, software, and information are provided to computers and other devices on demand, just like the electricity grid existing today. In general, the concept of cloud computing can incorporate various computer technologies, including web infrastructure, Web 2.0 and many other emerging technologies.

The key technological hypes about cloud computing in higher education are:

- On-demand self-service: A consumer can unilaterally provision computing capabilities, such as server time and network storage, as needed automatically without requiring human interaction with each service provider.
- Broad network access: Capabilities are available over the network and accessed through standard mechanisms that promote use by heterogeneous thin or thick

client platforms (e.g., mobile phones, laptops, and PDAs).

- Resource pooling: The provider's computing resources are pooled to serve multiple consumers using a multi-tenant model, with different physical and virtual resources dynamically assigned and reassigned according to consumer demand. There is a sense of location independence in that the customer generally has no control or knowledge over the exact location of the provided resources but may be able to specify location at a higher level of abstraction (e.g., country, place, or datacenter). Examples of resources include storage, processing, memory, network bandwidth, and virtual machines.
- Rapid elasticity: Capabilities can be rapidly and elastically provisioned, in some cases automatically, to quickly scale out and rapidly released to quickly scale in. To the consumer, the capabilities available for provisioning often appear to be unlimited and can be purchased in any quantity at any time.
- Measured Service: Cloud systems automatically control and optimize resource use by leveraging a metering capability at some level of abstraction appropriate to the type of service (e.g., storage, processing, bandwidth, and active user accounts). Resource usage can be monitored, controlled, and reported providing transparency for both the provider and consumer of the utilized service.

McKinsey & Company (2010) presented a typology of software-as-a-service (SaaS) depicting it through Delivery Platforms like Managed hosting contracting with hosting providers to host or manage an infrastructure (for example, IBM, OpSource), Cloud computing using an on-demand cloud-based infrastructure to deploy an infrastructure or applications (for example,

Amazon Elastic Cloud), Development Platforms like Cloud computing—using an on-demand cloud-based development environment to provide a general purpose programming language (for example, Bungee Labs, Coghead), Application-Led Platforms like SaaS applications—using platforms of popular SaaS applications to develop and deploy application (for example, Salesforce. com, NetSuite, Cisco-WebEx).

It implies a service-oriented architecture, reduced information technology overhead for the end-user, greater flexibility, reduced total cost of ownership, on-demand services and many other things. As per Wikipedia (2010), cloud computing describes a new supplement, consumption and delivery model for IT services based on Internet, and it typically involves the provision of dynamically scalable and often virtualized resources as a service over the Internet. The evolution of on-demand information technology services, products based on virtualized resources have been around for some time now (Averitt et al., 2007), but the term became popular in October 2007 when IBM and Google announced a collaboration in that domain (Bell, 2008; Bulkeley, 2007). This was followed by IBM's announcement of the "Blue Cloud" effort (Kirkpatrick, 2007). Since then, everyone is talking about "Cloud Computing". Certainly, there are many ways to look at cloud computing but the benefits need to be qualified in order to be quantified. Recently the iPhone has become very popular since it is in essence a cloud computing oriented device.

CLOUD COMPUTING SERVICES IN HIGHER EDUCATION

Cloud Computing Services in higher education vary depending on the service level via the surrounding management layer. It may be Software as a Service (SaaS), Platform as a Service (PaaS), Infrastructure as a Service (IaaS), or Data Storage as a Service (DaaS).

Cloud Software as a Service (SaaS): The capability provided to the consumer is to use the provider's applications running on a cloud infrastructure. The applications are accessible from various client devices through a thin client interface such as a web browser (e.g., web-based email). The consumer does not manage or control the underlying cloud infrastructure including network, servers, operating systems, storage, or even individual application capabilities, with the possible exception of limited user-specific application configuration settings. This will lead to end of a traditional, on-premises software. The functional interface makes End user interaction with the Application's function management Metering and Billing based on number of users. E.g. Application services like SalesForce.com.

Cloud Platform as a Service (PaaS): The capability provided to the consumer is to deploy onto the cloud infrastructure consumer created or acquired applications created using programming languages and tools supported by the provider. The consumer does not manage or control the underlying cloud infrastructure including network, servers, operating systems, or storage, but has control over the deployed applications and possibly application hosting environment configurations. This provides an independent platform or middleware as a service on which developers can build and deploy customer application. Common solutions provided in this tier range from APIs and tools to database and business process management system, to security integration, allowing developers to build applications and run them on the infrastructure that cloud vendors owns and maintains. Examples - Microsoft windows azure platforms services, Google apps. The functional interface interacts with Application development and Deployment environment Management, Manage scale out of Application, Metering and Billing based on application QoS. E.g. Application Infrastructure services like Force.com.

Cloud Infrastructure as a Service (IaaS): The capability provided to the consumer is to provision processing, storage, networks, and other fundamental computing resources where the consumer is able to deploy and run arbitrary software, which can include operating systems and applications. The consumer does not manage or control the underlying cloud infrastructure but has control over operating systems, storage, deployed applications, and possibly limited control of select networking components (e.g., host firewalls). This primarily compasses the hardware and technology for computing power, storage, operating systems or other infrastructure, delivered as off premises, on-demand services rather than dedicated as on-site resources. Because customers can pay for exactly the amount of service they use, like for electricity or water, this service is also called utility computing. Examples – Amazon elastic compute cloud (Amazon EC 2) or Amazon simple storage service (Amazon S3), Eucalyptus open-source cloud computing system. The functional interface makes Virtual machine for hosting OS based stacks Management: Manage life cycle of guest machines, Metering and Billing based on infrastructure usage. E.g. System Infrastructure services like VMWARE V-CLOUD.

Cloud Data Storage as a Service (DaaS): Delivery of virtualized storage on demand. By abstracting data storage behind a set of service interfaces and delivering it on demand, a wide range of actual offerings and implementations are possible. The only type of storage that is excluded from this definition is that which is delivered, not based on demand, but on fixed capacity increments. Storage as a Service is a business model in which a large company rents space in their storage infrastructure to a smaller company or individual. Storage as a Service is generally seen as a good alternative for a small or mid-sized business that lacks the capital budget and/or technical personnel to implement and maintain their own storage infrastructure. The functional interface includes Data storage interfaces used by any of the other types Management, Data Requirements and Storage Usage.

CLOUD COMPUTING MODELS IN HIGHER EDUCATION

The Cloud Computing models are categorized based on the targeted group using the cloud service. They can be grouped as:

- **Private cloud:** The cloud infrastructure is operated solely for an organization. It may be managed by the organization or a third party and may exist on premise or off premise. (cloud enterprise owned or leased)
- **Community cloud:** The cloud infrastructure is shared by several organizations and supports a specific community that has shared concerns (e.g., mission, security requirements, policy, and compliance considerations). It may be managed by the organizations or a third party and may exist on premise or off premise.
- **Public cloud:** The cloud infrastructure is made available to the general public or a large industry group and is owned by an organization selling cloud services. The resources are dynamically provisioned on a fine-grained, self-service basis over the Internet, via web applications/web services, from an off-site third-party provider who shares resources and bills on a fine-grained utility computing basis.
- **Hybrid cloud:** The cloud infrastructure is a composition of two or more clouds (private, community, or public) that remain unique entities but are bound together by standardized or proprietary technology that enables data and application portability (e.g., cloud bursting for load-balancing between clouds). A hybrid cloud environment consisting of multiple internal and/or external providers "will be typical for most enterprises".

PERSPECTIVES OF CLOUD COMPUTING IN HIGHER EDUCATION

People may have different perspectives from different views. For example, from the view of end-user, the cloud computing service moves the application software and operation system from desktops to the cloud side, which makes users be able to plug-in anytime from anywhere and utilize large scale storage and computing resources. On the other hand, the cloud computing service provider may focus on how to distribute and schedule the computer resources. Enterprises will act as cloud providers and deliver application, information or business process services to customers and business partners.

A user of the service doesn't necessarily care about how it is implemented, what technologies are used or how it's managed. Only that there is access to it and has a level of reliability necessary to meet the application requirements. In essence this is distributed computing. An application is built using the resource from multiple services potentially from multiple locations. But the difference is, the endpoint to access the services has to be known to the user whereas in cloud it provides the user available resources. Behind this service interface is usually a grid of computers to provide the resources. The grid is typically hosted by one company and consists of a homogeneous environment of hardware and software making it easier to support and maintain. Once you start paying for the services and the resources utilized it becomes utility computing. Cloud computing really is accessing resources and services needed to perform functions with dynamically changing needs. An application or service developer requests access from the cloud rather than a specific endpoint or named resource. The cloud manages multiple infrastructures across multiple organizations and consists of one or more frameworks overlaid on top of the infrastructures tying them together.

The cloud is a virtualization of resources that maintains and manages itself. There are of course people to keep resources like hardware, operating systems and networking in proper order. But from the perspective of a user or application developer only the cloud is referenced.

STANDARDS REQUIRED IN HIGHER EDUCATION

In this section we explore the readiness of various standards, gaps and opportunities for improvement in higher education. The standards must cover many areas such as Interoperability, Security, Portability, Governance, Risk Management, Compliance, etc.

National Institute of Standard and Technology (NIST) (2010) USA has initiated activities to promote standards for cloud computing. To address the challenges and to enable cloud computing, several standards groups and industry consortia are developing specifications and test beds.

Some of the existing Standards and Test Bed Groups are:

- Cloud Security Alliance (CSA)
- Distributed Management Task Force (DMTF)
- Storage Networking Industry Association (SNIA)
- Open Grid Forum (OGF)
- Open Cloud Consortium (OCC)
- Organization for the Advancement of Structured Information Standards (OASIS)
- TM Forum
- Internet Engineering Task Force (IETF)
- International Telecommunications Union (ITU)
- European Telecommunications Standards Institute (ETSI)
- Object Management Group (OMG)

On the other side, a cloud API provides either a functional interface or a management interface (or both). Cloud management has multiple aspects that can be standardized for interoperability. Some Possible Standards are:

- Federated security (e.g. identity) across Clouds
- Metadata and data exchanges among Clouds
- Standards for moving applications between Cloud platforms
- Standards for describing resource/performance capabilities and requirements
- Standardized outputs for monitoring, auditing, billing, reports and notification for Cloud applications and services
- Common representations (abstract, APIs, protocols) for interfacing to Cloud resources
- Cloud-independent representation for policies and governance
- Portable tools for developing, deploying, and managing Cloud applications and services
- Orchestration and middleware tools for creating composite applications across Clouds

There is an urgent need to define minimal standards to enable cloud integration, application and data portability. It is required to avoid specifications that will inhibit innovation and need to separately address different cloud models.

CLOUD COMPUTING IN INDIA

India is globally known for its strengths in innovation in IT services and associated models and cloud computing is an emerging opportunity in this space. India has always been a playground and a test bed to pilot IT strategic adoption techniques. Indian Subcontinent is a very unique and a potent geography for platform vendors. No other geography will give the platform vendor access to the whole ecosystem. This market has a huge, untapped potential at every level, be it enterprise or public sector. System integrators such as Microsoft, IBM, Wipro, Infosys and TCS are busy assessing the opportunity and creating the relevant service offerings.

Opportunities

By 2030, the population of India will be largest in the world estimated to be around 1.53 billion. India's current population is about 1.15 billion and about 70% of it resides in the rural areas and villages. Thus India has a great potential to make it an economic as well as an IT superpower (India Online, 2010). Obama Administration recently termed India as a great and emerging global power. Also global economic fortunes and global ambitions make it a potential power. But the major hindrance in this direction is the lack of infrastructure for the development of the technical know-how amongst the people living in the rural areas and the villages. With the introduction of the new cloud computing paradigm these problems can be easily eliminated because it doesn't require the end users to have any type of infrastructure, as all of them are delivered as services (whether it be infrastructure as a service (IaaS), Platform as a service (PaaS), Software as a service (SaaS)) on a pay per-cycle basis (utility computing) virtually which makes it easier and cheaper for the people living in rural areas to actively involve themselves in the IT sector.

Present Status

Indian businesses are definitely adopting cloud computing, but it's still in a budding phase. The decision makers have to understand the need of IaaS, PaaS and SaaS for their organization and then adapt to public, private, or hybrid clouds. Cloud vendors take India seriously as India hasn't hit the saturation levels yet. It is understood that TCS, Infosys and Wipro amongst others are taking steps towards making cloud-based services available to their customers. With India poised to achieve massive growth in cloud computing, mature markets in the region are nurturing early adopters while developing markets are presenting many green field opportunities for cloud vendors. We need to work on Evaluating the business case for public, private and hybrid cloud models; Developing an enterprise integration and migration strategy towards cloud provisioning; Optimising the management of virtualized environment and cloud implementation; Tracking developments in cloud security, governance and standards; and Learning lessons from recent SaaS, PaaS and IaaS implementation.

Government and Enterprises

Cloud computing holds the potential for the Indian government to offer better services while adding a green touch to its e-governance enabled transformation. Cloud computing holds the promise to transform the functioning of governments. In www.apps.gov, the United States administration has taken a definitive stride to infuse cloud computing paradigm into its Enterprise Architecture. The government (both central as well as state) and the public sectors have to understand the benefits of the Cloud in a right direction. By setting up a private cloud, state governments can gain access to virtually unlimited, centralized computing. Through this, they can save cost by limiting the servers and maintenance in the local data centers. Cloud printing has qualitative advantages such as reduced worker frustration and productivity loss resulting from searches for enabled network printers; increased productivity, especially for mobile and remote workers; the ability to deliver anywhere and then print the latest file version at the last minute; enabling non-employees to print to selected corporate printers etc. The Small and Medium Enterprise (SME) may use public SaaS and public clouds and minimise growth of data centres; Large enterprise data centres may evolve to act as private clouds; Large enterprises may also

use hybrid cloud infrastructure software to leverage both internal and public clouds; and Public cloud may adopt standards in order to run workloads from competing hybrid cloud infrastructure.

Challenges

To realise the full potential of cloud computing and to be mainstream member of IT portfolio and choices, the challenges has to be met. There is a lot of challenges to be tackled related to privacy and security and associated regulations compliance, vendor lock-in and standards, interoperability, latency, performance and reliability concerns, besides supporting R&D and creating specific test beds in public-private partnership. It further enhances scientific and technological knowledge on all related foundation elements of cloud computing. The role of academic institutes to subscribe to Cloud Services that provide student / teacher / parent collaboration on subscription, is a massive and important transition needed at this hour. Cloud computing could add a new dimension to India's ongoing e-governance program. Certain preparatory steps could be initiated by the Government of India to launch cloud computing as a model for e-governance programs. These are as follows: Setup a nodal agency for cloud computing; Create pilot solutions and demonstrate their success; Develop a legal framework and risk management program; and Creating a solution portfolio for cloud migration. State governments and their departments are at varying levels of e-governance maturity. As a result, citizens and businesses get varying degrees of accessibility and quality of government services across India. Usage of cloud computing can ensure the reach of citizen services in all states irrespective of their present e-governance readiness.

CLOUD IN HIGHER EDUCATION

IT is a critical component of modern higher education. Despite miraculous improvements in price and performance, total IT costs in higher education seem destined to remain on an upward trajectory, in part because of the voracious demands of researchers for bandwidth and computing power and of students for sound and video-intensive applications. Equally to blame for higher education's IT cost management challenge may be higher education's long tradition of building its own systems and tendency to self-operate almost everything related to IT. Growing external expectations require higher education to sharpen its focus on its core mission and competencies. The unsustainable economics of higher education's traditional approaches to IT, increased expectations and scrutiny, and the growing complexity of institutional operations and governance call for a different modus operandi. So too does the mass consumerization of services, for which students and faculty are more likely to look outside the institution to address their IT needs and preferences. Cloud computing represents a real opportunity to rethink and re-craft services in higher education. Among the greatest benefits of scalable and elastic IT is the option to pay only for what is used. Robust networks coupled with virtualization technologies make less relevant where work happens or where data is stored. Cloud computing allows the flexibility for some enterprise activities to move above campus to providers that are faster, cheaper, or safer and for some activities to move off the institution's responsibility list to the "consumer" cloud (below campus), while still other activities can remain in-house, including those that differentiate and provide competitive advantage to an institution.

The cloud is no longer just a concept. Commercial cloud computing already encompasses an expanding array of on-demand, pay-as-you-go infrastructure, platform, application, and software services that are growing in complexity and capability. The flexibility the cloud offers coupled

with the mounting economic pressures, along with the massive unbundling and commoditization taking place in IT and a variety of industries, are prompting higher education leaders to consider new sourcing arrangements. While some leaders are acclimating to this new IT environment and testing the marketplace with ventures ranging from computing cycles and data storage to student e-mail, disaster recovery, or virtual computing labs, most remain cautious observers as they assess its potential impact.

The major hurdle for development of IT related education in the rural areas is the lack of institutes with proper infrastructure. To tap the maximum potential of the rural India it is very important that these IT study institutes be located in the rural areas itself with proper tools such as proper applications, infrastructure and development platforms. The difficulty lies in the huge amount of money spent on buying software licenses, setting up proper infrastructure required for computation, storage etc.

The evolution of Cloud Computing can change the facets of rural area, through the three fundamental concepts namely IaaS, PaaS and SaaS. Expenses on software licenses shall be reduced by pay-per cycle basis whether it would be software development packages or working platforms. Instead of setting up huge and expensive infrastructure such as high speed processing computers or huge data storage devices, they can use these resources from the cloud providers.

The cloud computing environment will lead to better skilled people in rural areas and villages. These technocrats from the rural areas will involve themselves in the IT sector to empower the technology development. Hence, the maximum potential of the rural India can be utilized. The spread of rural technologies will be facilitated if they are also employment generators. This scenario will surely lead to the increase in the standard of living of the rural people and convert a simple and poor economy into modern and high-income economy. Economic development is the social and technological progress of any nation. Cloud computing may give an extensive growth to the Economy of rural areas by providing IT opportunities to the people which will lead to efficient business management. The income from the business will provide better funds for the development of technical institutes in the rural areas which will result into more number of technical people. For academia, cloud computing lets students, faculty, staff, administrators, and other campus users access file storage, e-mail, databases, and other university applications anywhere, on-demand.

Cloud Class Room: Cloud Computing also finds applications in the classroom teaching-learning process. Tools like Google Docs, Microsoft Office Live Workspace, and Zoho Office Suite are already in usage. Online office suites like these typically include word processor, spreadsheet, and presentation functionalities which users can utilize to create and edit documents completely online with collaboration capabilities between geographically separated users. Cloud computing may just be a buzzword today, but classroom experience with Google Docs, it was shown that it offers a new and better tool for teaching and collaboration. On the other end IBM and Google are each shelling out between $20 million and $25 million to start college programs focused on cloud computing. The vision goes like this: Run multiple data centers in parallel and allow users to share resources. Microsoft, Sun Microsystems, Hewlett-Packard and others all have a similar vision of computing in the cloud. IBM and Google will at first offer 400 computers to teach cloud computing techniques. The duo plans to expand to 4,000. So far, six universities–University of Washington, Carnegie Mellon, MIT, Stanford, University of California at Berkeley and the University of Maryland–are participating (Young, 2008).

Cloud Library: We also view on these new services at increasing the value of the subscriptions it offers to library members. It eliminates many of the redundancies inherent in the current patterns of library automation and allows libraries to take advantage of Web-scale efficiencies. Visits to libraries, focus groups, and over a decade of engagement in the library automation world have convinced us that libraries require less complexity in their management systems. Libraries spend a great deal of time on repetitive tasks, such as cataloging best-sellers, while ignoring the most valuable aspects of their collections: the archives, the rare items, the unique collections. Libraries must transfer effort into higher value activity and embrace the web as the primary technology infrastructure. Some are already using the cloud in the form of GoogleDocs. Finally, Cloud computing will make the library anywhere and anytime for the user. The cloud has emerged and libraries need to start thinking about how they may need to adjust services in order to effectively adapt to how users are interacting with it.

Benefits of Cloud in Higher Education

The prospect of a maturing cloud of on-demand infrastructure, application, and support services is important as a possible means of driving down the capital and total costs of IT in higher education; facilitating the transparent matching of IT demand, costs, and funding; scaling IT; fostering further IT standardization; accelerating time to market by reducing IT supply bottlenecks; countering or channeling the ad hoc consumerization of enterprise IT services; increasing access to scarce IT talent; creating a pathway to a $24 \times 7 \times 365$ environment; enabling the sourcing of cycles and storage powered by renewable energy; increasing interoperability between disjointed technologies between and within institutions; and facilitate inter-institutional collaboration. In 2009, National Science Foundation (NSF), USA announced $ 5 million grants to 14 leading US universities through its Cluster Exploratory (CLuE) programme to participate in the IBM/Google Cloud Computing University Initiative. Indian Universities should also be given such grants to enable Cloud Computing in Higher Education. This may lead to usage of scarce resources as services by institutions. A higher education cloud might act as a repository for modular courses that institutions can use or build on, making it possible to reduce redundancies. We need to come together in groups to optimize our strength and not simply to determine how to bridge the gap. This is the right time for this conversation because of our necessity to take advantage of emerging technologies to change how we do business on campus, and that includes looking at a higher education solution for maximizing the benefits of cloud computing that is inevitable.

FUTURE OF CLOUD COMPUTING

Cloud computing, from becoming a significant technology trend in 2010, there is a wide spread consensus amongst industry observers that it is ready for noticeable deployment in 2011 and is expected to reshape IT processes and IT market places in the next 3 years. This implies that in the near future there would be a requirement for professionals in this field. As companies increasingly depend more on blogs/ online document storage or other web based applications, enterprising youngsters can actually set up a business to help people set-up these applications. Thus while there would be bigger players like Amazon, Google, IBM, Microsoft, Yahoo, who would need such professionals in the field of cloud computing, the smaller players too would need fresh talent. While these companies invest heavily to make cloud computing mainstream, it is the nimble startups like Nivio who rush to take advantage to ever cheaper cloud computing infrastructure to deliver innovative applications. The undeniable consensus is that cloud computing is going to be with us for a number of years. One thing that stands as a testament to the financial industry is that it is able

to work in every industry and translate problems and obstacles into bridges towards success.

Future will be filled with services either at Management level or at Functional level. Users will be at one end, Service Providers at others end and the Service Managers or the middle layer dealers will help out gluing both. Despite its possible security and privacy risks, Cloud Computing has six main benefits that the public sector and government IT organizations are certain to want to take the advantages. They are:

- Reduced Cost: Cloud technology is paid incrementally, saving organizations money.
- Increased Storage: Organizations can store more data than on private computer systems.
- Highly Automated: No longer do IT personnel need to worry about keeping software up to date.
- Flexibility: Cloud computing offers much more flexibility than past computing methods.
- More Mobility: Employees can access information wherever they are, rather than having to remain at their desks.
- Allows IT to Shift Focus: No longer having to worry about constant server updates and other computing issues, government organizations will be free to concentrate on innovation.

The major decisions facing successful implementation of cloud technologies is whether to use a solution providers cloud or bring the cloud inside and oversee the process internally. Most of the industries/companies will re-label their products as cloud computing resulting in a lot of marketing innovation on top of real innovation. A sudden transformational change is poised to succeed where so many other attempts to deliver on-demand computing to anyone with a network connection have failed. Some skepticism is warranted. Developing the business strategy and technical migration towards cloud services is the order of the day. We can alter the environment in which we operate, we can shape the higher education cloud, but we will begin to lose control to do so if we don't start now.

CONCLUSION

Cloud computing overlaps some of the concepts of distributed, grid and utility computing. However it does have its own meaning if contextually used correctly. The conceptual overlap is partly due to technology changes, usages and implementations over the years. Cloud Computing in higher education built on decades of research in virtualization, distributed computing, utility computing, and, more recently, networking, web and software services. It implies a service-oriented architecture, reduced information technology overhead for the end-user, great flexibility, reduced total cost of ownership, on-demand services and many other things. But Cloud computing in higher education has become the new buzz word driven largely by marketing and service offerings from big corporate players like Google, IBM and Amazon. As information becomes even more on-demand and mobile, cloud computing is likely to grow in higher education. Is Cloud Computing limited by the availability of internet? Will Cloud computing actually work for the 'unconnected'? This has to be made very well comprehensible. The clear concept and definition of Cloud Computing by the experts have paved the way for people to explore the Giant Transition in higher education. The services that shall be provided by cloud, the models on which it can be deployed, and the different dimensional requirement for the user and the service provider will lead to a contemporary era in higher education. The outlook for Cloud Computing in India and more particularly in Higher Education, needs a make shift to the cloud. Definitely the future implications in higher education are scalable and expansive, and this cheap, utility-supplied comput-

ing will ultimately change the higher education and the society as profoundly as cheap electricity did in the past. Ultimately, the cloud must help higher education consolidate and collaborate.

REFERENCES

Averitt, S., Bugaev, M., Peeler, A., Schaffer, H., Sills, E., Stein, S., et al. (2007, May 7-8). The virtual computing laboratory. In *Proceedings of the International Conference on Virtual Computing Initiative*, Triangle Park, NC.

Bell, M. (2008). *Introduction to service-oriented modeling, service-oriented modeling: Service analysis, design, and architecture*. New York, NY: John Wiley & Sons.

Bulkeley, W. M. (2007). IBM, Google, Universities combine 'Cloud' forces. Retrieved from http://online.wsj.com/article/SB119180611310551864.html

Burton Group. (2010). *Comprehensive research and advisory solution*. Retrieved from http://www.burtongroup.com/research/

Carr, N. (2008). *The big switch: Rewiring the world, from Edison to Google*. New York, NY: Norton & Company.

Cloud Portal. (2010). *Cloud computing portal*. Retrieved from http://cloudcomputing.qrimp.com/portal.aspx

Foster, I., & Kesselman, C. (2004). *The grid 2: Blueprint for a new computing infrastructure* (2nd ed.). San Francisco, CA: Morgan Kauffman.

Gartner. (2010). *Gartner research*. Retrieved from http://blogs.gartner.com/

Gartner. (2010). *Gartner's expertise in a variety of ways*. Retrieved from http://www.gartner.com/

Geelan, J. (2009, May 18-19). *Deploying virtualization in the enterprise.* Paper presented at the Virtualization Conference, Prague, Czech Republic.

Google. (2010). *Indian version of this popular search engine.* Retrieved from http://www.google.co.in/

India Online. (2010). *Population of India.* Retrieved from http://www.indiaonlinepages.com/population/

INTEROP. (2008). *Enterprise software customer survey.* Retrieved from http://www.interop.com/

Kirkpatrick, M. (2007). *IBM unveils Blue Cloud - what data would you like to crunch?* Retrieved from http://www.readwriteweb.com/archives/ibm_unveils_blue_cloud_what_da.php

Magoules, F., Pan, J., Tan, K.-A., & Kumar, A. (2009). *Introduction to computing, numerical analysis and scientific computation series.* Boca Raton, FL: CRC Press.

Mckinsey & Company. (2010). *Highlights and features.* Retrieved from http://www.mckinsey.com/

McKinsey & Company. (2010). *Management consulting & advising.* Retrieved from http://www.mckinsey.com/

National Institute of Standards and Technology. (2010). *NIST homepage.* Retrieved from http://www.nist.gov/

Rewatkar, L. R., & Lanjewar, U. A. (2010). Data management in market-oriented cloud computing. *Advances in Computer Science and Technology*, 3(2), 217–222.

Uptîme Institute. (2009). *Clearing the air on cloud computing.* Retrieved from http://uptimeinstitute.org

Wikipedia. (2010). *Welcome to Wikipedia.* Retrieved from http://en.wikipedia.org/wiki/

Young, J. (2008). 3 ways that web-based computing will change colleges and challenge them. *The Chronicle of Higher Education*, 55(10), 16.

This work was previously published in the International Journal of Cloud Applications and Computing (IJCAC), Volume 1, Issue 2, edited by Shadi Aljawarneh & Hong Cai, pp. 1-13, copyright 2011 by IGI Publishing (an imprint of IGI Global).

Chapter 7
Using Free Software for Elastic Web Hosting on a Private Cloud

Roland Kübert
University of Stuttgart, Germany

Gregory Katsaros
University of Stuttgart, Germany

ABSTRACT

Even though public cloud providers already exist and offer computing and storage services, cloud computing is still a buzzword for scientists in various fields such as engineering, finance, social sciences, etc. These technologies are currently mature enough to leave the experimental laboratory in order to be used in real-life scenarios. To this end, the authors consider that the prime example use case of cloud computing is a web hosting service. This paper presents the architectural approach as well as the technical solution for applying elastic web hosting onto a private cloud infrastructure using only free software. Through several available software applications and tools, anyone can build their own private cloud on top of a local infrastructure and benefit from the dynamicity and scalability provided by the cloud approach.

INTRODUCTION

In the past years, cloud computing has evolved to be one of the major trends in the computing industry. Cloud computing is, basically, the provisioning of IT resources on demand to customers over some kind of network, most probably the internet. Cloud computing is in this sense an evolution of utility computing, with the difference that in cloud computing one does not demand infrastructure but higher-level services (compute capacity, storage, software) and does not need the knowledge to work with infrastructure (Danielson, 2008). One of the most prominent cloud providers nowadays is Amazon's Elastic Compute Cloud (EC2) (Amazon, n. d.), which even predates the term "cloud computing" (Figure 1) and can be seen as the foundation of this type of computing.

DOI: 10.4018/978-1-4666-1879-4.ch007

Figure 1. Search volume of the terms "cloud computing" (blue) and "Amazon ec2" (red) according to Google Trends (Google, 2011a)

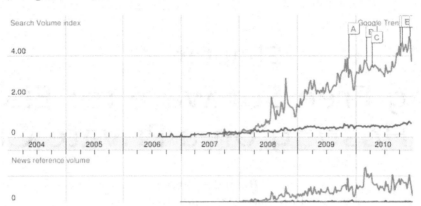

Migrating to public clouds is often said to lead to lower capital expenditure, as there is no up-front cost for buying infrastructure, having floor space for it etc. Instead, the costs incurred by cloud computing relate to operational expenditure, for example if using a cloud provider with pay as you go scheme. While it is questionable that the total cost of either buying and running a server or buying capacity on demand from a cloud provider can be compared directly with each other, the whole burden of operating a data center, controlling and managing the infrastructure etc. is removed if a cloud provider is used(Golden, 2009).

Exactly this lack of control over the infrastructure is what puts cloud users at risk. Richard Stallman, president of the Free Software Foundation, coined the term "carless computing", stating that users should rather keep control over their own data and not hand data over to providers that move it to unknown servers at unknown locations (Arthur, 2010). Additionally, with public cloud providers, the problem of vendor lock-in always exists. Vendor lock-in "is the situation in which customers are dependent on a single manufacturer or supplier for some product (i.e., a good or service), or products, and cannot move to another vendor without substantial costs and/or inconvenience" (The Linux Information Project, 2006). The costs of lock-in to a customer can be

severe and include, amongst others, "a substantial inconvenience and expense of converting data to other formats" and "a lack of bargaining ability to reduce prices and improve service" (The Linux Information Project, 2006).

Besides the lack of control over the placement of one's own data and vendor lock-in, other problems exist with private clouds, as with any business offer: the provider might decide that it is no longer interested in providing service to a customer, thereby disrupting the clients business, at least temporarily. This has happened, for example, to the non-profit organization WikiLeaks (Gross, 2010; MacAskill, 2010).

The question is then: is there a viable alternative to public cloud providers which retains some of the flexibility one gains by moving to the cloud? And the answer to this question, luckily, is positive: yes, there is an alternative and it is building a private cloud. A private cloud is essentially the same thing as a public cloud, only hosted on a private network on one's own physical infrastructure. Obviously, the host of a private cloud has to care about his own resources, which means that there are no upfront advantages for CAPEX. This, however, is less valid if a physical infrastructure already exists and is – either totally or partially – changed to a virtualized infrastructure. OPEX will probably be reduced when staying on one's

Figure 2. Search volume of the terms "public cloud" (red) and "private cloud" (blue) according to Google Trends(Google, 2011b)

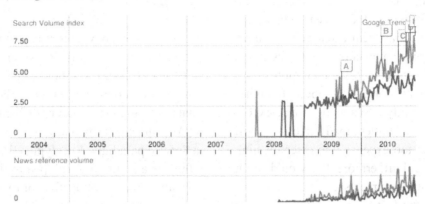

private cloud and not going to a public cloud, with the added advantage of having total control over the physical infrastructure, thereby avoiding the problems mentioned above. Private clouds seem to be of more and more interest, as can be seen in Figure 2.

But how can one turn private, non-virtualized physical resources on one's site into a private cloud? This transition is not too difficult, as we will demonstrate in this work. Building up this knowledge might be an up-front investment but can be cheaper in the long run, all the while keeping the advantages of a private cloud in mind. We will describe how one can turn an existing in-house cluster of physical hosts into a virtualized infrastructure and will demonstrate the usefulness of this by showing how a scalable web hosting solution can be built on top of this private cloud. We will describe an exemplary existing infrastructure, the software components we employ and demonstrate that we arrive at a solution that rivals public cloud offerings but has further advantages and is using free software exclusively.

The methodology that we have followed is to firstly explore the term "Elastic Web Hosting" and its significance and characteristics. After that, the identified capabilities have been transformed into architectural requirements and the high level design of our approach, based on the private cloud

paradigm, is presented. Further exploration of the realization of the solution is provided through specific details of the implementation and the software that has been used, before the approach is validated against a private cloud test-bed environment. Finally, the last section summarizes our findings and concludes our work.

Elastic Web Hosting

The term "elasticity" may have been used implicitly or explicitly before, but its mainstream usage in computing stems from Amazon's product "Elastic Compute Cloud" (EC2). In general, elasticity means "the ability to adapt" (Oxford Dictionaries, 2010b), therefore the fact of being "able to adjust to new conditions" (Oxford Dictionaries, 2010a). In the sense of computing, this means that resources are provisioned on demand. This feature is made easy through the use of virtualization technology: users do not access physical hardware but virtual hosts that seem like physical hosts and can run anywhere on sufficient hardware. It is understandable how the term elasticity became being used to describe these techniques as it has a connotation of dynamicity in contrast to the staticity of classical resource provisioning: if a users hosts a web site on a server and the capacity is maxed out, a new server needs to be bought, installed and configured. Elastic solutions, however, provide infrastructure that is not visible to the user and can scale depending on the current demand (of course, if the real infrastructure's capacity is maxed out, the same problem occurs, but it is assumed that the real infrastructure capacity is quite high). As important as the notion of adapting to higher capacity is the ability to scale down once excess capacity is detected. Often, load spikes are short and it would be too expensive to cater for high load all the time. Furthermore, this would result in general in low resource usage.

Apart from the fact the provisioning of bare virtual machines is a quite straight forward use case for Elasticity, the same goes for web hosting: as web servers are normally stateless, meaning that they store no data from one request to another, additional web servers can be provided if an increasing number of requests are incoming. If virtual machine images are prepared accordingly, for example with an installed web server and a deployed web site, once a virtual machine is up and running it can serve requests and it can be decommissioned once the load is decreasing again. In this paper we will propose and present a solution based on several free software tools that will support us in realizing elastic web hosting on a private cloud infrastructure.

ARCHITECTURE

In this section we will describe the architecture of the proposed solution using a three step approach: initially we will briefly discuss private cloud architectures starting from a generic point of view and finishing with the introduction of the elastic web hosting scenario into such a model, then we will elaborate on what role load balancing can play in the web hosting case and finally we will populate the mechanism with dynamic VM allocation which will realize the scalability of the proposed architecture.

Private Cloud: High-Level Architecture

As mentioned above, more and more people are interested in building their own private cloud infrastructure. As this solution gains ground, several open source toolkits and APIs have been developed that allow the management of Virtual Machines (VMs) and generally realize the Infrastructure as a Service paradigm. Eucalyptus (Baun & Kunze, 2009) is a solution developed at the University of California that offers the ability to deploy and control VM instances via different hypervisors (Xen or KVM) over physical resources. Nimbus (Freeman, LaBissoniere, Marshall, Bresnahan, & Keahey, 2009) is similar to Eucalyptus and provides the necessary interfaces to give users control over VMs. It is, however, running on top of a Globus Toolkit Java container (Foster, 2005). Another virtual infrastructure manager toolkit that allows to build private or hybrid clouds is OpenNebula (Sotomayor, Montero, Llorente, & Foster, 2009) mainly supported by the University of Madrid and being already used into several research initiatives (BonFIRE, 2010; Reservoir, 2010).

Regardless of the differences, all the aforementioned solutions are sharing a common, high level architecture for implementing a private cloud. An abstract architectural approach of a private cloud is shown in Figure 3.

The multilayered architecture consists of the physical (local) infrastructure, on top of that lies the hypervisor (Xen, KVM etc.) while over that layer the virtual management framework (OpenNebula, Eucalyptus etc.) is located. Finally, the latter exposes the proper interfaces in order for the users/administrators to access and control the private cloud. Based on this architectural model we can clearly identify two distinctive infrastructures: the physical infrastructure and the virtual infrastructure (Figure 4). With the introduction of virtualization technologies which have been exploited by cloud offers, the consumer is no

Figure 3. General architecture of a private cloud

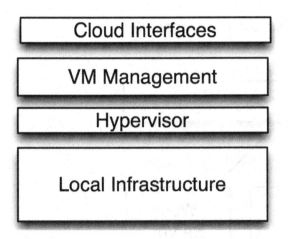

longer interested in the physical resources but in the VMs that have been instantiated. The physical infrastructure becomes transparent for him who now through Infrastructure as a Service and on-demand computing negotiates access to a virtual environment with certain specifications. This virtual infrastructure is dynamic, flexible and customizable according to the application or the service that every user wants to execute. We shall not elaborate on the advantages and disadvantages of cloud computing in general as this is not the main focus of this paper.

Figure 4. Virtual and physical infrastructure

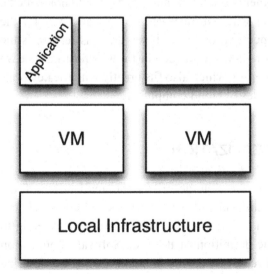

Coming back to our scenario and according to the private cloud architecture presented above, elastic web hosting can be realized when a web server container serves as the hosted application in a VM running atop a cloud infrastructure. It is clear, though, that this case is not innovative and new as such services have been provided by public clouds (e.g. Amazon EC2) almost from the begging of this cloud trend. The proposal that we elaborate in this paper is to apply this cloud-enabled web hosting on to private cloud infrastructures using free software solutions and APIs, in order to populate the system with certain new functionalities and features that will realize the elastic web hosting. In the next two sections we will analyze load balancing and scalability, two major capabilities of the proposed cloud-enabled, elastic web hosting.

Load Balancing

Using load balancing techniques over web hosting services is a fairly old topic (Cherkasova, 1999). In our proposed approach we will be hosting web servers on a private cloud infrastructure, benefiting from the flexibility of virtualization and of self-controlling and managing our private infrastructure. Furthermore, we have populated the architecture with an additional control interface: instead of having a single web server hosted in a VM we have multiple web server containers running on several VM instances deployed within the cloud. The distribution of the incoming requests is managed by a front-end component that is performing the load balancing. As shown in Figure 5none of the available web servers is directly accessible by the clients. The load balancer will take incoming requests on port 80 and distribute them equally towards the web servers. If a web server is deployed or shut down at run time, the load balancer needs to be informed of this and needs to balance requests accordingly.

The introduction of this component has a very important impact on the reliability of the pro-

Figure 5. Load balancing on web hosting

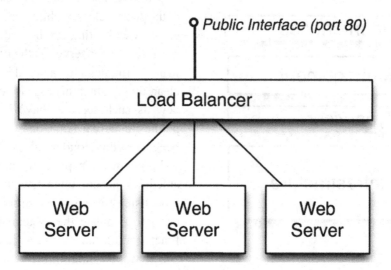

vided web hosting service. By having the control over the incoming request, through the load balancer we can optimize our resource utilization and overall the quality of the provided service (Quality of Service, QoS). In addition, this structure helps us to acquire increased capacity without investing to a single web server with extreme technical specifications. Furthermore, the proposed model realizes a fault tolerant system: in case a web server is unavailable or shut down the http requests will continue being served by the remaining web servers.

Implementing Scalability

The elasticity of the proposed solution is being achieved by instantiating a new VM that will host an extra web server when the capacity of the current system is reaching a limit. To this end, we populate the features of the front-end control component with monitoring and VM management functionalities. Overall the high level operation of the system when a new incoming request is arriving is described by the following pseudo code (Table 1):

Through this functionality we can guarantee that at all times the incoming requests will be

Table 1. Generic load balancing algorithm

```
If (web_server_capacity<limit)
{
forward(request)
} else {
instantiate_new_VM()
reset_load_balancer()
forward(request)
}
```

served and in the same time we will succeed in having good resource utilization. This dynamicity allows us to utilize the free resources with other applications and activities and not dedicate our whole infrastructure in a static way for the operation of the web server. This feature is the core concept of the elastic web hosting that we propose which also fits greatly with the key principles of cloud computing.

REALIZATION

This section describes the realization of the architecture described in the previous section. All software used is free software according to the definition of the Free Software Foundation (Free Software Foundation, 2010). We describe the actual physical infrastructure that has been used to realize the architecture as well. Existing infrastructure will most probably be different for other sites but nonetheless presenting our infrastructure will be helpful to others which want to implement a virtualized infrastructure for the same or a different use case.

Physical Infrastructure

The existing physical infrastructure is shown in Figure 6. As it can be seen, the actual physical resources, which are used as compute nodes, are running in an isolated network that can be accessed only via central frontend node. Jobs submitted to this frontend node are distributed to the compute nodes via some resource manager. Users can access the front-end either from the internet or another internal network, provided they are allowed to do so.

It is pretty straightforward to augment this physical infrastructure with a virtual infrastructure layer: the frontend node provides a virtual infrastructure manager and uses some or all of the physical hosts to deploy virtual machines to. The situation presented in Figure 7shows how the physical infrastructure can be partitioned in such a way that the original functionality – in our case, execution of computational jobs on the physical infrastructure – is still kept while allowing the parallel establishment of a virtualized infrastructure.

It might, however, be beneficial if the physical infrastructure is converted wholly to a virtualized one. As can be seen from the figure, this can be done step by step as more and more of the physical hosts are prepared to host virtual machines and the corresponding virtual machines for execution of compute jobs are prepared.

Figure 6. Existing, non-virtualized infrastructure

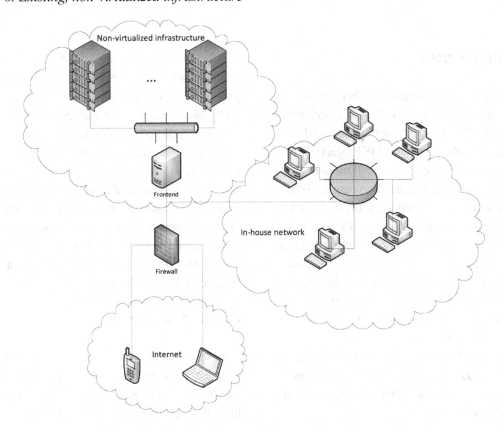

Figure 7. Parallel existing non-virtualized and virtualized infrastructure

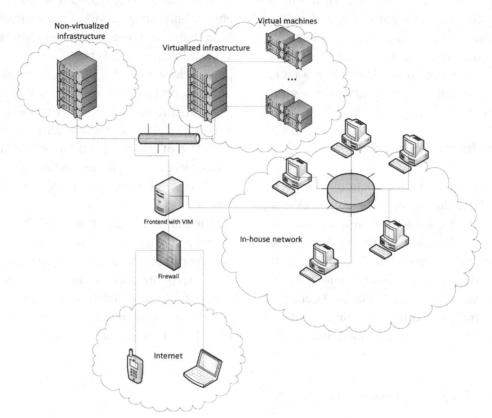

Operating System

CentOS (short for Community ENTerprise Operating System) is a GNU/Linux distribution based on the Red Hat Enterprise Linux (RHEL) distribution by Red Hat (CentOS Project). While RHEL is aimed at the commercial market, CentOS is free software. It is used on nearly 30% of all Linux web servers, making it the distribution used most often on web servers (Gelbman, 2010). CentOS also offers straightforward virtualization support out of the box: CentOS provides groups of packages that can be selected for installation and one package group provides everything necessary for virtualization (both "full virtualization", where unmodified guest systems can be run but require special hardware support on the host, and "para-virtualization", where a special component, a hypervisor, is introduced between the hardware layer and guest systems).

The main choice for CentOS was the out-of-the box support for both fully virtualized and para-virtualized kernels. This however, obviously only applies to CentOS as an operating system on the physical infrastructure. The fact that virtual machines running CentOS as well can be built very easily and that the process is very well documented (CentOS Wiki, 2009) led us to the decision to use CentOS as well as the operating system for the virtual machines. Other GNU/Linux or BSD distributions (for example FreeBSD (FreeBSD Wiki, 2010), would probably be an equally good choice.

Cloud Middleware

For running a private cloud, different choices of virtual infrastructure managers (VIM) exist: Eucalyptus, Nimbus and OpenNebula, just to name three popular ones. OpenStack, a VIM developed by Rackspace and NASA (OpenStack, 2010), is another promising project that, however, is still under heavy development while the other three VIMs can be already taken to production use. There are distinct advantages and disadvantages to each of these three VIMs and the choice of which to use is not fixed. For our case, as we are building a private cloud around a central head node and have a small group of machines which are, except for the web traffic, only accessed by trusted users, we decided to use OpenNebula (Sotomayor et al., 2009). The latest version of OpenNebula, version 2.0, released in October of 2010, provides support for Xen, KVM and VMWare, adds an image repository for the management of VM images and its basic installation on the frontend node is easy and quite small. The configuration of the physical hosts which will run the VMs is minimal as well.

Summarizing, the same solution we developed might be achieved with Eucalyptus or Nimbus, but for our case OpenNebula fit well with the requirements we had. It would be interesting to compare the same solution set up with the other VIMs as well.

Load Balancer

As with the operating system and the virtual infrastructure manager, there are multiple choices for load balancing. In our case, we do not have strong requirements on the load balancer: we want to run it on a dedicated VM, has it distribute incoming requests to multiple servers, be able to query the current status and interact with it during run-time. With *balance* (Inlab Software GmbH, 2010), we found a straightforward, easy to use software for this. Balance is a small (about 2k lines of code) generic TCP proxy with round-robin load balancing and failover mechanisms. It can easily be controlled at runtime through a simple command line interface. Its simplicity means one can build and deploy *balance* easily and can get started with simple load balancing tasks right away.

Setting *balance* up for load balancing incoming requests to port 80 to two servers is as easy as the following command line:

```
# balance www host1 host2
```

Balance will run in the background and will balance incoming requests dynamically between hosts *host1* and *host2*. For our demonstration setup, we decided to limit, without the loss of generality, the amount of connections that each server can take severely in order to facilitate testing. This is easily achievable with *balance*, as for each host it can be specified how much connections this host can mange, as shown in the following command line (Walcott, 2005):

```
# balance www host1::8 host2::8
```

This command line specifies that both *host1* and *host2* can manage 8 simultaneous connections. The configuration of balance needs to be in sync with the configuration of the web servers, as it is of no use to tell balance that each host can handle 250 requests if the web server is configured to only allow 150 requests.

One can connect to a running instance and get the connection status like this:

```
# balance http -i -c show
GRP Type# S ip-address portc totalc
maxc sentrcvd
0 RR 0 ENA 10.0.0.12 80 1 4 8 0 0
0 RR 1 ENA 10.0.0.13 80 0 4 8 0 0
```

Each line specifies a host, the "c" column specifies the number of current connections (1 to host 10.0.0.12 and none to 10.0.0.13). "totalc" specifies the total connections established until now and "maxc" the number of connections that a host can take as a maximum.

Adding a host dynamically to the setup as displayed with the *show* command above can be achieved by the following commands:

```
# balance http -i -c "create
10.0.0.14 80"
# balance -i -c "enable 2"
```

Finally, a host can be disabled with the following command:

```
# balance -i -c "disable 2"
```

As it can be seen, *balance* allows all the tasks necessary at runtime. It can be either controlled interactively – that is, by an administrator – or by dedicated component using the commands specified above.

HTTP SERVER

The sole purpose of the HTTP server we employ in this use case is to serve simple static web pages. We therefore do not have special requirements on the HTTP server and in fact users are free to choose from quite a number of popular HTTP servers that are free software: Apache HTTP Server, Lighttpd and nginx, just to name the most prominent ones. We decided to use the most popular one of these, Apache HTTP Server, but any of the others would be fine as well (Netcraft, 2010).

Monitoring and Management

While elasticity is one major characteristic of the proposed architecture, monitoring of the load and management of the virtual infrastructure are two very crucial operations of the system. To this end, we have created a mechanism based on the Nagios monitoring system (Nagios, 2009) along with the OpenNebula cloud middleware that automatically reserves a new virtual machine and re-distributes the load to all running VMs.

The proposed mechanism consists of three main components:

- Nagios API: this is an open source monitoring framework through which we can acquire the status of multiple hosts regarding their availability, load, memory and various other metrics. In our case, we have specifically implemented a Nagios service check in order to monitor the number of connections for the httpd service that is running in every host, a PING check for the availability and a CPU load service check.

- Event Broker Module: this is a custom module written in C code and using the Nagios Event Broker API (NEB) (Ingraham, 2006). The operation of this component is based on a notification that the Nagios API generates every time that a service check is being applied. The module receives notifications from Nagios and checks and forwards this information to the Load Manager.

- Load Manager: this component takes as input the monitoring data coming from the Nagios API regarding VM instances. In case that a web server in a VM is reaching a certain threshold regarding its capacity and/or other performance oriented parameters (e.g. CPU load, memory), then the Load Manager deploys a new virtual machine and reconfigures the Load Balancer to use an extra web server. To achieve this, it communicates with the VM Management (OpenNebula API) and the Load Balancer (balance API).

The monitored data (Figure 8) offered to the Load Manager is derived from the Load Balancer itself (e.g. number of connection) and as described before from the Event Broker Module. While the monitored data are customizable (especially those from Nagios) the logic that the manager implements can vary based on available and needed information. To this end, one might set multiple rules and respective actions to be applied using

Figure 8. Monitoring and management architecture

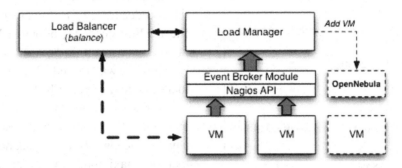

different input every time. This feature extends even more the dynamicity and elasticity of the architecture.

LOAD CREATION SOFTWARE

In order to test our solution, we used the open source tool curl-loader (Iakobashvili, 2007). Curl-loader, written in C, can simulate the behavior of a big number of protocols, for example HTTP(S)/FTP(S), and uses the client protocol stacks of libcurl (Stenberg, 2010).

Curl-loader is controlled by a configuration file. The configuration file we have used is listed below; it starts with 100 clients initially and adds each second another single client until 250 clients are running in total.

```
------------------------------------
########## GENERAL SECTION ##########
BATCH_NAME= 250-clients
CLIENTS_NUM_MAX=250
CLIENTS_NUM_START=100
CLIENTS_RAMPUP_INC=1
INTERFACE =eth0
NETMASK=255.255.0.0
IP_ADDR_MIN= 192.168.1.1
IP_ADDR_MAX= 192.168.53.255
CYCLES_NUM= 1
URLS_NUM= 1
```

```
########## URLs SECTION ##########

URL=http://balance-host/index.html
URL_SHORT_NAME="balance-index"
REQUEST_TYPE=GET
TIMER_URL_COMPLETION = 0
TIMER_AFTER_URL_SLEEP = 0
```

Each client obtains one single file, index.html, using an HTTP get request. This request is not time-limited.

Validation and Testing

The implementation of the load balancing system we have described above has been validated using curl-loader, described in the previous section, to simulate clients accessing web servers. This simple solution we described is sufficient to adapt dynamically to incoming load, provisioning new virtual machines when more requests are incoming than can be handled at the moment and shutting down VMs if excess capacity is provisioned at the moment.

In addition, our test-bed consisted of one front-end node and two workers with the following technical specifications (Table 2):

We used OpenNebula to start up 4 VMs on the workers and each one of those instances hosted an Apache HTTP web server. The testing scenario that we executed was to use curl-loader to apply load on the web servers already started and

Table 2. Technical specification of the validation testbed

```
Front-end
CPU
2x Intel(R) Xeon(TM) CPU 3.20GHz
16kiB L1 cache
1MiB L2 cache
RAM
8GiB system memory
Composed of 4 x 2GiB DIMM DDR Synchronous 333 MHz
(3.0 ns)
Disk (SATA)
2 x 250GB HDS722525VLSA80
Raid 1 setup (mirrored)
Network
2x 82546GB Gigabit Ethernet Controller
Workers
CPU
2x Intel(R) Xeon(TM) CPU 3.20GHz
16kiB L1 cache
1MiB L2 cache
RAM
8GiB system memory
Composed of 4 x 2GiB DIMM DDR Synchronous 333 MHz
(3.0 ns)
Disk (SATA)
2 x 250GB HDS722525VLSA80
Raid 1 setup (mirrored)
Network
82546GB Gigabit Ethernet Controller
```

force the manager to start a new VM when the connections monitored are reaching the defined limit of 100 connections.

For our tests, we used the SSH transfer manager, which copies VM images to the host on which they are to be executed using the Secure Shell network protocol (Ylonen & Lonvick, 2006). For a VM image with approximately 2GB size, transfer over a Gigabit Ethernet link using this transfer manager took in our case around 90 seconds. We decided to use the SSH transfer manager as this requires no further setup except configuring SSH on both the frontend node and the "worker" nodes; using other transfer managers – OpenNebula supports for example as well the network file system (NFS) (Shepler et al., 2003) – might give different results. The encryption overhead imposed by SSH is probably the limiting factor at this point.

As the 90 seconds mentioned above are the time needed for the pure transfer of the VM image file, the time needed to boot up a VM needs to be factored in as well. For our 2GB VM image, the boot up time was around 50 seconds, bringing the total time to deploy a VM to 140 seconds.

This is a quite good time. We have previously mentioned the fact that the threshold, at which a new VM needs to be deployed, needs to be lower than the maximum capacity at which VMs can operate. Now, it is immediately obvious why: assuming that an incoming connection means that all connections are used at the moment, just then deploying a new VM means that for around 140 seconds, no more new connections can be accepted. Depending on the number of connections allowed in total and the inter-arrival time of requests, the threshold can be set accordingly. Of course this does not mean that a single connection might not be served, as total contention situations might occur if the load increases suddenly. However, for slow rises in the number of connections this solution can be applied.

A good technique that requires slightly more implementation effort at the Load Manager is to deploy VMs at a certain threshold but put them into pause initially. The VM will already consume allocated resources like RAM but will not be scheduled to be run on the actual CPUs. When the load rises further, the VM can be unpaused and can immediately serve requests. The load manager then needs to take care of pausing and undeploying resources again. Depending on the situation, the algorithm for the load manager can be adapted; it can, for example, undeploy all VMs that have not been used for a certain time span, provided that the capacity is still high enough but can leave a fixed number of VMs in paused mode in order to react quickly on rising load.

CONCLUSION

In this paper we have discussed the advantages of using a private cloud in contrast to a public cloud. Furthermore, we have shown how this private cloud can be implemented using only free software. In this context, we presented how we can apply and realize elastic web hosting on a private cloud infrastructure using only free software as well. Thereby, we relied on a software stack that is completely free software, starting from the operating system, up to the web servers, monitoring software, load balancer and virtual infrastructure manager. The solution we developed has been validated through an elastic web hosting use case on a small-scale test-bed but the architecture is directly applicable to larger-scale infrastructures. This is only one usage of the virtualized infrastructure, which can be used for various purposes. It shows, however, how easy and efficient the setup and operation of a private cloud is. The virtualized infrastructure can co-exist with other uses of the physical infrastructure; we have shown how this can be easily achieved by partitioning the existing physical infrastructure in one part that is used physically and one that uses virtualization. Resources for providing the virtualized infrastructure do not need to be very expensive; in our case, surplus nodes from an old compute cluster have been used; due to their age, they might lack hardware virtualization but para-virtualization is of course possible. As many companies previously have acquired IT resources, there is no point in letting these go to waste by moving all services into a public cloud and we have shown they can be easily used as a basis for a virtualized infrastructure.

The advantage of this solution is that the complete infrastructure – physical and virtual – stays in-house but can, due to the ability of OpenNebula to manage hybrid clouds, even be augmented with private cloud resources if one so wishes. Having total control over the infrastructure is surely an advantage for an enterprise, which would partially,

been given up when using a hybrid model with a public and a private cloud at the same time. A public cloud can, however, be used if the existing physical infrastructure is not big enough anymore in order to bridge the gap from the time when this is realized until more physical resources can be acquired. Using a private cloud requires in-house knowledge of the employed techniques. As we have shown, this know-how is not difficult to acquire but it still has to be done. However, even the pure usage of an external cloud cannot be done without acquiring some know-how, so the question is which solution is more beneficial in the long-term and we think that the answer to this question surely is the in-house solution.

REFERENCES

Amazon. (n. d.). *Amazon elastic compute cloud (Amazon EC2)*. Retrieved from http://aws.amazon.com/de/ec2/

Arthur, C. (2010). *Google's Chrome OS means losing control of data, warns GNU founder Richard Stallman*. Retrieved from http://www.guardian.co.uk/technology/blog/2010/dec/14/chrome-os-richard-stallman-warning

Baun, C., & Kunze, M. (2009). Building a private cloud with Eucalyptus. In *Proceedings of the IEEE International Conference on E-Science Workshops* (pp. 33-38).

BonFire. (2010). *Building service testbeds on fire*. Retrieved from http://www.bonfire-project.eu/

Cent, O. S. (2009). *The community ENTerprise operating systems*. Retrieved from http://www.centos.org/

Cent, O. S. Wiki. (2009). *Creating and installing a CentOS 5 domU instance*. Retrieved from http://wiki.centos.org/HowTos

Cherkasova, L. (1999). *FLEX: Design and management strategy for scalable web hosting service (Tech. Rep. No. HPL 1999□64R1)*. Palo Alto, CA: Hewlett-Packard Laboratories.

Danielson, K. (2008). *Distinguishing cloud computing from utility computing*. Retrieved from http://www.ebizq.net/blogs/saasweek/2008/03/distinguishing_cloud_computing/

Foster, I. T. (2005). Globus toolkit version 4: Software for service-oriented systems. *Journal of Computer Science and Technology, 21*(4), 513–520. doi:10.1007/s11390-006-0513-y

Free Software Foundation. (2010). *The free software definition*. Retrieved from http://www.gnu.org/philosophy/free-sw.html

FreeBSD Wiki. (2010). *FreeBSD/Xen: FreeBSD/Xen port*. Retrieved from http://wiki.freebsd.org/FreeBSD/Xen

Freeman, T., LaBissoniere, D., Marshall, P., Bresnahan, J., & Keahey, K. (2009). *Nimbus elastic scaling in the clouds*. Retrieved from http://www.nimbusproject.org/files/epu_poster4.pdf

Gelbman, M. (2010). *Highlights of web technology surveys, July 2010: CentOS is now the most popular Linux distribution on web servers*. Retrieved from http://w3techs.com/blog/entry/highlights_of_web_technology_surveys_july_2010

Golden, B. (2009). *Capex vs. Opex: Most people miss the point about cloud economics*. Retrieved from http://www.cio.com/article/484429/Capex_vs._Opex_Most_People_Miss_the_Point_About_Cloud_Economics

Google. (2011a). *Google trends: Cloud computing, Amazon ec2*. Retrieved from http://www.google.de/trends?q=cloud+computing%2C+amazon+ec2&ctab=0&geo=all&date=all

Google. (2011b). *Google trends: Private cloud, public cloud*. Retrieved from http://www.google.de/trends?q=private+cloud%2C+public+cloud

Gross, D. (2010). *WikiLeaks cut off from Amazon servers*. Retrieved from http://articles.cnn.com/2010-12-01/us/wikileaks.amazon_1_julian-assange-wikileaks-amazon-officials?_s=PM:US

Iakobashvili, R. M. M. (2007). *Welcome to curl-loader*. Retrieved from http://curl-loader.sourceforge.net/

Ingraham, R. W. (2006). *The Nagios 2.X event broker module API*. Retrieved from http://nagios.sourceforge.net/download/contrib/documentation/misc/NEB%202x%20Module%20API.pdf

Inlab Software GmbH. (2010). *Balance*. Retrieved from http://www.inlab.de/balance.html

Linux Information Project. (2006). *Vendor lock-in definition*. Retrieved from http://www.linfo.org/vendor_lockin.html

MacAskill, E. (2010). *WikiLeaks website pulled by Amazon after US political pressure*. Retrieved from http://www.guardian.co.uk/media/2010/dec/01/wikileaks-website-cables-servers-amazon

Nagios. (2009). *The industry standard in IT infrastructure monitoring*. Retrieved from http://www.nagios.org/

Netcraft. (2010). *November 2010 web server survey*. Retrieved from http://news.netcraft.com/archives/2010/11/05/november-2010-web-server-survey.html

OpenStack. (2010). *OpenStack open source cloud computing software*. Retrieved from http://www.openstack.org/

Oxford Dictionaries. (2010a). *Adaptable*. Retrieved from http://oxforddictionaries.com/view/entry/m_en_gb0007570

Oxford Dictionaries. (2010b). *Elasticity*. Retrieved from http://oxforddictionaries.com/view/entry/m_en_gb0980800

Reservoir. (2010). *Reservoir fp7*. Retrieved from http://62.149.240.97/

Shepler, S., Callaghan, B., Robinson, D., Thurlow, R., Beame, C., Eisler, M., et al. (2003). *Network File System (NFS) version 4 protocol (request for comments no. 3530): IETF.* Retrieved from http://www.ietf.org/rfc/rfc3530.txt

Sotomayor, B., Montero, R. S., Llorente, I. M., & Foster, I. (2009). Virtual infrastructure management in private and hybrid clouds. *IEEE Internet Computing, 13,* 14–22. doi:10.1109/MIC.2009.119

Stenberg, D. (2010). *libcurl - the multiprotocol file transfer library.* Retrieved from http://curl.haxx.se/libcurl/

Walcott, C. (2005). *Taking a load off: Load balancing with balance.* Retrieved from http://www.linux.com/archive/feature/46735

Ylonen, T., & Lonvick, C. (2006). *The Secure Shell (SSH) protocol architecture (request for comments no. 4251):* IETF. Retrieved from http://www.ietf.org/rfc/rfc4251.txt

This work was previously published in the International Journal of Cloud Applications and Computing (IJCAC), Volume 1, Issue 2, edited by Shadi Aljawarneh & Hong Cai, pp. 14-28, copyright 2011 by IGI Publishing (an imprint of IGI Global).

Chapter 8

Applying Security Policies in Small Business Utilizing Cloud Computing Technologies

Louay Karadsheh
ECPI University, USA

Samer Alhawari
Applied Science Private University, Jordan

ABSTRACT

Over a decade ago, cloud computing became an important topic for small and large businesses alike. The new concept promises scalability, security, cost reduction, portability, and availability. While addressing this issue over the past several years, there have been intensive discussions about the importance of cloud computing technologies. Therefore, this paper reviews the transition from traditional computing to cloud computing and the benefit for businesses, cloud computing architecture, cloud computing services classification, and deployment models. Furthermore, this paper discusses the security policies and types of internal risks that a small business might encounter implementing cloud computing technologies. It addresses initiatives towards employing certain types of security policies in small businesses implementing cloud computing technologies to encourage small business to migrate to cloud computing by portraying what is needed to secure their infrastructure using traditional security policies without the complexity used in large corporations.

1. INTRODUCTION

At this time, organizations are expected to gain an increased in competitiveness and chances to focus their efforts and use their resources on their core competence. Therefore, cloud computing is defined as "a model for enabling convenient, on-demand network access to a shared pool of configurable computing resources (e.g., networks, servers, storage, applications, and services) that can be rapidly provisioned and released with

DOI: 10.4018/978-1-4666-1879-4.ch008

minimal management effort or service provider interaction" (Mell & Grance, 2010). Furthermore, cloud computing enables dynamic provisioning of resources based on the requirements of the user (Yogesh & Navonil, 2010). In a recent study, Cloud computing is not a new technology, but it is a new way of delivering technology. It is a new way of executing business applications by relying more on a third party's infrastructure, then local infrastructure (Srinivasan, 2010). In addition, the implementation of cloud computing is definitely accelerating (Cervone, 2010) and much of this is being motivated by new business requirements and enabled by information technology (IT).

Most importantly, Katharine and David (2010) explained that in concepts of the cloud, while widespread usage is not yet common, some governments are taking stages to guarantee their information remains authentic and accessible. For this viewpoint, cloud computing is increasingly being considered as a technology that has the possible of changing how the internet and the information systems are presently operated and used (Amir, 2010).

Lately, governments of several countries have realized the potential of cloud computing in offering enhanced services to its citizens. For example, UK Government is developing a secure cloud infrastructure called "G-Cloud" for public sector bodies. More significantly, the strategy will also provide some standardization for capabilities for the promotion of shared services with accredited cloud service providers (Heath, 2010).

However, a client computer on the Internet can communicate with many servers at the same time, some of which may also be exchanging information among themselves (Hayes, 2008). Furthermore, the cost of this service can be determined by several factors such as an hour of usage, software type, and storage space utilization (Srinivasan, 2010). Therefore, this can be translated into saving of the software license, number of support labor, maintenance, office space and utilities.

Recently, Wittow and Buller (2010) stated that traditional computing model is based on using hardware and software resources, which required on-site computing power and disk storage space, as well as the technical human expertise necessary to implement, maintain and secure those resources. Also, complicated and expensive upgrade procedures were necessary to take advantage of new developments and features available for software applications in the traditional computing model (Wittow & Buller, 2010). In addition, the upgraded software or/and hardware often required upgrading licenses and increasing backup and recovery capabilities to reduce the downtime that users would experience should a software or hardware failure occur. Furthermore, local administrators with specialized, technical skill-sets were historically responsible for application and hardware maintenance (Wittow & Buller, 2010). In addition, the "traditional model" often involved managing a large hardware infrastructure with dissimilar operating systems and applications that required individual backups, monitoring and software updates (Wittow & Buller, 2010). The traditional computing model required companies (and individuals) to make a significant financial commitment to set up software and hardware resources, and these were frequently difficult to expand when the needs of users changed (Wittow & Buller, 2010).

Furthermore, for small business, cloud computing can be a saving and reliability factors for relying increasingly on these technologies. In fact, clouds technologies may be very suitable for small businesses since clouds offer technical support and lower cost of service. Hence, an important issue in cloud computing allows for rapid increases in capacity or capability without the need to invest in additional infrastructure, personnel, or software licensing (Wittow & Buller, 2010). Furthermore, cloud computing free individuals and small businesses from worries about quick obsolescence and a lack of flexibility (Greengard, 2010). Therefore, small business will not need complicated infrastructure such as servers and lots of managed switches.

Another advantage for small business is the increasing popularity of internet notebooks, or "netbooks," (Wittow & Buller, 2010). Netbooks are typically low-cost, lightweight laptop computers with reduced hardware capacity and processing power that are primarily designed to provide the user with access to the Internet (Wittow & Buller, 2010). In this respect, netbooks provide users with vast resources because the cloud is fully accessible without requiring users to make a substantial investment in local hardware (Wittow & Buller, 2010). The virtually unlimited resources available in the cloud make the local system's limited hardware capabilities irrelevant (Wittow & Buller, 2010). Therefore, the user will not need to install Microsoft Office locally, which uses more CPU power and memory comparing to the internet browser which uses much fewer resources to run Microsoft Office as an example. In addressing these issues, Yogesh and Navonil (2010) noted that the high-speed communication networks are essential for cloud computing. As a result, cloud-based applications like Google Reader and some office productivity tools have an "offline" option. The purpose of this is to allow users to continue with their task even when they have intermittent access to the internet.

Additionally, risks can be categorized as internal risks and external risks. External risks is linked between the customer and the cloud provider and can range from resource exhaustion, isolation failure, interception of data in transit, ineffective deletion of data, DoS, loss of encryption keys, supply chain failure, cloud provider acquisition, cloud service termination, compliance challenges, lock-in and loss of governance (European Network and Information Security Agency, 2009). On the other hand, internal risks apply to the premises of the customer only and can range from malware, insiders, social engineering and theft. Furthermore, many small businesses don't have adequate or existed security policies to minimize risks. Also,

many small businesses do not have money to invest in security (Srinivasan, 2010). Furthermore, security incidents in small businesses are usually lectured by employees with no expertise in security (Srinivasan, 2010). With cloud computing model, the design and implementation of security policies might be easier especially for small businesses than the traditional computing model.

It is evident from the prior analysis that organization increasingly turns to IT security providers, in addressing this issue. Organizations are generally concerned with external security threats (such as viruses and hacking attempts) (D'Arcy & Hovav, 2007). Moreover, all research discusses the risks of using cloud computing by business with cloud providers without discussing the internal risks. Businesses should examine all risks associated with implementing new technologies such as cloud computing. In fact, a study by Vista Research in 2002 estimated that 70% of security breaches involving losses of more than $100,000 were internal, often perpetrated by disgruntled employees (Standage, 2002). Additionally, Cervone (2010) noted that one of the important advantages of cloud computing is the potential cost savings that can be gained. Usually cloud computing has little or no upfront capital costs. For the most part, operational responsibilities are shifted to the cloud provider, who is then responsible for the on-going maintenance of the hardware used by the cloud.

For this viewpoint, the authors have chosen a selection of topics to discuss how to mitigate risks inside the company implementing cloud computing technology model using security policies. In essence, this paper examines the reasons why some security policies are needed and how they fit into all elements of the small business that utilize cloud computing technology model. Furthermore, the paper discusses the possibilities of applying different types of security policies to enhance security of small business and reduces the risks to acceptable level. To achieve this goal, the authors will explore different security policies and how it can be mapped to cloud computing implemented within small businesses. The objective is to help small business understand what is needed to secure their internal infrastructure using security policies.

The rest of the paper is organized as follows: in the next section, we review relevant literature. Section three explains security policy for small business using cloud computing and proposed the model for the types of security policy applicable to small business; finally, the last section presents the overall conclusions and areas for further research.

2. LITERATURE REVIEW

2.1. Types of Internal Risks

There is a need to give secure and safe information security systems through the use of firewalls, intrusion detection and prevention systems, encryption, and authentication; therefore, this section discusses different types of internal risks that a small business might encounter. For example, client-side infrastructure embodies a vast array of vulnerabilities, particularly in the case of consumer-oriented devices and software, and even more so in the case of devices that support user profiles such as the browser using Active-X or Javascript (Clarke, 2010). Social engineering is the practice of using deception or persuasion to fraudulently obtain goods or information, and the term is often used in relation to computer systems or the information they contain (Twitchell, 2006). Malware is a variety of forms of hostile, intrusive, or annoying software or program code and a pervasive problem in distributed computer and network systems (Cesare & Xiang, 2010).

Furthermore, internal thefts of data are a major concern for all organizations regarding to the size of the business. In fact, employee theft at small companies can have more serious consequences because they do not have the resources of their larger counterparts. Often a lifetime of hard work can be lost because of a single unscrupulous em-

ployee (Morris, n.d.). Moreover, misuse of computer resources by surfing the internet or use the corporate email system for non-business related or using chat applications and downloading unauthorized software and unauthorized data access.

2.2. Cloud Computing

As mentioned earlier in the introduction section; in a cloud computing environment, the organization running an application does not typically own the physical hardware used for the applications. In fact, when running applications in the cloud, an organization does not usually know exactly where the computation work of the applications is being processed (Cervone, 2010). Therefore, cloud computing are the latest technology that is being feted by the IT industry as the next (potential) revolution to modify how the internet and information systems work and are used by the world at large (Amir, 2010).

However, Cervone (2010) noted that One of the important advantages of cloud computing is the potential cost savings that can be gained. Usually cloud computing has little or no upfront capital costs. For the most part, operational responsibilities are shifted to the cloud provider, who is then responsible for the on-going maintenance of the hardware used by the cloud. Correspondingly, Mark-Shane (2009) defines that cloud computing as simply the sharing and use of applications and resources of a network environment to get work done without concern about ownership and management of the network's resources and applications.

Consequently, Buyya, Yeo, and Venugopa (2008) noted that the cloud computing is defined as a type of parallel and distributed system consisting of a collection of interconnected and virtualized computers that are dynamically provisioned and presented as one or more unified computing resources based on service-level agreements established during the negotiation between the service provider and consumers. For this viewpoint, cloud computing is a relatively novel distributed computing technology that promises providing services that are scalable through on-demand provisioning of computing resources (Weiss, 2007).

National Institute of Standards and Technology (NIST) is a US federal technology agency that works with industry to develop and apply technology, measurements, and standards defines cloud computing architecture by portraying five essential characteristics (Mell & Grance, 2010):

1. On-demand self-service. A consumer can unilaterally provision computing capabilities, such as server time and network storage, as needed automatically without requiring human interaction with each service's provider.
2. Broad network access. Capabilities are available over the network and accessed through standard mechanisms that promote use by heterogeneous thin or thick client platforms (for example, mobile phones, laptops, and PDAs).
3. Resource pooling. The provider's computing resources are pooled to serve multiple consumers using a multitenant model, with different physical and virtual resources dynamically assigned and reassigned according to consumer demand.
4. Rapid elasticity. Capabilities can be rapidly and elastically provisioned, in some cases automatically, to quickly scale out and rapidly released to quickly scale in.
5. Measured service. Cloud systems automatically control and optimize resource use by leveraging a metering capability at some level of abstraction appropriate to the type of service (for example, storage, processing, bandwidth, and active user accounts). Resource usage can be monitored, controlled, and reported, providing transparency for both the provider and consumer of the utilized service.

Furthermore, cloud computing services is classified into three models, which is referred to as the "SPI Model," where 'SPI' refers to Software, Platform or Infrastructure (as a Service) (Cloud Security Alliance, 2009), respectively (European Network and Information Security Agency, 2009):

1. Software as a service (SaaS): is software offered by a third party provider, available on demand, usually via the Internet configurable remotely. Examples include online word processing and spreadsheet tools, CRM services and web content delivery services (Salesforce CRM, Google Docs, etc).
2. Platform as a service (PaaS): allows customers to develop new applications using APIs deployed and configurable remotely. The platforms offered include development tools, configuration management, and deployment platforms. Examples are Microsoft Azure, Force and Google App engine.
3. Infrastructure as service (IaaS): provides virtual machines and other abstracted hardware and operating systems which may be controlled through a service API. Examples include Amazon EC2 and S3, Terremark Enterprise Cloud, Windows Live Skydrive and Rackspace Cloud.

As well, clouds computing deployment can be divided into: 1) public: available publicly - any organization may subscribe; 2) private: services built according to cloud computing principles, but accessible only within a private network; 3) partner: cloud services offered by a provider to a limited and well-defined number of parties (European Network and Information Security Agency, 2009).

2.3. Security Policies

Security at the application level covers various aspects, including authentication, authorization, message integrity, confidentiality, and operational defense (Kannammal & Iyengar, 2007). Also, the transmission and storage of information in the digital form coupled with the widespread propagation of networked computers has created new concerns for policy (Bronk, 2008). An essential business tool and knowledge-sharing device, the networked computer is not without vulnerability, including the disruption of service and the theft, manipulation, and destruction of electronic data (Bronk, 2008). Therefore, the development of the information security policy is a critical activity (Kadam, 2007). Credibility of the entire information security program of an organization depends upon a well-drafted information security policy (Kadam, 2007). Some authors have studied the effectiveness of the information security policy. The success of the policy is dependent on the way the security contents are addressed in the policy document and how the content is communicated to users (Höne & Eloff, 2002).

In the security policy management, Huong Ngo (1999) noted that the security policy is to the security environment like the law is to a legal system. Without a policy, security practices will be developed without clear demarcation of objectives and responsibility, leading to increased weakness. Therefore, a policy is the start of security management. To integrate all related functions, the policy should be developed at both board and department levels, and executed in conjunction with the IT department and authorized users (Huong Ngo, 1999).

Information protection program should be part of any organization's overall asset protection program. Management is charged to ensure that adequate controls are in place to protect the assets of the enterprise. An information security program that includes policies, standards and procedures will allow management to demonstrate a standard of care (Peltier, 2004). Furthermore, management from all communities of interest, including general employees, IT personnel and information security specialist, must make policies the basis for all information security planning, designing

and deployment (Whitman & Mattord, 2009). Any quality security programs begin and end with policy and security policies are the least cost expensive control to execute, but the most difficult to implement properly (Whitman & Mattord, 2009). Similarly, Swanson and Guttman (1996) stated that policy is senior management's directives to create a computer security program, establish its goals and assign responsibilities. The term policy is also used to refer to the specific security rules for particular systems. Additionally, policy may refer to entirely different matters, such as the specific managerial decisions setting an organization's e-mail privacy policy or fax security policy.

Clearly, the complexity of issues involved means that the size and shape of information security policies may vary widely from a company to a company. This may depend on many factors, including the size of the company, the sensitivity of the business information they own and deal with in their marketplace and the numbers and types of information and computing systems they use (Diver, 2007). Therefore, for small size business the complexity and number of security policies needed are reduced than large size business if the small business implements cloud computing technology model.

The purpose of security policy is to protect people and information, set rules for expected behavior by users, define and authorize the consequences of violation, minimize risks and help to track compliances with regulation (Diver, 2007). Additionally, there are three different types of security policy: enterprise information security policy (EISP), Issue-Specific security policy (ISSP), and System Specific policy (SysSp) (Swanson & Guttman, 1996; Whitman & Mattord, 2009):

EISP is a general security policy and supports the mission, vision and direction of the organization and sets the strategic direction, scope for all security efforts (Whitman & Mattord, 2009). Furthermore, EISP should: 1) Create and define a computer security program; 2) Set organizational strategic directions; 3) Assign responsibilities and address compliance issues.

ISSP should 1) Address specific areas of technology such as e-mail usage, internet usage, privacy, corporate's network usage; 2) Requires consistent updates as changes in technology take place; 3) Contains issue statement, applicability, roles and responsibilities, compliance and point of contact.

SysSP should 1) Focus on decisions taken by management to protect a particular system, such as defining the extent to which individuals will be held accountable for their actions on the system, should be explicitly stated; 2) Be made by a management official. The decision management makes should be based on a technical analysis; 3) Vary From System to System. Variances will occur because each system needs defined security objectives based on the system's operational requirements, environment, and the From System to System. Variances will occur because each system needs defined security objectives based on the system's operational requirements, environment, and the manager's acceptance of risk. 4) Be Expressed as Rules. Who (by job category, organization placement, or name) can do what (e.g., modify, delete) to which specific classes and records of data, and under what conditions. Furthermore, SysSP can be divided into managerial guidance created by management to guide the implementation and configuration of technology and behavior of people and technical specification, which are an implementation of the managerial policy.

Hence, the real and fundamental success of security policy is actually needed to be maintained, distributed, read, understood, agreed and signed by employees and enforced by the organization in order to be effective (Whitman & Mattord, 2009). The main focus of small business is the ISSP, with less emphasis on EISP and not as much of on SysSP. Part of EISP can be incorporated into ISSP since small business doesn't have a complicated infrastructure.

3. SECURITY POLICY FOR SMALL BUSINESS USING CLOUD COMPUTING

Obviously, as a small business relying on cloud computing to execute business transactions this can be translated into saving of IT infrastructures, servers, labors and complexity. Therefore, a small business implementing cloud computing will need fewer numbers of security policies comparing to a large enterprise to manage the network infrastructures and servers; especially that security policies are the least cost expensive control to execute (Whitman & Mattord, 2009).

However, for this to be realized, a small business infrastructure can consist of a thin client interface as a web browser using a thin-client computer or laptop or desktop, switch, network cables and internet line as presented in Figure 1.

From literature reviews, many small businesses don't have the resources or capital to develop in-house applications, or hire technical. Therefore, for small business the SaaS might be feasible and easier to implement than any other kind of cloud service. For example, if a small business employed cloud SaaS, then the organization doesn't manage or control the fundamental cloud infrastructure, including network, servers, operating systems, storage, even individual application capabilities, with the possible exception of limited user specific application configuration settings (Cloud Security Alliance, 2009). On the other hand, in PaaS service model, the consumer does not manage or control the underlying cloud infrastructure including network, servers, operating systems, or storage, but has control over the deployed applications and possibly application hosting environment configurations (Cloud Security Alliance, 2009) and in this model it gets more complicated for small businesses. Also, for cloud IaaS, the consumer does not manage or control the underlying cloud infrastructure but has control over operating systems, storage, deployed applications, and possibly limited control

Figure 1. Typical network architecture for small business

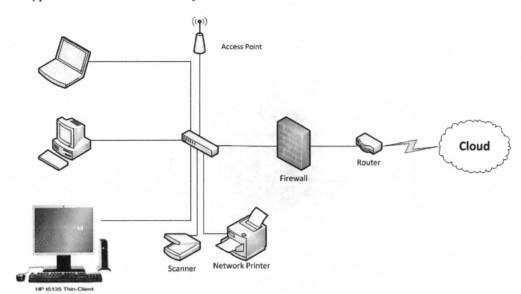

of select networking components (e.g., host fire-walls) (Cloud Security Alliance, 2009) and this is considered as the most complicated setup for small business, since this might require personnel with information technology background.

Any small business utilizing SaaS to conduct their day-to-day operation will need a certain kind of security policies for using information systems. In addition to many factors that must be taken into account, including audience type and company business and size (Diver, 2007). The key to ensuring that your company's security policy is useful and useable is to develop a suite of policy documents that match your audience and marry with existing company policies. Policies must be useable, workable and realistic (Diver, 2007). A small business might not have the internal expertise to create security policies; therefore, it is better and feasible to consult a third party to study the current infrastructure and requirements. The successful deployment of a security policy is closely related not only to the complexity of the security requirements but also to the capabilities/functionalities of the security devices (Preda et al., 2009).

In this regard, this paper has proposed the authors' viewpoint on the security policies for small business considering implementing cloud computing technologies, which requires fewer number of security policies as proposed in Figure 2.

Based on the above figure, are recommended security policies that can be implemented for small business with a brief description for each. Some of these policies can be combined together or might be not applicable due to usage of thin client computer versus laptop or desktop as an example:

- **Internet Usage Policy (IUP):** it applies to all internet users (individuals working for the Company, including permanent full-time and part-time employees, contract workers, temporary agency workers, business partners, and vendors) who access the Internet through the computing or networking resources. The Company's Internet users are expected to be familiar with and to comply with this policy, and are also required to use their common sense and exercise their good judgment while using Internet services (SANS,

Figure 2. Proposed the model for types of security policy applicable for small business

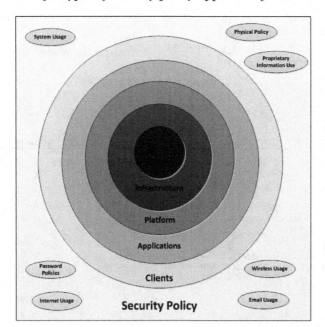

2006). Since the internet might create the possibilities of infection to the company's system via viruses, spyware, adware or Trojan. Therefore, the objective of IUP is to protect the internet resources from abusing by employees surfing the internet for non-business related or to obtain, view any pornographic or unethical. The IUP provides a clear distinction between personal use and work use related. Therefore, the depending of small business on the internet to conduct their tasks via cloud computing model requires implementing this policy to conserve the internet line usage and to protect the assets.

- **Email Usage Policy (EUP):** This policy covers the appropriate use of any email sent from a Company's email address and applies to all employees, vendors, and agents operating on behalf of the Company. The email system shall not to be used for the creation or distribution of any disruptive or offensive messages (SANS, 2006). This policy protects against unauthorized

email usage, distribution of non-work related emails, malware infections, and lost productivity. Also, this policy prohibits employees to send any confidential information using email to outside and to send pornographic jokes or stories which might be considered as a sexual harassment. Furthermore, this policy is directed against employees by virtue of any protected classification including race, gender, nationality, religion, and so forth, will be dealt with according to the harassment policy. Therefore, when small business decides to host their email system into the clouds, the EUP can protect the small business's email assets.

- **System usage policy (SUP):** it protects against program installation without authorization such as: no Instant Messaging, no file sharing software. Furthermore, SUP specifies the restrictions on use of your account or password (not to be given away) (Computer Technology Documentation Project, 2010). Information systems such as VoIP phones, email, web, software, printers, network, computers, computer accounts, video system and smart phones are for use of the employees to support the business. Therefore, SUP helps to ensure the system used to support the business is protected against unauthorized activities.

- **Wireless Security Policy (WSP):** The purpose of this policy is to secure and protect the information assets owned by the Company. This policy applies to all wireless infrastructure devices that connect to a Company's network or reside on a Company's site that provides wireless connectivity to endpoint devices including, but not limited to, laptops, desktops, cellular phones, and personal digital assistants (PDAs) (SANS, 2006). WSP protects against users attempting to modify or extend the network, any effort to break into or gain unauthorized access to the system and running data packet collection programs. Therefore, WSP ensures the protection of company's assets while using this technology to reduce the usage of cables and switches installation to facilitate the implementation of the cloud computing model.

- **Physical Policy (PhyP):** The purpose of this document is to provide guidance for Visitors to Company's premises, as well as for employees sponsoring Visitors to the Company (SANS, 2006). Also, PhyP can be used for securing network switches, access points and cables against unauthorized usage. Therefore, PhyP protects company's assets against theft, modification and unauthorized usage.

- **Password Policies (PassP):** This policy is to help keep user accounts secure. It defines how often users must change their passwords, how long they must be, complexity rules (types of characters used such as lower case letters, uppercase letters, numbers,

and special characters), and other items (Computer Technology Documentation Project, 2010). Furthermore, PassP protects against sharing of passwords among employees and disclose of important passwords to unauthorized personnel.

- **Proprietary Information Use (PIU):** Acceptable use of any proprietary information owned by the company. It defines where it can be stored and where it may be taken, how and where it can be transmitted (Computer Technology Documentation Project, 2010).

Security policies need tools to support it such as anti-virus, web security, URL filtering, data loss prevention. These tools will add cost, but some cloud providers provide protection against viruses, spyware, botnets and thousands of other advanced internet based threats with low monthly cost based on SaaS service without the need to deploy and manage security appliances and PC-based anti-virus/firewall utilities (Moran, 2010). This works by simply point all their internet traffic to certain web security cloud application. This cloud application protects small businesses against viruses, spyware, botnets and thousands of other advanced internet based threats. Furthermore, this cloud application provides a full suite of URL filtering capabilities. Therefore, unwanted URL categories can be blocked, while productivity draining categories can be controlled. Moreover, overall internet usage can be easily monitored with a simple cloud based reporting portal.

The implementation of security policy requires training and awareness sessions to ensure that employees understand their responsibilities toward company's assets. Therefore, the implementation can only be done by a major drive to educate everyone. Furthermore, the right message should reach the right people. The training programs have to be designed keeping in mind the actual groups being addressed. The trainer has to use easy language to the audience without technical jargon (Kadam, 2007). Moreover, it is recommended to cover only the relevant policies for each group depending on their job role with possible of customizing training program (Kadam, 2007). Finally, each policy document should be updated regularly. At a minimum, an annual review strikes a good balance, ensuring that policy does not become out of date due to changes in technology or implementation (SANS, 2006).

4. CONCLUSION

Cloud computing model is a relatively a new topic which has not been adequately researched up to now. Thus, this paper presents a suggestion for securing the internal infrastructure for small business using SaaS cloud computing service. To the best of our knowledge, this is the first theoretical study to provide a comprehensive identify internal risks to small business using cloud computing, Most importantly, this research describes the benefit of security policies and the benefit of cloud computing for small business in terms of reduction to cost and risk. The motivation for this will primarily be cost savings since the greater the success. The aim is to encourage small business to migrate to cloud computing model by portraying what is needed to secure their infrastructure using traditional security policies without the complexity used in large corporations.

This research suggests using specific applicable security policies to reduce the risks and encourage small business to migrate to cloud computing technologies.

Through an empirical study, the findings of this study can provide a foundation that can facilitate further study in cloud computing for small business to reduce cost and risk. Hopefully, the research is not intended to describe the process of designing security policy, on the other hand, the intention is this research is to contribute to the understanding of risks and security requirement for small businesses implementing or plan to implement cloud computing model. The research has succeeded in proposing security reduction techniques against internal risks for small business by implementing certain types of security policies.

The paper has confirmed the proposed model which satisfies the research aim. Also, the paper revealed a considerable number of interesting issues that would require future study such as: investigate and enhance the predictive power of the model proposed in this research. One major direction for further research would be geared towards to improve risk reduction by automated security policy creations based on small business requirements. Also, a future research can be amid on developing the appropriate security policies for small business with in-house technical capabilities using PaaS cloud service model.

REFERENCES

Amir, M. S. (2010). It's written in the cloud: the hype and promise of cloud computing. *Journal of Enterprise Information Management, 23*(2), 131–134. doi:10.1108/17410391011019732

Bronk, C. (2008). Hacking the nation-state: Security, information technology and policies of assurance. *Information Security Journal: A Global Perspective, 17*(3), 132-142.

Buyya, R., Yeo, C. S., & Venugopal, S. (2008). *Market-oriented cloud computing: Vision, hype, and reality for delivering IT Services as computing utilities*. In *Proceedings of the 10th IEEE International Conference on High Performance Computing and Communications* (p. 1).

Cervone, H. F. (2010). An overview of virtual and cloud computing. *OCLC Systems & Services, 26*(3), 162–165. doi:10.1108/10650751011073607

Cesare, S., & Xiang, Y. (2010). Classification of malware using structured control flow. In *Proceedings of the Eighth Australasian Symposium on Parallel and Distributed Computing*, Brisbane, Australia. (Vol. 107).

Clarke, R. (2010). User requirements for cloud computing architecture. In *Proceedings of the 10th IEEE/ACM International Conference on Cluster, Cloud and Grid Computing* (pp. 623-630).

Cloud Computing World. (2010). *Why your small business needs cloud computing.* Retrieved from http://www.cloudcomputingworld.org/cloud-computing-for-businesses/why-your-small-business-needs-cloud-computing.html

Cloud Security Alliance. (2009). *Security guidance for critical areas of focus in cloud computing V2.1.* Retrieved from http://www.privatecloud.com/2010/01/26/security-guidance-for-critical-areas-of-focus-in-cloud-computing-v2-1/?fbid=uCRTvs9w3Cs

Computer Technology Documentation Project. (2010). *Network and computer security tutorial 2010.* Retrieved from http://www.comptechdoc.org/

D'Arcy, J., & Hovav, A. (2007). Deterring internal information systems misuse. *Communications of the ACM, 50*(10), 113–117. doi:10.1145/1290958.1290971

Diver, S. (2007). *Information security policy – a development guide for large and small companies* (p. 43). Reston, VA: SANS Institute.

European Network and Information Security Agency. (2009). *Cloud computing - benefits, risks and recommendations for information security.* Retrieved from http://itlaw.wikia.com/wiki/Cloud_Computing:_Benefits,_Risks,_and_Recommendations_for_Information_Security

Greengard, S. (2010). Cloud computing and developing nations. *Communications of the ACM, 53*(5), 18–20. doi:10.1145/1735223.1735232

Hayes, B. (2008). Cloud computing. *Communications of the ACM, 51*(7), 9–11. doi:10.1145/1364782.1364786

Heath, N. (2010). *How cloud computing will help government save taxpayer £3.2bn.* Retrieved from http://www.silicon.com/management/public-sector/2010/01/27/how-cloud-computing-will-help-government-save-taxpayer-32bn-39745389/

Höne, K., & Eloff, J. H. O. (2002). Information security policy — what do international information security standards say? *Computers & Security, 21*(5), 382–475. doi:10.1016/S0167-4048(02)00504-7

Huong Ngo, H. (1999). Corporate system security: towards an integrated management approach. *Information Management & Computer Security, 7*(5), 217–222. doi:10.1108/09685229910292817

Kadam, A. W. (2007). Information security policy development and implementation. *Information Systems Security, 16*(5), 246–256. doi:10.1080/10658980701744861

Kannammal, A., & Iyengar, N. C. S. N. (2007). A model for mobile agent security in e-business applications. *International Journal of Business and Information, 2*(2), 185–198.

Katharine, S., & David, B. (2010). Current state of play: Records management and the cloud. *Records Management Journal, 20*(2), 217–225. doi:10.1108/09565691011064340

Mark-Shane, E. S. (2009). Cloud computing and collaboration. *Library Hi Tech News, 26*(9), 10–13. doi:10.1108/07419050911010741

Mell, P., & Grance, T. (2010). The NIST definition of cloud computing. *Communications of the ACM, 53*(6), 50–50.

Moran, J. (2010). *ZScaler web security cloud for small business.* Retrieved from http://www.smallbusinesscomputing.com/webmaster/article.php/3918716/ZScaler-Web-Security-Cloud-for-Small-Business.htm

Morris, M. (n. d.). *Employee theft schemes.* Retrieved from http://www.cowangunteski.com/documents/EmployeeTheftSchemes_001.pdf

Peltier, T. R. (2004). Developing an enterprisewide policy structure. *Information Systems Security, 13*(1), 44–50. doi:10.1201/1086/44119.13.1.20040301/80433.6

Preda, S., Cuppens, F., & Cuppens-Boulahia, N. Alfaro.J., Toutain, L., & Elrakaiby, Y. (2009). Semantic context aware security policy deployment. In *Proceedings of the 4th International Symposium on Information, Computer, and Communications Security*, Sydney, Australia.

SANS. (2006). *Information security policy templates.* Reston, VA: SANS Institute.

Srinivasan, M. (2010). Cloud security for small businesses. In Proceedings of the *Allied Academies International Conference of the Academy of Information &. Management Science, 14,* 72–73.

Standage, T. (2002). The weakest link. *The Economist,* 11-14.

Swanson, M., & Guttman, B. (1996). *Generally accepted principles and practices for securing information technology systems.* Retrieved from http://csrc.nist.gov/publications/nistpubs/800-14/800-14.pdf

Twitchell, D. P. (2006). Social engineering in information assurance curricula. In *Proceedings of the 3rd Annual Conference on Information Security Curriculum Development*, Kennesaw, GA.

Weiss, A. (2007). Computing in the clouds. *netWorker, 11*(4), 16–25. doi:10.1145/1327512.1327513

Whitman, M., & Mattord, H. (2009). *Principles of information security* (3rd ed.). Boston, MA: Course Technology.

Wittow, M. H., & Buller, D. J. (2010). Cloud computing: Emerging legal issues for access to data, anywhere, anytime. *Journal of Internet Law, 14*(1), 1–10.

Yogesh, K. D., & Navonil, M. (2010). It's unwritten in the Cloud: The technology enablers for realising the promise of cloud computing. *Journal of Enterprise Information Management, 23*(6), 673–679. doi:10.1108/17410391011088583

Zscaler. (2010). *Zscaler web security cloud for small business.* Retrieved from http://www.zscaler.com/pdf/brochures/ds_zscalerforsmb.pdf

This work was previously published in the International Journal of Cloud Applications and Computing (IJCAC), Volume 1, Issue 2, edited by Shadi Aljawarneh & Hong Cai, pp. 29-40, copyright 2011 by IGI Publishing (an imprint of IGI Global).

Chapter 9
The Financial Clouds Review

Victor Chang
University of Southampton and University of Greenwich, UK

Chung-Sheng Li
IBM Thomas J. Watson Research Center, USA

David De Roure
University of Oxford, UK

Gary Wills
University of Southampton, UK

Robert John Walters
University of Southampton, UK

Clinton Chee
Commonwealth Bank, Australia

ABSTRACT

This paper demonstrates financial enterprise portability, which involves moving entire application services from desktops to clouds and between different clouds, and is transparent to users who can work as if on their familiar systems. To demonstrate portability, reviews for several financial models are studied, where Monte Carlo Methods (MCM) and Black Scholes Model (BSM) are chosen. A special technique in MCM, Least Square Methods, is used to reduce errors while performing accurate calculations. Simulations for MCM are performed on different types of Clouds. Benchmark and experimental results are presented for discussion. 3D Black Scholes are used to explain the impacts and added values for risk analysis. Implications for banking are also discussed, as well as ways to track risks in order to improve accuracy. A conceptual Cloud platform is used to explain the contributions in Financial Software as a Service (FSaaS) and the IBM Fined Grained Security Framework. This study demonstrates portability, speed, accuracy, and reliability of applications in the clouds, while demonstrating portability for FSaaS and the Cloud Computing Business Framework (CCBF).

1. INTRODUCTION

The Global economic downturn triggered by the finance sector is an interdisciplinary research question that expertise from different sectors needs to work on altogether. There are different interpretations for the cause of the problem. Firstly,

DOI: 10.4018/978-1-4666-1879-4.ch009

Hamnett (2009) conducted a study to investigate the cause, and concluded unsustainable mortgage lending leads to out of control status and that the housing bubble and subsequent collapse were result of these. Irresponsible mortgage lending was the cause for Lehman Brother collapse that has triggered global financial crisis. Secondly, Lord Turner, Chair of the Financial Service Authority (FSA), is quoted as follows: "The problem, he said, was that banks' mathematical models assumed a 'normal' or 'Gaussian' distribution of events, represented by the bell curve, which dangerously underestimated the risk of something going seriously wrong" (Financial Times, 2009). Thirdly, there are reports showing the lack of regulations on financial practice. Currently there are remedies proposed by several governments to improve on this (City A.M., 2010). All the above suggested possibilities contribute to complexity that caused global downturn. However, Cloud Computing (CC) offers a good solution to deal with challenges in risk analysis and financial modelling. The use of Cloud resources can improve accuracy of risk analysis, and knowledge sharing in an open and professional platform (Chang, Wills, & De Roure, 2010a, 2010c). Rationales are explained as follows. The Clouds provide a common platform to run different modelling and simulations based on Gaussian and non-Gaussian models, including less conventional models. The Clouds offer distributed high-performing resources for experts in different areas within and outside financial services to study and review the modelling jointly, so that other models with Monte Carlo Methods and Black Scholes Models can be investigated and results compared. The Clouds allow regulations to be taken with ease while establishing and reminding security and regulation within the Clouds resources.

2. LITERATURE REVIEW

Literature review is presented as follows. Three challenges in business context and Software as a Service (SaaS) are explained. This paper is focused on the third issue, enterprise portability, and how financial SaaS is achieved with portability. Financial models with Monte Carlo methods and Black Scholes models are also explained.

2.1. Three Challenges in Business Context

There are three Cloud Computing problems experienced in the current business context (Chang, Wills, & De Roure, 2010b, 2010c). Firstly, all cloud business models and frameworks proposed by several leading researchers are either qualitative (Briscoe & Marinos, 2009; Chou, 2009; Weinhardt et al., 2009; Schubert, Jeffery, & Neidecker-Lutz, 2010) or quantitative (Brandic et al., 2009; Buyya et al., 2009; Patterson et al., 2009). Each framework is self-contained, and not related to others' work. There are few frameworks or models which demonstrate linking both quantitative and qualitative aspects, and when they do, the work is still at an early stage.

Secondly, there is no accurate method for analysing cloud business performance other than the stock market. A drawback with the stock market is that it is subject to accuracy and reliability issues (Chang, Wills, & De Doure, 2010a, 2010c). There are researchers focusing on business model classifications and justifications for which cloud business can be successful (Chou, 2009; Weinhardt et al., 2009). But these business model classifications need more cases to support them and more data modelling to validate them for sustainability. Ideally, a structured framework is required to review cloud business performance and sustainability in systematic ways.

Thirdly, communications between different types of clouds from different vendors are often difficult to implement. Often work-arounds require writing additional layers of APIs, or an interface or portal to allow communications. This brings interesting research questions such as portability, as portability of some applications from desktop to cloud is challenging (Beaty et al., 2009; Patterson et al., 2009). Portability refers to moving enterprise applications and services, and not just files or VM over clouds.

2.2. Financial Models

Gaussian-based mathematical models have been frequently used in financial modelling (Birge & Massart, 2001). As the FSA has pointed out, many banks' mathematical models assumed normal (Gaussian) distribution as an expected outcome, and might underestimate the risk for something going wrong. To address this, other non-Gaussian financial models need to be investigated and demonstrated for how financial SaaS can be successfully calculated and executed on Clouds. Based on various studies (Feiman & Cearley, 2009; Hull, 2009), one model for pricing and one model for risk analysis should be selected respectively. A number of methods for calculating prices include Monte Carlo Methods (MCM), Capital Asset Pricing Models and Binomial Model. However, the most commonly used method is MCM since MCM is commonly used in stochastic and probabilistic financial models, and provides data for investors' decision-making (Hull, 2009). MCM is thus chosen for pricing. On the other hand, methods such as Fourier series, stochastic volatility and Black Scholes Model (BSM) are used for volatility. As a main stream option, BSM is selected for risk analysis, since BSM has finite difference equations to approximate derivatives. Origins in literature and mathematical formulas in relation to MCM and BSM are presented in the next two sections.

2.2.1. Monte Carlo Methods in Theory

Monte Carlo Simulation (MCS), originated from mathematical Monte Carlo Methods, is a computational technique used to calculate risk analysis and the probability of an event or investment to happen. MCS is based on probability distributions, so that uncertain variables can be described and simulated with controlled variables (Hull 2009; Waters 2008). Originated from Physics, Brownian Motions follow underlying random variables can influence the Black-Scholes models, where stock price becomes

$$dS = \mu S dt + \sigma S dW_t \qquad (1)$$

where W is Brownian the dW term here stands in for any and all sources of uncertainty in the price history of the stock. The time interval is divided into M units of length δt from time 0 to T in a sampling path, and the Brownian motion over the interval dt are approximated by a single normal variable of mean 0 and variance δt, and leading to

$$S(k\delta t) = S(0) \exp\left(\sum_{i=1}^{k}\left[\left(\mu - \frac{\sigma^2}{2}\right)\delta t + \sigma\varepsilon_i\sqrt{\delta t}\right]\right) \qquad (2)$$

for each k between 1 and M, and each ε_i is drawn from a standard normal distribution. If a derivative H pays the average value of S between 0 and T then a sample path ω corresponds to a set $\{\varepsilon_1, ..., \varepsilon_M\}$ and hence,

$$H(\omega) = \frac{1}{M+1}\sum_{k=0}^{M} S(k\delta t) \qquad (3)$$

The Monte Carlo value of this derivative is obtained by generating N lots of M normal variables, creating N sample paths and so N values of H, and then taking the mean. The error has order $\varepsilon = O(N^{-1/2})$ convergence in standard deviation based on the central limit theorem.

2.2.2. Black Scholes Model (BSM)

The BSM is commonly used for financial markets and derivatives calculations. It is also an extension from Brownian motion. The BSM formula calculates call and put prices of European options (a financial model) (Hull, 2009). The value of a call option for the BSM is

$$C(S,t) = SN(d_1) - Ke^{-r(T-t)}N(d_2) \qquad (4)$$

where

$$d_1 = \frac{\ln(\frac{S}{K}) + (r + \frac{\sigma^2}{2})(T-t)}{\sigma\sqrt{T-t}}$$

and

$$d_2 = d_1 - \sigma\sqrt{T-t}$$

The price for the put option is

$$P(S,t) = Ke^{-r(T-t)} - S + (SN(d_1) - Ke^{-r(T-t)}N(d_2))$$
$$= Ke^{-r(T-t)} - S + C(S,t)$$

$$(5)$$

For both formulas (Hull, 2009),

- $N(\bullet)$ is the cumulative distribution function of the standard normal distribution
- T - t is the time to maturity
- S is the spot price of the underlying asset
- K is the strike price
- r is the risk free rate
- σ is the volatility in the log-returns of the underlying asset.

2.3. Least Square Methods (LSM) for Monte Carlo Simulations (MCS)

Variance Gamma Processes are used in our previous papers (Chang, Wills, & De Roure, 2010a, 2010c), and although it reduces errors while calculating pricing and risk analysis on Clouds, it can only go up to 20,000 simulations in one go before performance drops off. In addition, it takes approximately 10 seconds for error correction due to stratification of sampling, although it takes less than 1 second for 5,000 simulations per attempt for executing financial applications with Octave 3.2.4 on Clouds. This leads us to investigate other methodology that can offer much more simulations to be executed in one go, in other words, improvements in performance on Clouds while maintaining accuracy and quality of our simulations. Monte Carlo Methods (MCM)

are used in our simulations, and this means other methods supporting MCM are required to meet our objectives. Various methods such as stochastic simulation, Terms Structure Models (Piazzesi, 2010), Triangular Methods (Mullen et al., 1988; Mullen & Ennis, 1991), and Least Square Methods (LSM) are studied (Longstaff & Schwartz, 2001; Moreno & Navas, 2001; Choudhury et al., 2008). LSM is chosen because of the following advantages. Firstly, LSM provides a direct method for problem solving, and is extremely useful for linear regressions. LSM only needs a short starting time, and is therefore a good choice. Secondly, Terms Structure Models and Triangular Methods are not necessarily used in the Clouds. LSM can be used in the Clouds, because often jobs that require high computations in the Clouds, need extensive resources and computational powers to run. LSM is suitable if a large problem is divided into several sections where each section can be calculated swiftly and independently. This also allows improvements in efficiency.

Here is the explanation for the LSM. There is a data set $(x_1, y_1), (x_2, y_2), ..., (x_n, y_n)$ and the fitting curve $f(x)$ has the deviation $d_1, d_1, ..., d_n$ which are caused from each data point, the least square method produces the best fitting curve with the property as follows

$$Minimum\ Least\ Square\ Error(\prod) =$$
$$d_1^2 + d_2^2 + ... + d_{x-1}^2 = \sum_{i-1}^{n} d_i^2 = \sum_{i=1}^{n} [y_i - f(x_i)]^2$$

$$(6)$$

The least squares line method uses an equation $f(x) = a + bx$ which is a line graph and describes the trend of the raw data set $(x_1, y_1), (x_2, y_2), ..., (x_n, y_n)$. The n should be greater or equal to 2 $(n \geq 2)$ in order to find the unknowns a and b. So the equation for the least square line is

$$\prod = d_1^2 + d_2^2 + ... + d_n^2 = \sum_{i=1}^{n} d_1^2 = \sum_{i=1}^{n} [y_i - (a + bx_i)]^2$$

$$(7)$$

The least squares line method uses an equation $f(x) = a + bx + cx^2$ which is a parabola graph. The n should be greater or equal to 3 $(n \geq 3)$ in order to find the unknowns a, b, and c. When you get the first derivatives of $\prod$ in parabola, you will have

$$\prod = d_1^2 + d_2^2 + \ldots + d_{n-1}^2 + d_n^2 =$$
$$\sum_{i=1}^{n} d_i^2 = \sum_{i=1}^{n} [y_i - (\sum_{i=1}^{n} a + \sum_{i=1}^{n} bx_i + \sum_{i=1}^{n} cx_i^2)]^2 \tag{8}$$

The LSM has been mathematically proven, and allows advanced calculations of complex systems. The LSM is the most suitable for a complex problem divided into several sections where each section runs its own calculations. These complex systems include robot, financial modelling and medical engineering. Longstaff and Schwartz (2001) have developed an algorithm based on LSM Monte Carlo simulations (MCS) to estimate best values precisely. Moreno and Navas (2001) have adopted a similar approach, and demonstrate their algorithm and robustness of LSM MCS for pricing American derivatives. Choudhury (2008) used an approach presented Longstaff and Schwartz, except they focused on code algorithms and performance optimisation. These three papers have demonstrated how LSM can be used for financial computing to achieve accurate estimation and optimisation. Abdi (2009) demonstrate that LSM is very useful for regression and explain why LSM is popular and versatile for calculations. He also states the drawback is that LSM does not cope well with extreme calculations, but such volatile calculations will be handled by 3D Black Scholes (Section 4).

2.4. The Cloud Computing Business Framework

To address the three challenges in business context earlier, the Cloud Computing Business Framework (CCBF) is proposed. The core concept of CCBF is an improved version from Weinhardt's et al. (2009) Cloud Business Model Framework (CBMF) where they demonstrate how technical solutions and Business Models fit into their CBMF. The CCBF is proposed to deal with four research problems:

1. Classification of business models with consolidation and explanations of its strategic relations to IaaS, PaaS and SaaS.
2. Accurate measurement of cloud business performance and ROI.
3. Dealing with communications between desktops and clouds, and between different clouds offered by different vendors, which focus on enterprise portability.
4. Providing linkage and relationships between different cloud research methodologies, and between IaaS, PaaS, SaaS and Business Models.

The Cloud Computing Business Framework is a highly-structured conceptual and architectural framework to allow a series of conceptual methodologies to apply and fit into Cloud Architecture and Business Models. Based on the summary in Section 2.1, our research questions can be summed up as: (1) Classification; (2) Sustainability; (3) Portability and (4) Linkage. This paper focuses on the third research question, Portability, which is described as follows.

Portability*: This refers to enterprise portability, which involves moving the entire application services from desktops to clouds and between different clouds. For financial services and organisations that are not yet using clouds, portability involves a lot of investment in terms of outsourcing, time and effort, including rewriting APIs and additional costs. This is regarded as a business challenge. Portability deals with IaaS, PaaS and SaaS. Examples in Grid, Health and Finance will be demonstrated. Financial SaaS (FSaaS) Portability is the focus for this paper.*

2.5. Financial Software as a Service (FSaaS)

In relation to finance, portability is highly relevant. This is because a large number of financial applications are written for desktops. There are financial applications for Grid but not all of them are portable onto clouds. Portability often requires rewrites in software design and the API suitable for clouds. Apart from portability, factors such as accuracy, speed, reliability and security of financial models from desktop to clouds must be taken into consideration. The second problem related to finance is there are few financial clouds as described in opening section. Salesforce offers CRM but it is not directly related to financial modelling (FM). Paypal is a payment system and not dealing with financial modelling. Enterprise portability from desktops to clouds, and between different clouds, is useful for businesses and financial services, as they cannot afford to spend time and money migrating the entire applications, API libraries and resources to clouds. Portability must be made as easy as possible. However, there are more advantages in moving all applications and resources to clouds. These added values include the following benefits:

- The community cloud: this encourages groups of financial services to form an alliance to analyse complex problems.
- Risk reduction: financial computing results can be compared and jointly studied together to reduce risks. This includes running other less conventional models (non-Gaussians) to exploit causes of errors and uncertainties. Excessive risk taking can be minimised with the aid of stricter regulations.

Financial Software as a Service (FSaaS) is the proposal for dealing with two finance-specific problems. FSaaS is designed to improve the accuracy and quality of both pricing and risk analysis. This is essential because incorrect analysis or excessive risk taking might cause adverse impacts such as financial loss or severe damage in credibility or credit crunch. Research demonstration is on SaaS, which means it can calculate best prices or risks based on different values in volatility, maturity, risk free rate and so forth on cloud applications. Different models for FSaaS are presented and explained from Section 2.3 onwards, in which Monte Carlo Methods (MCM) and Black Scholes Models (BSM) will be demonstrated as the core models used in FSaaS.

3. FaaS PORTABILITY: MONTE CARLO SIMULATIONS WITH LEAST SQUARE METHODS

This section describes how Financial SaaS portability on clouds can be achieved. This mainly involves Monte Carlo Methods (MCM) and Black Scholes Model (BSM). Before describing how they work and how validation and experiments are done, current practice in Finance is presented as follows. Mathematical models such as MCM are used in Risk Management area, where models are used to simulate the risk of exposures to various types of operational risks. Monte Carlo Simulations (MCS) in Commonwealth Bank Australia are written in Fortran and C#. Such simulations take several hours or over a day (Chang, Wills, & De Roure, 2010c). The results may be needed by the bank for the quarterly reporting period.

Monte Carlo Methods (MCM) are suitable to calculate best prices for buy and sell, and provides data for investors' decision-making (Waters, 2008). MATLAB is used due to its ease of use with relatively good speed. While the volatility is known and provided, prices for buy and sale can be calculated. Chang, Wills, and De Roure (2010b, 2010c) have demonstrated their examples on how to calculate both call and put prices, with their respective likely price, upper limit and lower limit.

3.1. Motivation for Using the Least Square Method

As discussed in Section 2.3, Variance-Gamma Processes (VGP) with Financial Clouds and FSaaS with error reductions are demonstrated by Chang, Wills, and De Roure (2010a, 2010b, 2010c). It has two drawbacks: (1) the program focuses on error correction, which takes time, and seems to make the program slow to start; and (2) 20,000 simulations per attempt is the optimum. This is perhaps because of the high amount of memory required for VGP. Improvements are necessary, including the use of another HPC language or a better method. Adopting a better methodology not only enhances performance but also resolves some aspects of challenges. Barnard et al. (2003) demonstrate that having the right method is more important than using a particular language.

The Least Square Methods (LSM) fits into the improvement plan with the following rationale. Firstly, LSM provides a quick execution time, more than 50% compared with VGP (as shown in Section 5). Secondly, it allows the number of simulations to be pushed to 100,000 in one go, before encountering issues such as stability and performance. By offering these two distinct advantages over VGP, LSM is therefore a more suitable method for FSaaS to achieve speed, accuracy and performance. In addition, LSM has been extensively used in robots, or intelligent systems where a major problem is divided into sections, and each section is performed with fast and accurate calculations.

3.2. Coding Algorithm for the Least Square Method

This section describes the coding algorithm for the Least Square Method. Table 1 shows the initial part of the code, where key figures such as maturity, volatility and risk free rate are given. This allows us to calculate and track call prices if variations for maturity, risk free rate and volatility change.

Table 1. The first part of coding algorithm for LSM

```
S=100; %underlying price
X=100; %strike
T=1; %maturity
r=0.04; %risk free rate
dividend=0;
v=0.2; % volatility
nsimulations=10000; % No of simulations, which can be
updated
nsteps=10; % 10 steps are taken. Can be changed to 50, 100,
150 and 200 steps.
CallPutFlag="p";
%%%%%%%%%%%%%%%%%%%%%%%%%%%%%%%%%%
%AnalyAmerPrice=BjerkPrice(CallPutFlag,S,X,r,dividend,v
,T)
r=r-dividend; %risk free rate is unchanged
%AnalyEquropeanPrice=BlackScholesPrice(CallPutFlag,S,X
,T,r,v)
if CallPutFlag=="c",
 z=1;
else
 z=-1;
end;
```

Similarly, we can modify our code to track volatility for risk analysis if other variables are changed.

Both American price and European price methods are commonly used in Monte Carlo Simulations (Hull, 2009). It is an added value to calculate both prices in one go, and so both options are included in our code.

The next step involves defining the three important variables for both American and European options, which include cash flow from continuation (CC), cash flow from exercise (CE) and exercise flag (EF), shown in Table 2. The 'for' loop is to start the LSM process. Table 3 shows how the three variables CC, CE and EF are updated.

Table 4 shows the main body of LSM calculations. The 'regrmat' function is used to perform regression of continuation value. This value is calculated, and fed into the 'ols' function, which is a built-in function offered by open-source Octave to calculate ordinary LSM estimation. The p value is the outcome of the 'ols' function, which is then used to determine final values of CC, EF and CE. In MATLAB, the equivalent function is 'lscov' for the LSM.

Table 2. The second part of coding algorithm for the LSM

```
smat=zeros(nsimulations,nsteps);
CC=zeros(nsimulations,nsteps); %cash flow from continuation
CE=zeros(nsimulations,nsteps); %cash flow from exercise
EF=zeros(nsimulations,nsteps); %Exercise flag
dt=T/(nsteps-1);
smat(:,1)=S;

drift=(r-v^2/2)*dt;
qrdt=v*dt^0.5;
for i=1:nsimulations,
 st=S;
 curtime=0;
 for k=2:nsteps,
 curtime=curtime+dt;
 st=st*exp(drift+qrdt*randn);
 smat(i,k)=st;
 end
end
```

Table 5 shows the last part of the algorithm for the LSM. EF, calculated in Table 4 is used to decide values of an important variable 'payoff_sum', which is then used to calculate the best price for American and European options.

Table 4. The fourth part of coding algorithm for the LSM

```
for k=nsteps-1:-1:2,
 st=smat(:,k);
 CE(:,k)=max(z*(st-X),0);

%Only the positive payoff points are input for regression
 idx=find(CE(:,k)>0);
 Xvec=smat(idx,k);
 Yvec=CC(idx,k+1)*exp(-r*dt);
 % Use regression - Regress discounted continuation value at the
 % next time step to S variables at current time step
 regrmat=[ones(size(Xvec,1),1),Xvec,Xvec.^2];

p=ols(Yvec,regrmat); %p = lscov(Yvec, regrmat) for MAT-
LAB CC(idx,k)=p(1)+p(2)*Xvec+p(3)*Xvec.^2;
 %If exercise value is more than continuation value, then
 %choose to exercise
 EF(idx,k)=CE(idx,k) > CC(idx,k);
 EF(find(EF(:,k)),k+1:nsteps)=0;
 paramat(:,k)=p;
 idx=find(EF(:,k) == 0);
 %No need to store regressed value of CC for next use
 CC(idx,k)=CC(idx,k+1)*exp(-r*dt);
 idx=find(EF(:,k) == 1);
 CC(idx,k)=CE(idx,k);
end
```

Table 3. The third part of coding algorithm for the LSM

```
CC=smat*0; %cash flow from continuation
CE=smat*0; %cash flow from continuation
EF=smat*0; %Exercise flag
st=smat(:,nsteps);
CE(:,nsteps)=max(z*(st-X),0);
CC(:,nsteps)=CE(:,nsteps);
EF(:,nsteps)=(CE(:,nsteps)>0);

paramat=zeros(3,nsteps); %coefficient of basis functions
```

Upon running the MATLAB application, 'lsm', it calculates the best pricing values for American and European options. The following shows the outcome of executing LSM code.

```
> lsm
MCAmericanPrice = 6.3168
MCEuropeanPrice = 5.9421
```

4. A PARTICULAR FSaaS: THE 3D BLACK SCHOLES MODEL BY MATHEMATICA

Black Scholes Model (BSM) has been extensively used in financial modelling and optimisation. Chang, Wills and De Roure (2010a, 2010c) have demonstrated their Black Scholes MATLAB applications running on Clouds for risk analysis.

Table 5. The fifth part of coding algorithm for the LSM

```
payoff_sum=0;
for i=1:nsteps,
 idx=find(EF(:,i) == 1);
 st=smat(idx,i);
 payoffvec=exp(-r*(i-1)*dt)*max(z*(st-X),0);
 payoff_sum=payoff_sum+sum(payoffvec);
end

MCAmericanPrice=payoff_sum/nsimulations
st=smat(:,nsteps);
payoffvec=exp(-r*(nsteps-1)*dt)*max(z*(st-X),0);
payoff_sum=sum(payoffvec);
MCEurpeanPrice=payoff_sum/nsimulations
```

Often risk analysis is presented in visualisation, so that it makes analysis easier to read and understand. MATLAB is useful for calculation and 3D computation, but its 3D computational performance tends to be more time-consuming than Mathematica, which offers commands to compute 3D diagrams swiftly. For this reason, Mathematica is used as the platform for demonstration.

Miller (2009) explain how Mathematica can be used for BSM, and he demonstrates that it is relatively complex to model BSM, so the Black Scholes formulas (BSF) are therefore the best to be expressed in terms of auxiliary function. His rationale is that BSM is based on an arbitrage argument in which any risk premium above the risk-free rate is cancelled out. Hence, both BSF and auxiliary functions take the same five variables as follows.

p = current price of the stock.
k = exercise price of the option.
sd = volatility of the stock (standard deviation of annual rate of return)
r = continuously compounded risk-free rate of return, e.g., the return on U.S. Treasury bills with very short maturities.
t = time (in years) until the expiration date

The first step is to define the auxiliary function, 'AuxBS', which is then used to define Black Scholes function. The code algorithm and formals are presented as follows:

AuxBS[p_,k_,sd_,r_,t_] = (Log[p/k]+r t)/(sd Sqrt[t])+.5 sd Sqrt[t]

This is equivalent to

$$0.5 \; sd \; \sqrt{t} \; +(r \; t+Log[p/k])/(sd \; \sqrt{t} \;) \qquad (9)$$

Similarly, Black Scholes can be defined as:

BlackScholes[p_,k_,sd_,r_,t_] = p Norm[AuxBS[p,k,sd,r,t]]- k Exp[-r t] (Norm[AuxBS[p,k,sd,r,t]-sd Sqrt[t]])

The formula is: -©$^{-rt}$ k Norm[-0.5 sd

$$\sqrt{t} \; +(rt+Log[p/k])/(sd \; \sqrt{t} \;)]+p \; Norm[0.5 \; sd \; \sqrt{t} \; +(r \; t+Log[p/k])/(sd \; \sqrt{t} \;)] \qquad (10)$$

'Norm' is a function in Mathematica to compute complex mathematical modelling such as Gaussian integers, vectors, matrices and so on. By using these two functions effectively, pricing and risks can be calculated and then presented in 3D Visualisation. The advantages are discussed in the next section.

4.1. 3D Black Scholes

Methods such as Fourier series, stochastic volatility and BSM are used for volatility. As a main stream option, BSM is selected for risk analysis in this paper, since BSM has finite difference equations to approximate derivatives. Our previous papers (Chang, Wills, & De Roure, 2010a, 2010c) have demonstrated risk and pricing calculations based on Black Scholes Model (BSM). Results are presented in numerical forms, and occasionally require users and collaborators to visualise some scenarios of numerical computation in their minds. In other papers by Chang, Wills, and De Roure (2010b, 2010c), they demonstrate that Cloud business performance can be presented by 3D Visualisation. Where computational applications can be presented using 3D Visualisation, this can improve usability and understanding (Pajorova & Hluchy, 2010). Currently the focus of MCM is to demonstrate portability on top of computational simulations and modelling in pricing on different Clouds, and this does not need results to be on 3D formats. However, BSM is used to investigate risk. Risk can be difficult to be accurately measured, and models may have possibilities to undermine or miss areas and probability of risk. It is difficult to keep track risks if extreme circumstances happen. The use of 3D Visualisation can help to exploit any hidden errors or missing calculations. Thus, it helps the quality of risk analysis.

4.1.1. Scenarios in Risk Analysis with 3D Visualisation

This section describes some scenarios to calculate and present risks. The first scenario involves investigations of profits/loss in relation to put price. The call price (buying price) for a particular investment is 60 per stock. The put price (selling price) to get zero profit/loss is 60. The risk-free rate, the guarantee rate that will not incur loss, is between 0 and 0.5%. However, the profit and loss will be varied due to impacts of volatility, which means selling price between 50 and 60 will get to a different extent of loss. Similarly, selling prices between 60 and 70 will get to a different extent of profits. The intent is to find out the percentage of profit and loss for massive sale, and the risk associated with it. While using auxiliary and Black Scholes function, the result can be computed in 3D swiftly and presented in Figure 1, which is similar to a 3D parabola.

The second scenario is to identify the best put price for a range of fluctuating volatilities. Volatility is used to quantify the risk of the financial instrument, and is subject to fluctuation that may result in different put prices. The volatility ranges between 0.20% and 0.40%, the best put price is between 6.5 and 9.2, and the risk-free rate is between 0 and 0.5%. The higher the risk is, the more the return will be. However, this situation is reversed when risk (volatility in this case) goes beyond cut-off volatility. Hence, the task is to keep track the risk pattern, and to identify the cut-off point for volatility. Similarly, auxiliary and Black Scholes functions are used to compute 3D Visualisation swiftly, and result is presented in Figure 2 which looks like an inverted-V and shows the best price is 9 while volatility is 0.30.

4.1.2. Delta and Theta: Scenarios in Risk Analysis with 3D Visualisation

In BSM, the partial derivative of an option value with respect to stock price is known as Delta. Hull (2009) and Millers (2009) assert that Delta is useful in risk measurement for an option because it indicates how much the price of an option will respond to a change of price of the stock. Delta is a useful tool in risk management where a portfolio contains more than one option of the stock. The derivative function, D, is built in Mathematica. This much simplifies coding for Delta, which can be presented as

Delta[p_,k_,sd_,r_,t_] = D[BlackScholes[p,k,sd,r,t],p]

Figure 1. The 3D risk analysis to investigate volatile percentage of profits and loss

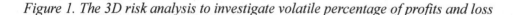

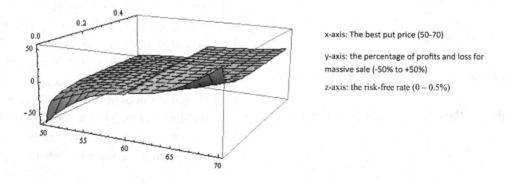

x-axis: The best put price (50-70)

y-axis: the percentage of profits and loss for massive sale (-50% to +50%)

z-axis: the risk-free rate (0 – 0.5%)

Figure 2. The 3D risk analysis to investigate the best put price in relations to fluctuating volatility

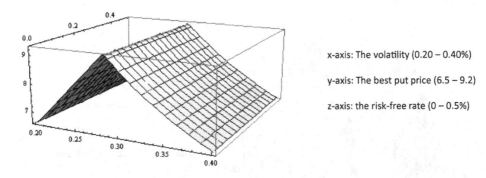

x-axis: The volatility (0.20 – 0.40%)

y-axis: The best put price (6.5 – 9.2)

z-axis: the risk-free rate (0 – 0.5%)

which corresponds to this formula

$$(0.398942\,\frac{1}{2}\,(0.5sd\sqrt{t}\,\frac{rt\,Log-\dfrac{p}{k}-}{sd\sqrt{t}})^2\,)/(sd\sqrt{t})$$

$$-(0.398942\quad rt\,\frac{1}{2}\,(0.5sd\sqrt{t}\,\frac{rt\,Log-\dfrac{p}{k}-}{sd\sqrt{t}})^2\quad k)/$$

$$(psd\sqrt{t})+Norm[0.5\ \ sd\ \ \sqrt{t}+(r\ \ t+Log[p/k])/$$

$$(sd\sqrt{t})]\qquad(11)$$

Delta computes positive derivatives in BSM, and to get an inverted Delta, a new function, Theta, is introduced.

**Theta[p_,k_,sd_,r_,t_] =
-D[BlackScholes[p,k,sd,r,t],t]**

which corresponds to this formula

$$0.398942\ rt\,\frac{1}{2}\,(0.5sd\sqrt{t}\,\frac{rt\,Log-\dfrac{p}{k}-}{sd\sqrt{t}})^2\ k\ (r/$$

$$(sd\sqrt{t})\,\text{-}(0.25\ sd)/\sqrt{t}\,\text{-}(r\ t+Log[p/k])/(2\ sd$$

$$t^{3/2}))\text{-}0.398942\,\frac{1}{2}\,(0.5sd\sqrt{t}\,\frac{rt\,Log-\dfrac{p}{k}-}{sd\sqrt{t}})^2\ p$$

$$(r/(sd\sqrt{t})+(0.25\ sd)/\sqrt{t}\,\text{-}(r\ t+Log[p/k])/(2\ sd$$

$$t^{3/2}))\text{-}\ e^{\text{-}rt}\ k\ r\ Norm[\text{-}0.5\ sd\ \sqrt{t}+(r\ t+Log[p/k])/$$

$$(sd\sqrt{t})]\qquad(12)$$

The third scenario is to investigate the extent of loss in an organisation during the financial crisis between 2008 and 2009, and to identify which put prices (in relations to volatility) will get the least extent of loss while keeping track of risks in 3D. This needs using Theta function to present the risk and pricing in relations high volatility. The put price is between 20 and 100, and the percentage of loss is between -5% and -25%, and the risk-free rate is 0 and 0.5%. In this case, risk-free rate means the percentage this organisation can get assistance from. The Theta function is used to compute the 3D risk swiftly and to get the result in Figure 3. This shows the percentage of loss gets better when the put prices are raised to approximately 55. However, when it gets to 60, this is the price that uncontrolled volatility (such as human speculation or natural disasters) takes hold and the percentage of loss goes down sharply at -25%. The percentage of loss is raised to -5%, and is slowly lowering its value to-25%. However, if the risk-free rate is improved up to 0.5%, the extent of loss is less, and stays nearly at -5%. It means credit guarantee from somewhere may help this organisation with the minimum impacts from loss. However, this is just a computer simulation and does not reflect the real difficulty faced by this organisation. Even so, our FSaaS simulations

Figure 3. The 3D risk analysis to explore the percentage of loss and the best put price in relations to the impact of economic downturn

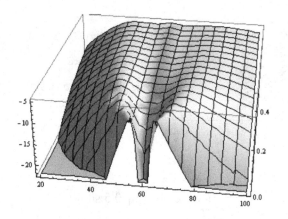

x-axis: The best put price (20 – 100)

y-axis: The percentage of loss (-5 and -25%)

z-axis: the risk-free rate (0 – 0.5%)

can produce a range of likely outcomes, which are valuable to decision-makers.

5. EXPERIMENT AND BENCHMARK IN THE CLOUDS

Monte Carlo Simulations with LSM can be used for FSaaS on Public, Private and Hybrid Clouds. This is further enhanced by the use of open source package, Octave 3.2.4, so that there is no need to write additional APIs to achieve enterprise portability. Applications written on the developer platform can be portable and executable on different desktops and Clouds of different hardware and software requirements, and execute as if they are on the same platform.

3D Black Scholes has fast execution time and only runs in Mathematica, which is not yet portable to different Clouds due to licensing issues and also there is no open source alternative to simplify the process of enterprise portability. At the time of writing, MATLAB licences on Private Clouds are still under development, and therefore results on MATLAB have only Private Cloud in Virtual Machines. Chang, Wills and De Roure (2010a, 2010c) have demonstrated the same FSaaS application running with Octave and

MATLAB on different Clouds, and their results demonstrate that the execution speed on MATLAB is approximately five times quicker than Octave, though MATLAB is more expensive and needs to deal with licensing issues regularly.

5.1. Experiments with Octave in Running the LSM on Different Clouds

Code written for LSM in Section 3.2 has been used for experimenting and benchmarking in the Clouds. 10,000 to 100,000 simulations (increase with an additional 10,000 simulations each time) of Monte Carlo Methods (MCM) adopting LSM are performed and the time taken at each of a desktop, private clouds and Amazon EC2 public clouds are recorded and averaged with three attempts. Hardware specifications for desktop, public cloud and private clouds are described as follows.

The desktop has 2.67 GHz Intel Xeon Quad Core and 4 GB of memory (800 MHz) with installed. One Amazon EC2 public cloud is used. The first virtual server is a 64-bit Ubuntu 8.04 with large resource instance of dual core CPU, with 2.33 GHz speed and 7.5GB of memory. There are two private clouds set up. The first private cloud is hosted on a Windows virtual server, which is created by a VMware Server on top of a rack server, and its network is in a network translated and secure domain. The virtual server has 2 cores of 2.67 GHz and 4GB of memory at 800 MHz. The second private cloud is a 64-bit Windows server installed on a rack, with 2.8GHz Six Core Opteron, 16 GB of memory. All these five settings have installed Octave 3.2.4, an open source compiler equivalent to MATLAB. The experiment began by running the FSaaS code (in Section 3.2) on desktop, private clouds and public cloud and started one at a time. Three attempts for each set of simulations are done, and the result is the average of three attempts. Benchmark is execution time, since it is a common benchmark used in several financial applications. Figure 4 shows the complete result of running FSaaS code on different Clouds.

Figure 4 shows the execution time for FSaaS application on desktop, public cloud and two private clouds. Experiments confirm with the followings. Firstly, enterprise portability is achieved and the FSaaS application can be executed on different platforms. Secondly, the improved FSaaS application can go for 100,000 simulations in one go on Clouds. Although above 100,000 simulations in one go, factors such as performance and stability must be balanced, before

Figure 4. Timing benchmark comparison for desktop, public cloud and two private clouds for time steps = 10

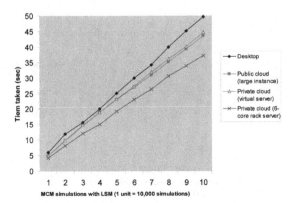

tuning up the capabilities of our FSaaS. The six-core processing rack server has the most advanced CPU, disk, memory, 64-bit operating system and networking hardware, and is not surprising that it is always the quickest. Although the desktop has similar hardware specification to server, it comes out slowest in all experiments. The difference between the Public Cloud (large instance) and Private Cloud (virtual server) is minimal. Although the large instance of a public cloud has the edge on hardware specification against the Virtual Private Cloud (VPC), the networking speed within the VPC is faster than the Public Cloud, and this explains the small differences between them.

Benchmark results show pricing and risk analysis can be calculated rapidly with accurate outcomes. Portability is achieved with a good reliable performance in clouds. These experiments demonstrate portability, speed, accuracy and reliability from desktop to clouds. Figure 4 shows the benchmark graph.

5.2. Experiments with MATLAB in Running the LSM on Desktop and one Private Cloud

MATLAB is used for high performance Cloud computation, since it allows faster calculations than Octave. The drawback of using MATLAB 2009 is license, which means all desktops and Cloud resources must be licensed prior setting up experiments. For this reason, only desktop and a Private Cloud (virtual machine) are used for experiments. The use of MATLAB 2009 reduces execution time for FSaaS, and also allows experiments to proceed with a higher number of time steps. The more the time steps used, the more accurate the outcome is, although higher numbers of time steps need more computing resources.

Five different sets of experiments are designed, and each set of experiments count execution time from 10,000 to 100,000 simulations as described in Section 5.2. The only difference is time step.

The first experiment gets time step equals to 10, and second experiment has time step equals to 50, and the third experiment sets time steps equal to 100, and the fourth experiment has time steps equal to 150, and finally, the last experiment gets time step equal to 200. The time step can be increased up to 1,000, but performance seems to drop off, particularly for experiments running high numbers of simulations. For this reason, the maximum time step in the experiments is limited to 200. Results for each set of experiments are recorded and shown in Figure 5, 6, 7, 8 and 9.

The results presented in Figures 5, 6, 7, 8 and 9 have the following implications. Firstly, the execution time and number of simulations are directly proportional to each other. It means the higher the number of simulations to be computed,

Figure 5. MATLAB timing benchmark for time step = 10

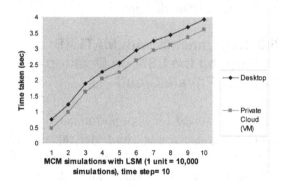

Figure 6. MATLAB timing benchmark for time step = 50

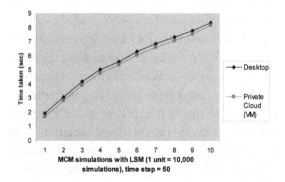

Figure 7. MATLAB timing benchmark for time step = 100

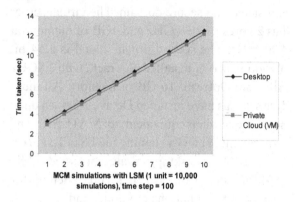

Figure 8. MATLAB timing benchmark for time step = 150

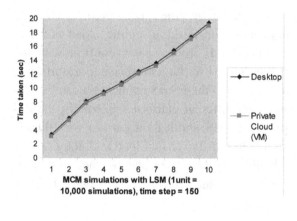

Figure 9. MATLAB timing benchmark for time step = 200

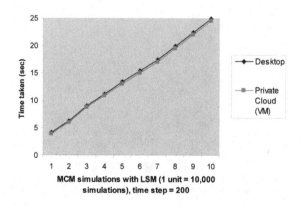

the longer the execution time on desktop and Private Cloud. It is not so obvious to identify linear relationship with a lower time step involved. This is likely because that execution time is so quick to complete that the range of errors and uncertainties are higher. When the time steps increases, it is easier to identify the linear relationship. This linear relationship also confirms what LSM suggests and recommends. Secondly, MATLAB 2009 offers quick execution time for the portability to Cloud, and the significant time reduction is experienced. However, the licensing issue still prevents from a large scale of adoptions to different Clouds. This means portability should be made as easy as possible, and not only includes technical implementations but also licensing issues. However, this paper will not go into details about licensing.

6. A CONCEPTUAL CLOUD PLATFORM: IMPLEMENTATIONS AND WORK-IN-PROGRESS

As discussed in previous sections, the primary objective for optimal provisioning and runtime management of cloud infrastructures at the infrastructure, platform, and software as a service levels is to optimise the delivery of the overall business outcome of the user. Improved business outcome in general refers to the increased revenue or reduced cost or both. Uncertainties of outcome, measured in terms of variance, are often regarded as negative impacts (or *risk*) and must be accounted for in the pricing calculations of the service delivery.

There are many types of risks that might impact the variance of the business outcome – including market risk, credit risk, liquidity risk, legal/reputation risk and operational risk. (Risk taxonomy was previously established in the context of various banking regulations such as Basil II.) Among these types of risks, operational risk is considered to be most directly related to the IT infrastructure

as it might impact the business through internal and external fraud, workplace safety, business practice, damage to physical assets, business disruption and system failures, and execution delivery and process management. In particular, over and under capacity, general availability of the system, failed transactions, loss of data due to virus or intrusion, poor business decision due to poor data, and failure of communication systems are all considered as part of the business disruption and system failures and need to be considered as part of the operational risk.

Behaviour models of systems are often constructed to predict the likely outcome under different context and scenarios. Both analytical and simulations methodologies have been applied to these behaviour models to predict the likely outcomes, and our demonstrations in MCM and BSM present some of these predictability features. Maximize the outcome requires minimizing the risk, cost, and maximize the performance.

In regard to all possible causes, "Poor business decision due to poor data quality" is the one that we address. The proposal of FSaaS can track and display risks in 3D Visualisation, so that there is no hidden area or missing data not covered within simulations. Accurate results can be computed quickly for 100,000 simulations in one go, and this greatly helps directors to make the right business decisions.

Apart from MCM and BSM simulations, other technologies such as workflows are used to present risks in business processes and help making the right business decision. This includes Risk Tolerance, which is commonly associated with the industry framework and business processes and have to be established top down. Figure 10 shows a business process-based behaviour model of a typical e-commerce operation. The customer interacts with the web site through web server for placing a new order or initiates a return/exchange. Either of the two scenarios will require interaction with the customer order system and accessing the customer records. A new order might also involve preparing the billing, sending the request to the warehouse for fulfillment. This business process based behaviour model clearly illustrates different types of operational risk involved during various stage of the business process. In Figure 10, the types of operational risk identified from the front end part of the business process includes Business Reputation Natural Disaster, System failure/ System Capacity Security, and other system failures and security issues. Business Risk includes Business Reputation and other system failures and security issues.

6.1. Contributions from Southampton: The Financial Software as a Service (FSaaS)

Figure 11 shows a conceptual architecture based on Operational Risk Exchange (www.orx.org), which currently includes 53 banks from 18 countries for sharing the operational risk data (a total of 177K loss incidents with a total of 62B Euros of loss as of the end of 2010), and demonstrated

how financial clouds could be implemented successfully for aggregating and sharing operational risk data. One of the main contributions from the University of Southampton is the use of MCM (MATLAB) for pricing and BSM (Mathematica) for risk analysis. This cloud platform offers calculation for risk modelling, fraud detection, pricing analysis and a critical analysis with warning over risk-taking. It reports back to participating banks and bankers about their calculations, and provides useful feedback for their potential investment.

Risk data computed by different models such as MCM, BSM and other models can be simulated and shared within the secure platform that offers anonymisation and data encryption. It also allows bank clients to double check with mortgage lending interests and calculations whether they are fit for purpose. This platform also works closely with regulations and risk control, thus risks are managed and monitored in the Financial Cloud platform. Our FSaaS is one part of the platform (as shown in the red arrow) to demonstrate accuracy, performance and enterprise

Figure 10. The operational risk and business risk analysis by workflow

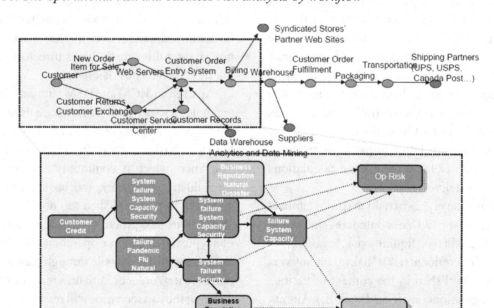

Figure 11. A conceptual financial cloud platform [using orx.org as an example] and contributions from Southampton in relations to this platform

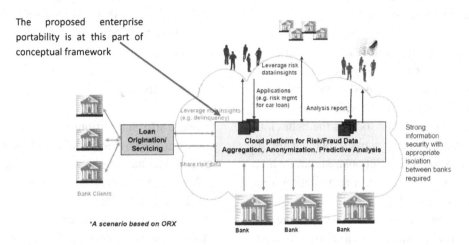

portability over Clouds, and is not only conceptual but has been implemented.

6.2. The IBM Fined Grained Security Framework

Figure 12 shows the Fined Grained Security Framework currently being developed at IBM Research Division. The framework consists of layers of security technologies to consolidate security infrastructure used by financial services. In additional to the traditional perimeter defence mechanisms such as access control, intrusion

detection (IDS) and intrusion prevention (IPS), this fine-grained security framework introduced fine-grained perimeter defence at a much finer granularity such as a virtual machine, a database, a JVM, or a web service container.

Starting with the more traditional approach side, the first layer of defence is Access Control and firewalls, which only allow restricted members to access. The second layer consists of Intrusion Detection System (IDS) and Prevent System (IPS), which detect attack, intrusion and penetration, and also provide up-to-date technologies to prevent attack such as DoS, anti-spoofing, port scanning,

Figure 12. The IBM Fine-Grained Security Framework (Li, 2010)

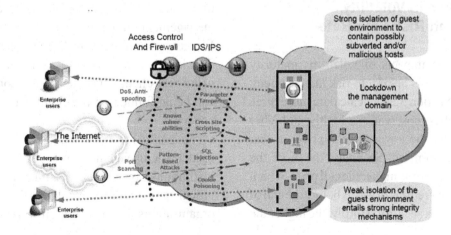

known vulnerabilities, pattern-based attacks, parameter tampering, cross site scripting, SQL injection and cookie poisoning.

The novel approach in the proposed fine-grained approach imposes the additional protection in terms of isolation management – which enforces top down policy based security management; integrity management – which monitors and provides early warning as soon as the behaviour of the fine-grained entity starts to behave abnormally; and end-to-end continuous assurance which includes the investigation and remediation after abnormally is detected. This environment intends to provide strong isolation of guest environment in an infrastructure or platform as a service environment and to contain possibly subverted and malicious hosts for security. Weak isolation can also be provided when multiple guest environments need to collaborate and work closely – such as in a three tier architecture among web server, application server, and database environment. Weak isolation usually focuses more on monitoring and captures end-to-end provenance so that investigation and remediation can be greatly facilitated. Strong isolation and integrity management is also required for the cloud management infrastructure – as this is often among the first few vulnerabilities of the cloud are exposed. See Figure 12 for details.

7. DISCUSSIONS

7.1. Variance in Volatility, Maturity and Risk Free Rate

Calculating the impacts of volatility, maturity and risk free rate is helpful to risk management. Our code in Section 3.2 can calculate these three aspects with these observations. Firstly, the higher the volatility is, the lower the call price, so that risk can be minimised. Secondly, the more the maturity becomes, the higher the call price, which improves higher returns of assets before the end of life in a bond or a security. Thirdly, the higher the

risk free rate, the higher the call price, as high risk free rate has reduced risk and boosts on investors' confidence level. Both Monte Carlo Methods and Black Scholes models are able to calculate these three aspects.

7.2. Accuracy

Monte Carlo Simulations are suitable to analyse pricing and provide reliable calculations up to several decimal numbers. In addition, the use of LSM reduces errors and thus improves the quality of calculation. New and existing ways to improve error corrections are under further investigation while achieving enterprise SaaS portability onto Clouds. In addition, the use of 3D Black Scholes will ensure the accuracy and quality of risk analysis. Risks can be quantified and also presented in 3D Visualisation, so that risks can be tracked and checked with the ease.

7.3. Implications for Banking

There are implications for banking. Firstly, security is a main concern. This is in particular when Cloud vendors tend to mitigate this risk technically by segregating different parts of the Clouds but still need to convince clients about the locality of their data, and data protection and security. Security concerns for banks in terms of using Cloud Computing, may be limited to cases where data need to be transferred (even for a moment) to the cloud infrastructure. However, certain risk management simulations, such as those involving Monte Carlo, where input data are usually random data based on statistical distribution (instead of using real client data), then these computations can be performed on the cloud without security concerns.

Secondly, financial regulators are imposing tighter risk management controls. Thus, financial institutions are involved in running more analytical simulations to calculate risks to the client organisations. This may present a greater need for the use of the Cloud computation and resources. Thirdly, portability of the Cloud can imply letting clients install their own libraries. Users who run MATLAB on the Cloud may only need the MATLAB application script or executable and to install the MATLAB Runtime once on the Cloud. For financial simulations written in Fortran or C++, users may also need Mathematical libraries to be installed in the Cloud. The Cloud must facilitate an easy way to install and configure user required libraries, without the need to write additional APIs like several practices do.

Portability would be important since bank personnel who run the simulations, should be able to install the necessary software infrastructure such as 'dlls'. One key benefit offered by Cloud is the cost. In Risk Management where mathematical models are always changing and becoming more advanced, the hardware requirement changes with it. Using the cloud service such as FSaaS would reduce upgrade costs. Hence greater hardware requirement may be facilitated by upgrading cloud subscription to a higher level, instead of decommissioning the company's own servers and replaced by new ones.

7.4. Enterprise Portability to the Clouds

Enterprise portability involves moving entire application services from desktops to clouds and between different Clouds, so that users need not worry about complexity and work as if on their familiar systems. This paper demonstrates financial clouds that modelling and simulations can take place on the Clouds, where users can connect and compute. This has the following advantages:

- Performance and speed: Calculations can be completed in a short time.
- Accuracy: The improved models based on LSM provide a more accurate range of prices comparing to traditional computation in normal distribution.

- Usability: users need not worry about complexity. This includes using iPhone or other user-friendly resources to compute. However, this is not the focus of this paper.

However, the drawback for portability is that additional APIs need to be written (Chang, Wills, & De Roure, 2010c). Clouds must facilitate an easy way to install and configure user required libraries, without the need to write additional APIs like several practices do. If writing APIs is required for portability, an alternative is to make APIs as easy and user-friendly as Facebook and iPhone do. In our demonstration, there is no need to write additional APIs to execute financial clouds.

7.5. Other Alternatives Such as Parallel Computing

In parallel computing, one way to speed up is to divide the data up into chunks and compute on different machines. However, there is an overhead in designing the problem (requiring human design effort) also there is machine overhead in sending the chunks of data to different machines and having a host machine to keep track of it. In a cloud, this may involve sending to different parts of the cloud and depending on how busy the cloud is; perhaps it will take longer in waiting time than when it actually takes to compute the chunk of data.

MCM is used for simulating losses due to Operational Risks, and there are plans in the Commonwealth Bank, Australia, to perform experiments in parallel computing with virtual machines, which have been recently set up.

8. CONCLUSION AND FUTURE WORK

FSaaS including MCM and BSM are used to demonstrate how portability, speed, accuracy and reliability can be achieved while demonstrating financial enterprise portability on different Clouds.

This fits into the third objective in the CCBF to allow portability on top of, secure, fast, accurate and reliable clouds. Financial SaaS provides a useful example to provide pricing and risk analysis while maintaining a high level of reliability and security. Our research purpose is to port and test financial applications to run on the Clouds, and ensure enterprise level of portability is workable, thus users can work on Clouds as they work on their desktops or familiar environments. Six areas of discussions are presented to support our cases and demonstration.

Benchmark is regarded as time execution to complete calculations after portability is achieved. Timing is essential since less time with accuracy is expected in using Financial SaaS on Clouds. The LSM provides added values and improvements. Firstly, it has a short starting and execution time to complete pricing calculations, and secondly, it allows 100,000 simulations in one go in different Clouds. This confirms enterprise portability can be delivered with LSM application on Octave 3.2.4 and MATLAB 2009. Five sets of experiments with MATLAB in running the LSM are performed, where the time steps have been increased for each set. The results confirm the linear relationship and also fast execution time for up to 100,000 simulations in one go on the Private Cloud. Portability should be made as easy as possible, and not only includes technical implementations but also licensing issues. In addition, the 3D Black Scholes presentation can enhance the quality of risk analysis, since risks are not easy to track down in real-time. The 3D Black Scholes improve the risk analysis so that risks can be presented in BSM formulas and are easier to be checked and understood. Three different scenarios of risk analysis are illustrated, and 3D simulations can provide a range of likely outcomes, so that decision makers can avoid potential pitfalls.

Implementations and work-in-progress for a conceptual Cloud Platform have been demonstrated. This includes the use of workflow to present risks in business processes, including the operational risk and business risk, so that risk tolerance can be established and the analysis can help making the right decision. Contributions from Southampton are the implementation of FSaaS, which allow pricing calculations and risk modelling to be computed fast and accurately to meet the research and business demands. Technical implementation in enterprise portability also meets challenges in business context: reduce time and cost with better performance. The IBM Fined Grained Security Framework provides a comprehensive model to consolidate security, which impose the additional protection in terms of isolation management and integrity management. This ensures trading, transaction and any financial related activities on Clouds are further protected and safeguarded.

Future work may include the following. HPC languages such as Visual C++ and/or.NET Framework 3.5 (or 4.0) will be used for the next stages. Other methods such as parallelism in MCM are potentially possible for further investigations. New error correction methods related to MCM will be investigated, and any useful outcomes will be discussed in the future work. New techniques to improve current 3D Black Scholes Visualisation will be investigated. There are plans to investigate Financial SaaS and its enterprise portability over clouds with Commonwealth Bank Australia, IBM US and other institutions, so that better platforms, solutions and techniques may be demonstrated. We hope to present different perspectives, recommendations and solutions for risk analysis, pricing calculations, security and financial modelling on Clouds, and to deliver improved prototypes, proof of concepts, advanced simulations and visualisation.

ACKNOWLEDGMENT

We greatly thank Howard Lee, a Researcher from Deakin University in Australia, for his effort to inspect our code and improve overall quality.

REFERENCES

Abdi, H. (2009). *The methods of least squares.* Dallas, TX: The University of Texas.

Assuncao, M. D., Costanzo, A., & Buyya, R. (2010). A cost-benefit analysis of using cloud computing to extend the capacity of clusters. *Journal of Cluster Computing, 13,* 335–347. doi:10.1007/s10586-010-0131-x

Barnard, K., Duygulu, P., Forsyth, D., De Freitas, N., Blei, D. M., & Jordan, M. I. (2003). Matching words and pictures. *Journal of Machine Learning Research,* 1107–1135. doi:10.1162/153244303322533214

Beaty, K., Kochut, A., & Shaikh, H. (2009, May 23-29). Desktop to cloud transformation planning. In *Proceedings of the IEEE International Symposium on Parallel and Distributed Processing,* Rome, Italy (pp. 1-8).

Birge, L., & Massart, P. (2001). Gaussian model selection. *Journal of the European Mathematical Society, 3*(3), 203–268. doi:10.1007/s100970100031

Brandic, I., Music, D., Leitner, P., & Dustdar, S. (2009, August 25-28). VieSLAF framework: Enabling adaptive and versatile SLA-management. In *Proceedings of the 6th International Workshop on Grid Economics and Business Models,* Delft, The Netherlands.

Briscoe, G., & Marinos, A. (2009, June 1-3). Digital ecosystems in the clouds: Towards community cloud computing. In *Proceedings of the 3rd IEEE International Conference on Digital Ecosystems and Technologies,* New York, NY (pp. 103-108).

Buyya, R., Yeo, C. S., Venugopal, S., Broberg, J., & Brandic, I. (2009). Cloud computing and emerging IT platforms: Vision, hype, and reality for delivering computing as the 5th utility. *Journal of Future Generation Computer Systems, 25*(6), 559–616. doi:10.1016/j.future.2008.12.001

Chang, V., Mills, H., & Newhouse, S. (2007, September). *From open source to long-term sustainability: Review of business models and case studies.* Paper presented at the UK e-Science All Hands Meeting, Nottingham, UK.

Chang, V., Wills, G., & De Roure, D. (2010a, July 5-10). A review of cloud business models and sustainability. In *Proceedings of the Third IEEE International Conference on Cloud Computing,* Miami, FL.

Chang, V., Wills, G., & De Roure, D. (2010b). Case studies and sustainability modelling presented by cloud computing business framework. *International Journal of Web Services Research.*

Chang, V., Wills, G., & De Roure, D. (2010c, September 13-16). *Cloud business models and sustainability: Impacts for businesses and e-research.* Paper presented at the UK e-Science All Hands Meeting Software Sustainability Workshop, Cardiff, UK.

Chang, V., Wills, G., De Roure, D., & Chee, C. (2010, September 13-16). *Investigating the cloud computing business framework - modelling and benchmarking of financial assets and job submissions in clouds.* Paper presented at the UK e-Science All Hands Meeting on Research Clouds: Hype or Reality Workshop, Cardiff, UK.

Chou, T. (2009). *Seven clear business models.* Active Book Press.

Choudhury, A. R., King, A., Kumar, S., & Sabharwal, Y. (2008). Optimisations in financial engineering: The least-squares Monte Carlo method of Longstaff and Schwarts. In *Proceedings of the IEEE International Symposium on Parallel and Distributed Computing* (pp. 1-11).

City, A. M. (2010). *Business with personality.* Retrieved from http://www.cityam.com

Feiman, J., & Cearley, D. W. (2009). *Economics of the cloud: Business value assessments.* Stamford, CT: Gartner RAS Core Research.

Financial Times. (2009). *Interview with Lord Turner, Chair of Financial Services Authority.* Retrieved from http://www.ft.com/cms/s/0/d76d0250-9c1f-11dd-a42e-000077b07658.html#axzz1Iqssz7az

Hamnett, C. (2009). *The madness of mortgage lenders: Housing finance and the financial crisis.* London, UK: King's College.

Hull, J. C. (2009). *Options, futures, and other derivatives* (7th ed.). Upper Saddle River, NJ: Pearson/Prentice Hall.

Li, C. S. (2010, July 5-10). Cloud computing in an outcome centric world. In *Proceedings of the IEEE International Conference on Cloud Computing*, Miami, FL.

Longstaff, F. A., & Schwartz, E. S. (2001). Valuing American options by simulations: A simple least-squares approach. *Review of Financial Studies, 14*(1), 113–147. doi:10.1093/rfs/14.1.113

Millers, R. M. (2011). *Option valuation.* Niskayuna, NY: Miller Risk Advisor.

Moreno, M., & Navas, J. F. (2001). On the robustness of least-square Monte Carlo (LSM) for pricing American derivatives. *Journal of Economic Literature Classification.*

Mullen, K., & Ennis, D. M. (1991). A simple multivariate probabilistic model for preferential and triadic choices. *Journal of Psychometrika, 56*(1), 69–75. doi:10.1007/BF02294586

Mullen, K., Ennis, D. M., de Doncker, E., & Kapenga, J. A. (1988). Models for the duo-trio and triangular methods. *Journal of Bioethics, 44*, 1169–1175.

Pajorova, E., & Hluchy, L. (2010, May 5-7). 3D visualization the results of complicated grid and cloud-based applications. In *Proceedings of the 14th International Conference on Intelligent Engineering Systems*, Las Palmas, Spain.

Patterson, D., Armbrust, M., Fox, A., Griffith, R., Jseph, A. D., Katz, R. H., et al. (2009). *Above the clouds: A Berkeley view of cloud computing* (Tech. Rep. No. UCB/EECS-2009-28). Berkeley, CA: University of California.

Piazzesi, M. (2010). *Affine term structure models.* Amsterdam, The Netherlands: Elsevier.

Schubert, L., Jeffery, K., & Neidecker-Lutz, B. (2010). *The future for cloud computing: Opportunities for European cloud computing beyond 2010 (public version 1.0).* Retrieved from http://cordis.europa.eu/fp7/ict/ssai/docs/cloud-report-final.pdf

Waters, D. (2008). *Quantitative methods for business* (4th ed.). Upper Saddle River, NJ: Prentice Hall.

Weinhardt, C., Anandasivam, A., & Blau, B., & StoBer, J. (2009). Business models in the service world. *IT Professional, 11*(2). doi:10.1109/MITP.2009.21

Weinhardt, C., Anandasivam, A., Blau, B., Borissov, N., Meinl, T., & Michalk, W. (2009). Cloud computing – a classification, business models, and research directions. *Journal of Business and Information Systems Engineering, 1*(5), 391–399. doi:10.1007/s12599-009-0071-2

This work was previously published in the International Journal of Cloud Applications and Computing (IJCAC), Volume 1, Issue 2, edited by Shadi Aljawarneh & Hong Cai, pp. 41-63, copyright 2011 by IGI Publishing (an imprint of IGI Global).

Chapter 10
Cloud Security Engineering:
Avoiding Security Threats the Right Way

Shadi Aljawarneh
Isra University, Jordan

ABSTRACT

Information security is a key challenge in the Cloud because the data will be virtualized across different host machines, hosted on the Web. Cloud provides a channel to the service or platform in which it operates. However, the owners of data will be worried because their data and software are not under their control. In addition, the data owner may not recognize where data is geographically located at any particular time. So there is still a question mark over how data will be more secure if the owner does not control its data and software. Indeed, due to shortage of control over the Cloud infrastructure, use of ad-hoc security tools is not sufficient to protect the data in the Cloud; this paper discusses this security. Furthermore, a vision and strategy is proposed to mitigate or avoid the security threats in the Cloud. This broad vision is based on software engineering principles to secure the Cloud applications and services. In this vision, security is built into all phases of Service Development Life Cycle (SDLC), Platform Development Life Cycle (PDLC) or Infrastructure Development Life Cycle (IDLC).

INTRODUCTION

Due to lack of control over the Cloud software, platform and/or infrastructure, several researchers stated that a security is a major challenge in the Cloud. In Cloud computing, the data will be virtualized across different distributed machines, hosted on the Web (Taylor, 2010; Marchany,

2010). In business respective, the cloud introduces a channel to the service or platform in which it could operate (Taylor, 2010).

Thus, the security issue is the main risk that Cloud environment might be faced. This risk comes from the shortage of control over the Cloud environment. A number of practitioners described this point. For example, Stallman (Arthur, 2010) from the Free Software Foundation re-called the Cloud computing with Careless Computing be-

DOI: 10.4018/978-1-4666-1879-4.ch010

cause the Cloud customers will not control their own data and software and then there is no monitoring over the Cloud providers and subsequently the data owner may not recognize where data is geographically located at any particular time.

Threats in the Cloud computing might be resulted from the generic Cloud infrastructure which is available to the public; while it is possessed by organization selling Cloud services (Marchany, 2010; Chow et al.,2009).

In Cloud computing, software and its data is created and managed virtually from its users and might only accessible via a certian cloud's software, platform or infrastructure. As shown in Figure 1, there are three Cloud models that describe the Cloud architecture for applications and services (Taylor, 2010; Marchany, 2010):

1. The Software as a Service (SaaS) model: The Cloud user rents/uses software for use on a paid subscription (Pay-As-You-Go).

Figure 1. Models of Cloud environment-taken from (Taylor, 2010)

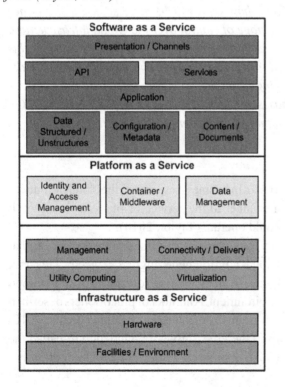

2. The Platform as a Service (PaaS) model: The user rents a development environment for application developers.
3. The Infrastructure as a Service (IaaS) model: The user uses the hardware infrastructure on pay-per-use model, and the service can be expanded in relation to demands from customers.

In spite of this significant growth, a little attention has been given to the issue of Cloud security both in research and in practice. Today, academia requires sharing, distributing, merging, changing information, linking applications and other resources within and among organizations. Due to openness, virtualization, distribution interconnection, security becomes critical challenge in order to ensure the integrity and authenticity of digitized data (Cárdenas et al., 2005; Wang et al., 2005).

Cloud opts to use scalable architecture. Scalability means that hardware units that are added bringing more resources to the Cloud architecture (Taylor, 2010). However, this feature is in tradeoff with the security. Therefore, scalability eases to expose the Cloud environment and it will increase the criminals who would access illegally to the Cloud storage and Cloud Datacenters as illustrated in Figure 2.

Availability is another characteristic for Cloud. So the services, platform, data can be accessible at any time and place. Cloud is candidate to expose to greater security threats, particularly when the cloud is based on the Internet rather than an organization's own platform (Taylor, 2010).

Although the security is a risk in the Cloud environment, several companies are offering now Cloud services including Microsoft Azure Services Platform, Amazon Web Services, Google and open source Cloud systems such as Sun Open Cloud Platform for academic, customers and administrative purposes (Taylor, 2010). Yet, some organizations have not realized the importance of security for the Cloud systems. These organiza-

Figure 2. Cloud computing Security - taken from (Marchany, 2010)

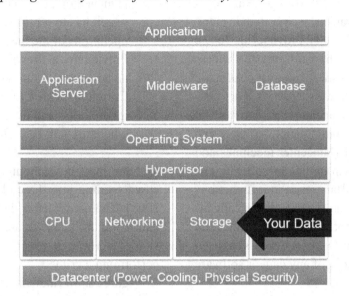

tions adopted some ready security and protection tools to secure their systems and platforms.

RELATED WORK

In this section, the Amazon Web Services (AWS) is only discussed. Amazon uses the Cloud services for introducing a number of web services for customers.

Amazon constructed Amazon Web Services (AWS) platform to secure the access for web services (Amazon, 2010). The AWS platform introduces a protection against traditional security issues in the Cloud network.

Physical access to AWS Datacenters is limited controlled both at the perimeter and at building ingress nodes by security experts to raise Video Surveillance (VS), Intrusion Detection Systems (IDS), and other electronic means. Authorized staff has to log in two authentication phases with restricted number of time for accessing to Amazon Web Services and AWS Datacenters at maximum (Amazon, 2010).

Note that Amazon only offers restricted Datacenter access and information to people who have a legal business need for these privileges. If the business need for these privileges is revoked, then the access is stopped, even though if employees continue to be an employee of Amazon or Amazon Web Services (Amazon, 2010).

However, one of the weakness of the AWS is the dynamic data which is generated from the AWS could be listened and penetrated from hackers or professional criminals.

Basically there are six areas for security vulnerabilities in cloud computing (Trusted Computing Group, 2010): (a) data at end-to-end points, (b) data in the communication channel, (c) authentication, (d) separation between clients, (e) legal issues and (f) incident response.

This article is organized as follows: a study shows that the Cloud threats and an overview of existing cloud computing concerns are described. Next, the proposed vision and strategies that might be improved to mitigate or avoid some of the concerns outlined. Finally conclusions future works are offered.

CLOUD THREATS

However, security principles (such as data integrity, and confidentiality) in the Cloud environment could be lost (Amazon, 2010). For example, a criminal might penetrate the web system in many forms (Snodgrass et al., 2004; Provos et al., 2007). An insider adversary, who gains physical access to Datacenters, is able to destroy any type of static content in the root of a web server. It is not only physical access to Datacenter that can corrupt data. Malicious web manipulation software can penetrate servers and Datacenter machines and once located them such malicious software can monitor, intercept, and tamper online transactions in a trusted organization. The result typically allows a criminal full root access to Datacenter and web server application. Once such access has been established, the integrity of any data or software is in question.

There are several security products (such as Antivirus, Firewalls, gateways, and scanners) to secure the Cloud systems but they are not sufficient because each one has only specific purpose and hence, they are called ad-hoc security tool. For example, Network firewalls provide protection at the host and network level (Gehling et al., 2005). There are, however, five reasons why these security defences cannot be only used to secure the systems (Gehling et al., 2005):

- They cannot stop malicious attacks that perform illegal transactions, because they are designed to prevent vulnerabilities of signatures and specific ports.
- They cannot manipulate form operations such as asking the user to submit certain information or validate false data because they cannot distinguish between the original request-response conversation and the tampered conversation.
- They do not track conversations and do not secure the session information. For example, they cannot track when session information in cookies is exchanged over an HTTP request-response model.
- They provide no protection against web application/services attacks since these are

launched on port 80 (default for web sites) which has to remain open to allow normal operations of the business.
- Previously, a firewall could suppose that an adversary could only be on the outside. Currently, with Cloud systems, an attack might originate from the inside as well, where firewall can offer no protection.

Note that the computer forensics has classified e-crime into three classes (Mohay et al., 2003): The computer is the target of the crime; data storage which is created during the commission of a crime; or a tool or scheme that used in performing a crime.

Figure 2 shows the data storage and Datacenters which are possibly targeted by the criminals. According to the computer forensics, the distrusted servers and Datacenters are the target of crime. Therefore, question should be attempted to answer is that whether data is safe and secure?

Data confidentiality might be exposed either from insider user threats or outsider user threats from (CPNI, 2010). For instance, Insider user threats might maliciously form from: cloud operator/provider, cloud customer, or malicious third party. The threat of insiders accessing customer data take place within the cloud is larger as each of models can offer the need for multiple users:

- **SaaS:** Cloud clients and administrators
- **PaaS:** Application developers and testers
- **IaaS:** Third party consultants

A VISION AND STRATEGY TO MITIGATE OR AVOID CLOUD SECURITY CONCERNS

In this article, a vision is proposed to avoid the Cloud security threat at the SaS level. Our vision is that SaS is based on Service-oriented architecture. A service is a standard approach to make a reusable component available and could be accessible across the web or possible technology. Thus, service provision is independent of the application that using the service.

In reference to the article (Aljawarneh, 2011) a case study has been described to mention a number of significant threat vulnerabilities that can be introduced during all phases of the software (service) development life cycle.

For instance, number security vulnerabilities might be occurred at the requirements specification process cycle (Bono et al., 2005; Cappelli et al., 2006):

- Ignoring to declare authentication and role-based access control requirements eased the insider and or outsider attacks.
- Ignoring to declare security requirements of duties for automated business processes provided a simplified method for attack.
- Ignoring to declare requirements for data integrity checks gave insiders the secu-

rity of knowing their actions would not be detected.

The existing Cloud services face some security issues because a security design is not integrated into the Cloud architecture development process (Glisson et al., 2005). Thus, organizations should pay more attention to the insider threats to operational systems; it turns out that vulnerabilities can be accidentally and intentionally introduced throughout the development life cycle – during requirements definition, design, implementation, deployment, and maintenance (Cappelli et al., 2006). Once business leaders are aware of these, they can implement practices that will aid in mitigating these vulnerabilities.

As illustrated in Figure 3, security should be built in all steps of the service development process to identify what a customer and an organization need for every stage of the software engineering principles. This proposed vision or strategy could

Figure 3. The proposed strategy

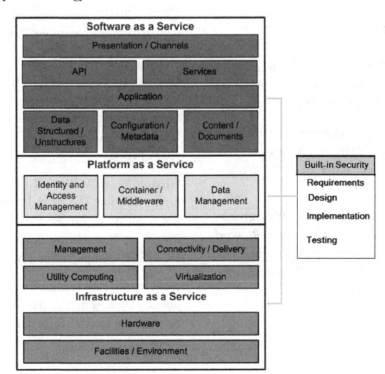

help to detect the threats and concerns at each stage instead of processing them at the implementation stage. Consequently, our vision and strategies might help the Cloud developers, providers and administrators to eliminate the attacks or mitigate them if possible in the design stage not waiting for actual attacks to occur.

CONCLUSION

Cloud faces some security issues at the SaS, PaS, IaS models. One main reason is that the lack of control over the Cloud Datacenters and distrubted servers. Furthermore, security is not integrated into the service development process.

Indeed, the traditional security tools alone will not solve current security issues and so it will be effective to incorporate security component upfront into the development methodology of Cloud system phases. In the next part of this article, we will propose a methodology that could help to mitigate the security concerns on the Cloud models.

REFERENCES

Aljawarneh, S. (2011). A web engineering security methodology for e-learning systems. *Network Security Journal, 2011*(3), 12-16.

Amazon. (2010). *Amazon web services: Overview of security processes*. Retrieved from awsmedia. s3.amazonaws.com/pdf/AWS_Security_Whitepaper.pdf

Arthur, C. (2010). *Google's ChromeOS means losing control of data, warns GNU founder Richard Stallman*. Retrieved from http://www.guardian. co.uk/technology/blog/2010/dec/14/chrome-os-richard-stallman-warning

Bono, S. C., Green, M., Stubblefield, A., Juels, A., Rubin, A. D., & Szydlo, M. (2005). Security analysis of a cryptographically-enabled RFID device. In *Proceedings of the 14th Conference on USENIX Security*, Berkeley, CA.

Cappelli, D. M., Trzeciak, R. F., & Moore, A. B. (2006). *Insider threats in the SLDC: Lessons learned from actual incidents of fraud: Theft of sensitive information, and IT sabotage*. Pittsburgh, PA: Carnegie Mellon University.

Cárdenas, R. G., & Sanchez, E. (2005). Security challenges of distributed e-learning systems. In F. F. Ramos, V. A. Rosillo, & H. Unger (Eds.), *Proceedings of the 5th International School and Symposium on Advanced Distributed Systems* (LNCS 3563, pp. 538-544).

Chow, R., Golle, P., Jakobsson, M., Shi, E., Staddon, J., Masuoka, R., et al. (2009). Controlling data in the cloud: Outsourcing computation without outsourcing control. In *Proceedings of the ACM Workshop on Cloud Computing Security* (pp. 85-90). New York, NY: ACM Press.

CPNI. (2010). *Information security briefing 01/2010 cloud computing*. Retrieved from www. cpni.gov.uk/Documents

Gehling, B., & Stankard, D. (2005). eCommerce security. In *Proceedings of the Information Security Curriculum Development Conference*, Kennesaw, GA (pp. 32-37). New York, NY: ACM Press.

Glisson, W., & Welland, R. (2005). Web development evolution: The assimilation of web engineering security. In *Proceedings of the Third Latin American Web Congress* (p. 49). Washington, DC: IEEE Computer Society.

Google. (2011b). *Google trends: private cloud, public cloud*. Retrieved from http://www.google. de/trends?q=private+cloud%2C+public+cloud

Marchany, R. (2010). *Cloud computing security issues: VA Tech IT security.* Retrieved from http://www.issa-centralva.org

Mohay, G., Anderson, A., Collie, B., & del Vel, O. (2003). *Computer and intrusion forensics* (p. 9). Boston, MA: Artech House.

Provos, N., McNamee, D., Mavrommatis, P., Wang, K., & Modadugu, N. (2007). The ghost in the browser analysis of web-based malware. In *Proceedings of the RST Conference on First Workshop on Hot Topics in Understanding Botnets,* Berkeley, CA (p. 4).

Ramim, M., & Levy, Y. (2006). Securing e-learning systems: A case of insider cyber attacks and novice IT management in a small university. *Journal of Cases on Information Technology, 8*(4), 24–34. doi:10.4018/jcit.2006100103

Snodgrass, R. T., Yao, S. S., & Collberg, C. (2004). Tamper detection in audit logs. In *Proceedings of the Thirtieth International Conference on Very Large Data Bases* (pp. 504-515).

Taylor, M. (2010). *Enterprise architecture – architectural strategies for cloud computing: Oracle.* Retrieved from http://www.techrepublic.com/whitepapers/oracle-white-paper-in-enterprise-architecture-architecture-strategies-for-cloud-computing/2319999

Trusted Computing Group. (2010). *Cloud computing and security –a natural match.* Retrieved from http://www.infosec.co.uk/

Wang, H., Zhang, Y., & Cao, J. (2005). Effective collaboration with information sharing in virtual universities. *IEEE Transactions, 21*(6), 840–853.

Chapter 11
Security Issues in Cloud Computing:
A Survey of Risks, Threats and Vulnerabilities

Kamal Dahbur
New York Institute of Technology, Jordan

Bassil Mohammad
New York Institute of Technology, Jordan

Ahmad Bisher Tarakji
New York Institute of Technology, Jordan

ABSTRACT

Cloud Computing (CC) is revolutionizing the methodology by which IT services are being utilized. It is being introduced and marketed with many attractive promises that are enticing to many companies and managers, such as reduced capital costs and relief from managing complex information technology infrastructure. However, along with desirable benefits come risks and security concerns that must be considered and addressed correctly. Thus, security issues must be considered as a major issue when considering Cloud Computing. This paper discusses Cloud Computing and its related concepts; highlights and categorizes many of the security issues introduced by the "cloud"; surveys the risks, threats, and vulnerabilities; and makes the necessary recommendations that can help promote the benefits and mitigate the risks associated with Cloud Computing.

INTRODUCTION

Cloud computing is arguably one of the most significant technological shifts of our time. The mere idea of being able to use computing in a similar manner to using a utility, such as electricity, is revolutionizing the IT services world and holds great potential. Customers, whether large enterprises or small businesses, are drawn toward the cloud's promises of agility, reduced capital costs, and enhanced IT resources. IT companies are

DOI: 10.4018/978-1-4666-1879-4.ch011

shifting from providing their own IT infrastructure to utilizing the computation services provided by the cloud for their information technology needs (Carr, 2008).

Cloud computing introduces a level of abstraction between the physical infrastructure and the owner of the information being stored and processed. Such indirect control of the physical environment introduces vulnerabilities unknown in previous settings. Such a radical change is of course not risk free. As IT services are contracted outside of the enterprise, the dependency on third party providers compels companies to rethink their risk management techniques and adapt accordingly.

After this brief introduction, the remainder of this paper is organized as follows: First, we provide an overview of cloud computing, its services, and core technologies; we provide a survey of known general risks, vulnerabilities and threats, and then explore the additional risks introduced by (or relevant to) cloud computing; some real world examples of vulnerabilities of cloud computing that have been reported in the literature are discussed; and we provide our conclusion and recommendations.

OVERVIEW OF CLOUD COMPUTING

As with any new technology, the definition of cloud computing is changing with the evolution of technology and its services. No standard definition for cloud computing has yet been agreed upon, especially since it encompasses so many different models and potential markets, depending on venders and services. In the simplest of terms, cloud computing is basically internet-based computing. The term "cloud" is used as a metaphor for the Internet, and came from the well known cloud drawing that was used in network diagrams to depict the Internet's underlying networking infrastructure. The computation in the internet is done by groups of shared servers that provide on demand hardware resources, data and software to devices connected to the net.

The National Institute of Standards and Technology NIST, gives a more formal definition: "Cloud computing is a model for enabling convenient, on-demand network access to a shared pool of configurable computing resources (e.g., networks, servers, storage, applications, and services) that can be rapidly provisioned and released with minimal management effort or service provider interaction" (NIST, 2010) also notes that this definition will probably change over time.

In this sense, users of cloud computing are raised to a level of abstraction where they are hidden and relived from the details of the hardware or software infrastructures that supports their computations. This greatly simplifies the costs involved in establishing and managing the IT that is needed to meet the requirements of any business. And since businesses will pay only for the required IT resources when and as they are needed, more and much more powerful resources can be provided at a fraction of the price of the real value for such resources.

Core Technologies

To better understand the security issues that are associated with CC, it is important to discuss the core concepts and technologies in cloud computing. CC is based on the general principle of utility computing – providing metered services of computing resources in a similar manner to the other utilities such as electricity. The measured service-oriented perspective for computing resources can be easily understood for the hardware resources. But this perspective can also be extended to software systems because they are designed and built in the form of autonomous interoperable services (Wei & Blake, 2010).

The large variety of devices that can connect to the internet, such as PDAs, mobile phones and handheld and static devices, all expanded the number of ways the cloud can be accessed.

Coupled with acceptance of the browser as some sort of universal interface for even very complex systems, the potential of cloud computing could be tapped using basically any device that can load a browser. High speed broadband networks, data centers, and server farms are also critical components.

But perhaps the most influential concept in cloud computing would be virtualization. In computing, virtualization is the creation of a virtual (rather than actual) version of computers or operating systems. In this sense, the physical traits of the computing platform are hidden from the users and instead another abstract computing platform is presented. The software that creates this virtual environment is usually called a hypervisor. In case of server consolidation, many small servers with different OSs are substituted by one powerful server, and the previous operating systems are run in virtual environments on this server. The large server can "host" many such "guest" virtual machines which can be more easily configured and controlled. And this will allow the same hardware to present itself in different capabilities as the users need, a key concept in cloud computing.

Cloud Computing Main Characteristics

Another way of defining cloud computing is to examine its characteristics, especially the ones that have been agreed upon and are generally accepted by different groups:

- Shared resources: or what NIST calls resource pooling, where no resources are dedicated to one user but instead are pooled together to serve multiple consumers. Resources whether on the application, host or network level, are assigned and reassigned as needed to these consumers. This creates a sense of location independence where users cannot pinpoint exactly where there computations are being executed (NIST, 2010).

- On-demand self-service: the users can assign themselves additional resources such as storage or processing power automatically without human intervention. This is comparable with autonomic computing where the computer system is capable of self management.

- Elasticity: along with self provisioning of resources, cloud computing is characterized with the ability to locate and release resources rapidly. This will allow the consumers to scale up the resources they need at any time to address heavy loads and usage spikes, and then scale down by returning the resources to the pool when finished.

- Pay as you go: or what is known as measured service. Computing in the cloud is offered as a utility which users pay for on a consumption basis, not unlike any other utility enterprises pay for such as electricity, gas and water.

It could be argued that the main feature of cloud computing is that the computation is done in the "cloud" and remaining characteristics stem from or complement this simple fact. Additional characteristics have also been reported in the literature but most of these are, in our opinion, complementary to the main characteristics we reported (Mather, Kumaraswamy, & Latif, 2009).

Cloud Computing Service Models

The services provided by cloud computing can be categorized into three service models, Software as a Service (SaaS), Platform as a Service (PaaS), and Infrastructure as a Service (IaaS). These three models often abbreviated as the SPI Service framework (i.e. SPI is short for Software, Platform and Infrastructure) are the basis of all services provided by cloud computing:

- Software as a Service (SaaS): in this model software is provided by the vendor over the net as a one-to-many model (single instance, multi-tenant architecture) as a substitute of the one-to-one typical model. Instead of users buying the software and installing it on their systems, they rent the software using pay-per-use, or subscription fee. Thus the user exchanges the capital expense acquiring software licenses for operation expenses renting software usage. Since the application is provided over the net, usually the package includes the usage of the software itself and the utilization of the hardware it runs on, in addition to some level of support.

Additional benefits of this model is centralized updating, so users don't need to worry about patching and versioning. Examples of SaaS would be Google Docs and Salesforce.com customer relationship Management CRM software.

- Platform as a Service (PaaS): the sophistication needed to create software that can run in the cloud entails the providers to create a development environment or platform on which these applications can be executed. The second service provided in the cloud is the utilization of the development environment itself. Users can create custom applications that target a certain platform, with tools offered by the platform provider. They then can deploy and run these applications on this platform, with full control over the applications and their configuration. Such applications may also be acquired from third parties. When using this service, users don't need, or even have the ability to manage the underlying cloud infrastructure, including servers, storage mediums and network configuration.

The benefits of such a service are large, since startup companies and small teams can start developing and deploying their own software without the need to acquire servers and teams to manage them. Examples of PaaS would be Google's Apps Engine and Microsoft's Azure Platform.

- Infrastructure as a Service (IaaS): in the third service, the users are given access to elements of the computing infrastructure itself. Using internet technologies, users can utilize the processing power, storage mediums and necessary networking components provided by the vendor. Users then can run arbitrary software and operating systems that best meets their requirements, with full control and management. This is much like traditional hosting services except when done in the cloud it is possible to scale the service to conform to the changing requirements, and to offer the pay per use model.

This model is very similar to utility computing where users pay for the consumption of disk space, processing power, or bandwidth they use. Examples would be Amazon.com, EC2 and S3.

CLOUD COMPUTING RISKS, THREATS AND VULNERABILITIES

Definitions of Terms

Before we discuss the security risks associated with doing business on the cloud, we need to define the terms "Risks", "Threats", and "Vulnerabilities". These terms, in addition to the term "Exposure", are often used to represent the same thing even though they have different meanings and relationships to each other. It is important to understand each word's definition and, more importantly, understand the relationship of these terms to the other concepts.

"Vulnerability" refers to a software, hardware, or procedural weakness that may provide an attacker the open door to enter a computer or network and have unauthorized access to resources within the environment. Vulnerability characterizes the absence or weakness of a safeguard that could be exploited. This vulnerability may be a service running on a server, unpatched applications or operating system software, or an unsecured physical entrance.

A "Threat" is any potential danger to information or systems. The threat is that someone, or something, will identify a specific vulnerability and use it against the company or individual. Threats exploit existing vulnerabilities in an attempt to cause damage or destruct a resource. A "Threat Agent" is the entity that takes advantage of vulnerability. A threat agent could be an intruder, a process, or an employee making an unintentional mistake that could expose confidential information or destroy a file's integrity.

A "Risk" is the likelihood of a threat agent taking advantage of vulnerability and the corresponding business impact. For example, if users are not educated on processes and procedures, there is a higher likelihood that an employee will make an intentional or unintentional mistake that may destroy data. Risk ties the vulnerability, threat, and likelihood of exploitation to the resulting business impact (Figure 1).

An "Exposure" is an instance of being exposed to losses from a threat agent. Vulnerability exposes an organization to possible damages. If a bank does not properly patch its servers then it may be exposed to possible breaches in relation to the open holes resulting from the missing patches.

A countermeasure (also referred to as a safeguard) is generally put into place to mitigate the potential risk. A countermeasure may be a policy, procedure, a software configuration, or a hardware device that eliminates vulnerability or reduces the likelihood that a threat agent will be able to exploit a vulnerability. Strong authentication mechanisms, computer anti-virus software and information security awareness are some examples of proper countermeasures.

Accepting Some Risk

In any enterprise, information security risks must be identified, evaluated, analyzed, treated and properly reported. Businesses that fail in identifying the risks associated with the technology they use, the people they employ, or the environment where they operate usually subject their business to unforeseen consequences that might result in severe damage to the business. Because risks cannot be completely eliminated, they need to be lowered into acceptable levels. Acceptable risks are risks that the business decides to live with,

Figure 1. Risk = Vulnerability x Threat x Impact x Likelihood

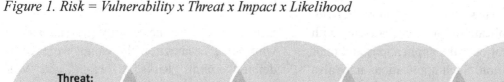

Safeguard: A control or countermeasure to a vulnerability

given that proper assessment for these risks was done and the cost of treating these risks might outweigh the benefits.

As we discussed above, CC is in reality a business model in which companies with limited resources to build their own IT infrastructure elect to outsource this to third parties who will in turn lease their infrastructure and probably applications as services. Thus, the risks, threats and vulnerabilities that are usually associated with technology in general exist in this environment, because outsourcing the technological service only transfers the liability. In addition, this new model introduces more security challenges due to mainly the fact that the data is on the cloud (i.e. hosted somewhere in the Internet space), being transferred across countries with different regulations, and most importantly might reside on a machine that hosts other data instances of other enterprises. In some instances, the data for the same enterprise might even be stored across multiple data centres. We will concentrate on the cloud-specific security issues (Figure 2).

Risks and Threats in the Cloud

Information security risks in Cloud Computing (CC) were subject for detailed analysis and assessment. One of the best efforts in this direction was realized by the European Network & Information Systems Agency (ENISA) whom developed a comprehensive detailed research in this regards (ENISA, 2010). Other groups such as Cloud Security Alliance (CSA) who specialize

Figure 2. Cloud computing risks, threats and vulnerabilities

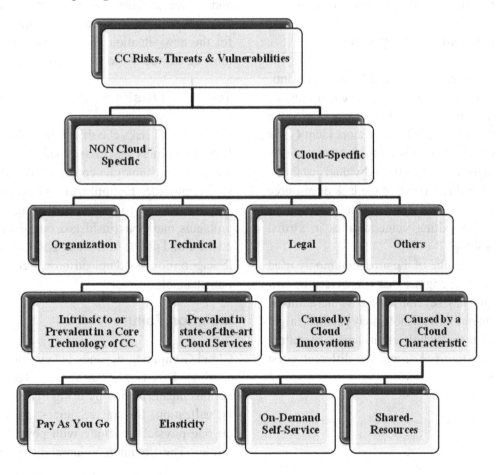

in cloud computing technology and information security matters also have significant publications (CSA, 2010). ENISA classifies Cloud Computing (CC) risks into three categories: Organizational, Technical and Legal. CSA threats model avoids classifying CC's risks but yet introduce a detailed list of considerable issues that need to be properly addressed.

Organizational Risks

The organizational risks classification includes all risks that may impact the structure of the organization or the business as an entity. Examples of such risks include, but not limited to, loss of business reputation due to co-tenant activities (or the tenants sharing the same resource), and any organizational change that can happen to the cloud provider (as a business organization) including provider failure, termination or acquisition.

Loss of Business Reputation

Because resources are shared by many tenants, offensive activities done by one tenant may affect the reputation of the other tenants sharing the same cloud service. When detectors identify the origin of the malicious activity, they may not really distinguish between the individual instances. This risk is a direct result of the lack of resource isolation existing in some cloud implementations coupled with exiting vulnerabilities in virtual machines setup.

VMWare is one of the most commonly used virtual machine engines that is provided by Microsoft and has several known vulnerabilities. These vulnerabilities, if not timely patched, might compromise that whole environment including both the hosting server and the different guest instances.

Cloud Service Termination or Failure

Cloud Computing service providers are subject for the same circumstances and pressures posed by the market. Under certain conditions and possible economic crises these CC providers might go out of business or alternatively be acquired by larger ones. If this happens, hosted guests will not be able to meet their contractual commitments with their clients, or even worse, might become subject to fines and legal obligations. Vulnerabilities that might lead to this risk include poor provider selection and lack of supplier redundancy.

Cloud Provider Acquisition

As mentioned in the previous point, the CC provider might be acquired by a different CC. The new CC may decide to change the nature business and services provided. As a result of these changes new threats might be introduced. Complete review for the new situation and the existing contacts must be carried out to assess the risks.

Technical Risks

The technical risks classification includes problems or failures associated with the provided services or technologies contacted from the cloud service provider. Examples of such risks include, but not limited to, resource-sharing isolation problems, malicious (insiders or outsiders) attacks on the cloud provider, and any possibility of data leakage on download/upload through communication channels.

Isolation Failure

Cloud Computing depends heavily on resource sharing. This is practically achieved through various means mainly by using hypervisors and virtualization. Several resources are mounted on one physical machine with powerful capabilities. The Common Vulnerabilities Exposure library (CVE) shows several vulnerabilities that are realized in VM environments that would lead to malicious activity spreading across the various hosted guests. In this sense, if one host was exposed to remote access vulnerability, this vulnerability will extend the capabilities of the attacker to reach other hosted instances including the physical main hosting server. Given this setup, the impact of known attacks would be considerably amplified. Selecting to host services on CC in the Private model will considerably lower the impact. Vulnerabilities resulting to this risk will include hypervisor vulnerabilities and lack of resource isolation amongst others.

Cloud Provider Insiders

The CC service provider is supposed to conduct proper background checks and screening before hiring people and to apply proper policies and procedures for access control. They also should have clear roles and responsibilities and not have a vulnerable environment that does not have adequate physical security. Failure to do any of these measures will result in malicious activities being easily carried out without detection. This will expose guest instances to major risks and high possibility of devastating breaches.

Data Leakage on Upload/Download

When the data is being transferred across the cloud unencrypted it is subject for traffic sniffing, spoofing, and man-in–the-middle attacks amongst others. Moreover, in some cases CC providers do not offer confidentiality or non-disclosure agreements which will in turn make liabilities unclear and subject the client's reputation at major risk. The lack of proper logical access controls will result in insiders being able to copy and move sensitive content on mass storage media away from the premises.

Legal Risks

The legal risks classification refers to issues that surround data being exchanged across multiple countries that have different laws and regulations concerning data traversal, protection requirements and privacy laws. Examples of such risks include, but not limited to, risks resulting from possible changes of jurisdiction and the liability or obligation of the vendor in case of loss of data and/or business interruption.

Risks from Changes of Jurisdiction

In Cloud Computing data is exchanged across multiple countries that have different laws and regulations concerning data traversal, protection requirements and privacy laws. Some countries might require data to be encrypted and others might not. For clients doing business on the cloud they are exposed to risks associated with not being able to address the laws that they need to comply with to properly protect their data. Fines and penalties might result from not being able to do so which would accordingly result in reputation damage.

Data Protection Risks

The data owner putting his clients' data on the cloud is exposed to the risks resulting from the fact that he does not really know who at the cloud service provider side handles the data. This data might be misused or even destructed. Given the various jurisdictions across different areas the obligations are not clear according leaving the tenant subject for fines.

Other Risks and Risk Categories

Cloud Computing is based on a new utilization of technology and many risks that used to be present in other technological implementations do still exist, and are realized as not cloud specific. Risks like social engineering, physical security, lost or stolen backups, and loss or compromise of security logs are just a few examples of such general security risks.

The Cloud Security Alliance (CSA) lists the following threats as the top risks associated with CC based on their recent research: malicious insiders, data loss/leakage, abuse and nefarious use of CC and shared technology vulnerabilities (CSA, 2010). Even though CSA prefers to prioritize risks, it easy to see that each of the listed threats can be included in the ENISA categories or as non-cloud specific, or general, security risk.

Cloud Specific Vulnerabilities

Other researchers prefer to focus on cloud specific vulnerabilities, without much focus on threats and risks (Grobauer, Walloscheck, & Stocker, 2010). According to such research, a particular vulnerability can be considered specific to cloud computing if it meets any of the following criteria:

- It is intrinsic to or prevalent in a core technology of cloud computing
- It has its root cause in one of essential cloud characteristics
- It is caused by cloud innovations making exiting (tried and tested) security controls hard or impossible to implement
- It is prevalent in established state-of-the-art cloud services

The first indicator is concerned with the essential technologies of cloud computing such as virtualization, web services and service oriented architecture SOA, and cryptography. Vulnerabilities such as obsolete cryptographic algorithms and virtual machine escapes are extremely relevant to cloud computing.

The second indicator checks if one of the characteristics that define the cloud is the main reason behind this vulnerability. For example data leakage or data loss caused by the cloud's rapid elasticity or by resource pooling is very cloud pertinent. Another example would be billing evasion or any weakness in the pay-as-you-go model.

The third indicator to check if the vulnerability is cloud specific checks whether such a weakness/vulnerability cannot be countered with well known security controls. For example IaaS offers virtualized networking capabilities to consumers without the full fledged networking controls such as IP zoning. Also Mather, Kumaraswamy, and Latif (2009) notes that key management procedures that were created initially for a fixed hardware structure do not port correctly to virtual machines that the cloud consists of.

The fourth indicator states that the origin of the vulnerability is in the current services that leading cloud computing companies are providing. Problems with any sort of code injection (command, SQL, XSS) or weak authentication schemes based on simple usernames and passwords would fall under this category.

REAL WORLD EXAMPLES

Many of the cloud specific vulnerabilities, the threats that might exploit them, and the risks associated with them have appeared as actual incidents. The following section will present some real world examples of cloud specific vulnerabilities and whether they have been exploited or not. This is by no means a comprehensive list, but is primarily intended to highlight the main security issues that might accompany cloud computing technologies.

Using IaaS to Host Crimeware

In its low level offering, cloud computing rents out storage space, processing cycles, and network components to consumers, allowing them to utilize them in whatever manner they wish within certain constraints. In the case of the certain cybercriminals, the cloud's IaaS was used as a platform to control a malicious botnet derived from the crimeware Zeus.

The Zeus crimeware toolkit is well established in the underground economy as being an easy to use and powerful tool for stealing personal data from remote systems. Based on the do-it-yourself DIY model, the crimeware allows entry level hackers to create their own versions of botnets. As in any botnet, there is a need to be able to communicate and command all the computers infected with it. In 2009, security experts uncovered a variant of Zeus using Amazon's EC2 IaaS to command and control their botnets (Danchev, 2009).

Even though using an ISP might offer better anonymity, using a cloud can provide traffic camouflaging, where it would be harder to detect and blacklist harmful activity that is hiding in traffic disguised as a valid cloud service.

The Blue Pill Rootkit

As mentioned before, cloud computing is primarily based on the concept of virtualization (i.e. several compartments in the same unit), Therefore, strong compartmentalization should be employed to ensure that individual customers do not impact the operations of other tenants running on the same cloud provider (ENISA, 2010). But virtual machine escape would not be the only problem stemming from virtual environments.

In 2006 Joanna Rutkowska, a security researcher for IT security firm COSEINC, claims to have developed a program that can trick a software to think that it is running on a certain system, where in reality it is running on a virtual version of this system (Blogspot, 2008). This rootkit which is coined the "blue pill" creates a fake reality for an entire operating system and all of the applications running on it including anti-malware sensors. The risk of such a program is that it could easily intercept all hardware requests from any software running on the system.

The creators of the blue pill claim it to be completely undetectable, although some have disputed this claim (Naraine, 2007). Whether detectable or not, it would hard not to acknowledge that it demonstrates how exploits can be developed based on virtualization technologies.

Cloud Computing Outage and Data Loss

Leading providers of cloud computing services have suffered, and in some cases more than once, from data loss or suspension of service. The following are just a few examples of such incidents:

In 2009 Salesforce.com suffered an outage that locked more than 900,000 subscribers out of crucial CC applications and data needed to transact business with customers. Such an outage has even greater impact on companies with most of their operations conducted within the cloud (Ferguson, 2009).

In September 2009 an estimated 800,000 users of a smart phone known as "Sidekick" temporary lost personal data that they could access from their smart phones. The outage lasted almost two weeks and some losses might have been were permanent. The data at the time was stored on servers owned by Microsoft, and accessed as a cloud service. At the time it was described as the biggest disaster in cloud computing history (BBC, 2009).

Rackspace was forced to pay out between $2.5 million and $3.5 million in service credits to customers in the wake of a power outage that hit its Dallas data center in late June 2009.

CONCLUSION

Cloud Computing is an exciting new IT frontier which holds great potential and introduces many benefits for many organizations. The need to understand what CC is, its capabilities and associated vulnerabilities and risks are imperative if enterprises are to make the shift towards outsourcing, and trusting, their computations to the cloud. Many security measures should be taken such as conducting security assessments and having deep understanding of related laws, regulations and best practices to ensure that the right cloud provider is selected. The selection of cloud service provider is very crucial; cloud computing providers should be selected carefully with focus on solid reputation and clear and comprehensive contracts. Companies should also develop solid Business Continuity (BCP) plans and have them tested, and have continuous monitoring software solutions.

Cloud computing is great to use but needs to be considered very carefully. Several risks need to be accounted for and addressed through proper controls to deal with its legal, technical and organizational risks. Deciding to do business on the cloud is a shared responsibility between business and IT. Therefore, proper alignment between business objectives and IT needs to be carefully well thought-out. Transferring risk through outsourcing to a cloud-service provider should not eliminate responsibility, and due diligence and due care must always be practiced.

REFERENCES

Armbrust, M., Fox, A., Griffith, R., Joseph, A. D., Katz, R., & Konwinski, A. (2010). A view of cloud computing. *Communications of the ACM, 53*, 4. doi:10.1145/1721654.1721672

BBC. (2009). *The sidekick cloud disaster*. Retrieved from http://bbc.co.uk./blogs/technology/2009/10/the_sidekick_cloud_disaster.html

Blogspot. (2008). *The blue pill*. Retrieved from http://theinvesiblethings.blogspot.com/2008/07/0wning-xen-invegas.html

Carr, N. (2008). *The big switch: Our new destiney*. New York, NY: W.W. Norton.

Chow, G. J. (2009). Controlling data in the cloud: Outsourcing computation without outsourcing control. In *Proceedings of the ACM Workshop on Cloud Computing and Security*, Chicago, IL (pp. 85-90).

CSA. (2010). *Top threats to cloud computing, v1.0.* Retrieved from https://cloudsecurityalliance.org/topthreats/csathreats.v1.0.pdf

Danchev, D. (2009). *Zeus crimeware using Amazon's EC2 as command and control server*. Retrieved from http://www.zdnet.com/blog/security/zeus-crimeware-using-amazons-ec2-as-command-and-control-server/5010

ENISA. (2010). *Cloud computing: Benefits, risks and recommendations for information security*. Retrieved from http://www.coe.int/t/dghl/cooperation/economiccrime/cybercrime/cy-activity-interface-2010/presentations/Outlook/Udo%20Helmbrecht_ENISA_Cloud%20Computing_Outlook.pdf

Ferguson, T. (2009). *Outage hits thousands of businesses*. Retrieved from http://news.cnet.com/8301-1001_3-10136540-92.html

Ferreiro, D. S. (2010). *Guidance on managing records in cloud computing environments*. Retrieved from http://www.egov.vic.gov.au/focus-on-countries/north-and-south-america-and-the-caribbean/united-states/government-initiatives-united-states/culture-sport-and-recreation-united-states/archives-and-public-records-united-states/guidance-on-managing-records-in-cloud-computing-environments.html

Grobauer, B., Walloscheck, T., & Stocker, E. (2010). Understanding cloud-computing vulnerabilities. *IEEE Security and Privacy, 9*(2), 50–57. doi:10.1109/MSP.2010.115

Hobson, D. (2009). *Global secure systems: Into the Cloud we go....have we thought about security issues?* Retrieved from http://www.globalsecuritymag.com/David-Hobson-Global-Secure-Systems,20090122,7110

Jensen, M., Schwenk, J., Gruschka, N., & Lo Iacono, L. (2009). On technical security issues in cloud computing. In *Proceedings of the IEEE International Conference on Cloud Computing*, Bangalore, India.

Mather, T., Kumaraswamy, S., & Latif, S. (2009). *Cloud security and privacy.* Sebastopol, CA: O'Reilly Media.

Naraine, R. (2007). *100% undetectable malware' challenge.* Retrieved from http://zdnet.com/blog/security/rutkowska-faces-100-undetectable-malware-challenge/334

NIST. (2010). *Cloud computing forum and workshop.* Retrieved from http://csrc.nist.gov/groups/SNS/cloud-computing/

Pearson, S. (2009). Taking account of privacy wehn designing cloud computing services. In *Proceedings of the International Workshop on Software Engineering Challenges of Cloud Computing.* Vancouver, BC, Canada.

Provos, N., Abu Rajab, M., & Mavrommatis, P. (2009). Cybercrime 2.0: When the cloud turns dark. *ACM Queue; Tomorrow's Computing Today, 7*(2), 46–47. doi:10.1145/1515964.1517412

Rutkowska, J. (2007). *Security challenges in virtualaized environments.* Paper presented at the Nordic Virtualization Forum.

Vieira, K., Schulter, A., Westphall, C., & Westphall, C. (2009). Intrusion detection techniques in grid and cloud computing environment. *IT Professional, 12*(4), 38–43. doi:10.1109/MITP.2009.89

Wei, J., Zhang, X., Ammons, G., Bala, V., & Ning, P. (2009). Managing security of virtual machine images in a cloud environment. In *Proceedings of the ACM Cloud Computing Security Workshop* (pp. 91-96).

Wei, Y., & Blake, B. M. (2010). Service-oriented computing and cloud computing: Challenges and opportunities. *IEEE Internet Computing, 14*(6), 72–75. doi:10.1109/MIC.2010.147

Youseff, L., Butrico, M., & Da Silva, D. (2008). *Towards a unified ontology of cloud computing.* Retrieved from http://www.cs.ucsb.edu/~lyouseff/CCOntology/CloudOntology.pdf

This work was previously published in the International Journal of Cloud Applications and Computing (IJCAC), Volume 1, Issue 3, edited by Shadi Aljawarneh & Hong Cai, pp. 1-11, copyright 2011 by IGI Publishing (an imprint of IGI Global).

Chapter 12
Cloud Computing Applied for Numerical Study of Thermal Characteristics of SIP

S. K. Maharana
MVJ College of Engineering, Bangalore, India

Praveen B. Mali
MVJ College of Engineering, Bangalore, India

Ganesh Prabhakar P
MVJ College of Engineering, Bangalore, India

Sunil J
MVJ College of Engineering, Bangalore, India

Vignesh Kumar V
MVJ College of Engineering, Bangalore, India

ABSTRACT

Thermal management of integrated circuit (IC) and system-in-package (SIP) has gained importance as the power density and requirement for IC design have increased and need exists to analyse the heat dissipation performance characteristics of IC under use. In this paper, the authors examine the thermal characteristics of materials of IC. The authors leverage Cloud Computing architecture to remotely compute the dissipation performance parameters. Understanding thermal dissipation performance, which explains the thermal management of IC, is important for chip performance, as well as power and energy consumption in a chip or SIP. Using architectural understanding of Software as a Service (SaaS), the authors develop an efficient, fast, and secure simulation technique by leveraging control volume method (CVM) of linearization of relevant equations. Three chips are kept in tandem to make it a multi-chip module (MCM) to realise it as a smaller and lighter package. The findings of the study are presented for different dimensions of chips inside the package.

DOI: 10.4018/978-1-4666-1879-4.ch012

1. INTRODUCTION

Thermal management of integrated circuit (IC) and of a system-in-package (SIP) has gained importance in the recent past since the power density and power requirement for IC design have gone up. With increasing integrated circuit (IC) power densities and performance requirements, thermal issues have become critical challenges in IC design. If not properly addressed, increased IC temperature affects other design metrics including performance, power and energy consumption, reliability and price. It is thus critical to consider thermal issues during IC design and synthesis. When determining the impact of each decision in the synthesis or design process, the impacts of changed thermal profile on performance, power, price, and reliability must be considered. This requires repeated use of detailed chip-package thermal analysis. This analysis is generally based on computationally expensive numerical methods. In order to support IC synthesis, a thermal simulator must be capable of accurately analyzing models containing tens of thousands of discrete elements. The study will be focused upon analysing thermal characteristic of multi-chip module (MCM).

There will be a two dimensional modelling of the entire module and then the computation of thermal characteristics will be done through MATLAB. In this study we have considered three chips which shall be kept either in series or in a stack.

We shall be computing the thermal characteristics of these chips in series or in stack through governing equation of heat flow.

Here we are attempting to compute the heat dissipation rate using conduction mode of heat transfer happening inside the system-in-package. The governing equations are approximated using control volume method (CVM) and then a code is scripted in MATLAB. Using the concept of Cloud computing architecture the entire computation is achieved through MATLAB. In particular we have leveraged SaaS, which is one of the Cloud

computing architectures, in our study. The motivation behind taking such an initiative to use Cloud Computing fundamentals is to understand how it could be beneficial to an engineering problem in the design of IC that we have considered here. As a first steps toward achieving this goal we have tried to use the resources, computing power and other ancillary applications in Cloud Server through a communication mechanism inside the research institute. The same computation has also been tested outside the private cloud network.

The details of the discretization of equations are described in the paper. The authors have designed a Java-based user interface through which we control the computing resources available in the server. The entire application designed for the above computational purpose is available to users through cloud network called as Software as a Service (SaaS) as shown in Figure 1.

Due to the flexible manufacturing processes for producing Ball Grid Array packages, there has been an explosion of new package designs, many of which contain more than one integrated circuit chip. Existing industry standards for IC package thermal characterizations apply only to single-chip packages. This basic calculation will present some techniques for calculating the junction temperature of a multi-chip module (MCM). Existing JEDEC standards all assume that there is one heat source in a package, characterized by a single junction

Figure 1. SaaS architecture

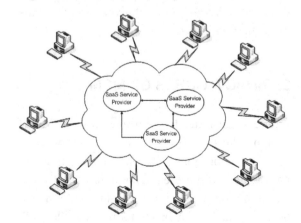

temperature, T_J. This allows the use of thermal metrics that can be employed in calculations in a manner analogous to electrical resistors and that can be easily extracted from thermal measurements. The format of such a metric is

$$\Theta_{J,REF} = \frac{T_J - T_{REF}}{P} = \frac{\Delta T_{J,REF}}{P} \tag{1}$$

Where $\mathbf{T_{REF}}$ is a reference temperature, measured in a specified location in a JEDEC-standard test environment, $\mathbf{\Theta_{J,REF}}$ is the thermal resistance of the chip and $\mathbf{P}$ is the total dissipated power in the chip. Examples of reference temperatures are those of the ambient air, the test board, and the package case.

Clearly, this methodology breaks down when dealing with an MCM. First, it only accounts for a single junction temperature. Second, even when the power applied to one chip is zero, $\Delta T_{J, REF}$ can be non-zero due to power applied to another chip in the package. Applying Equation 1 to the non-powered chip would involve dividing a finite value of $\Delta T_{J,REF}$ by zero power. This would lead to a value of infinity for $\mathbf{\Theta_{J,REF}}$ – not a very useful result. Intuition tells us that the temperature of any chip in an MCM would be affected by the particular combination of power levels applied to all the chips. This, clearly, represents a much more complicated situation than encountered with single-chip packages. We shall be analysing these situations in our study. Below given are two basic types of MCMs and some relatively simple means of dealing with them.

2. BASIC MODELS OF MCM

Figure 2 illustrates two limiting cases of MCM design. The first figure illustrates the so-called "stacked" configuration. (Of course, a given package could represent a combination of these configurations). The second figure is described here (for lack of a better term) as a "lateral"

configuration. Each chip attaches directly to the substrate.

3. METHODOLOGY

Governing Equations

Our study of heat transfer begins with an energy balance and Fourier's law of heat conduction. The first working equation we derive is a partial differential equation (PDE). In the analysis of a heat transfer system, as in all engineering systems, our first step should be to write out the appropriate two dimensional PDE equations. From the basic law of heat conduction and the conservation of thermal energy, the general differential equation can be written as

$$\frac{\partial}{\partial x}\left(k\frac{\partial T}{\partial x}\right) + \frac{\partial}{\partial y}\left(k\frac{\partial T}{\partial y}\right) + Q = \rho c\frac{\partial T}{\partial t} \tag{2}$$

Where, 'ρ' is the density constant, 'Q' is the internal heat generation per unit volume and 'k' refers to thermal conductivity.

Divide the Equation 2 by ρc.

$$\frac{1}{\rho c}\frac{\partial}{\partial x}\left(k\frac{\partial T}{\partial x}\right) + \frac{1}{\rho c}\frac{\partial}{\partial y}\left(k\frac{\partial T}{\partial y}\right) + \frac{Q}{\rho c} = \frac{\partial T}{\partial t} \tag{3}$$

Figure 2. Stacked and Lateral configuration

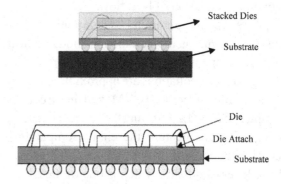

$$\frac{k}{\rho c}\left(\frac{\partial^2 T}{\partial x^2}\right) + \frac{k}{\rho c}\left(\frac{\partial^2 T}{\partial y^2}\right) + \frac{Q}{\rho c} = \frac{\partial T}{\partial t} \tag{4}$$

Let $Q = f(x,y,z,t)$

$$\alpha\left[\left(\frac{\partial^2 T}{\partial x^2}\right) + \left(\frac{\partial^2 T}{\partial y^2}\right)\right] + \frac{f(x,y,z,t)}{\rho c} = \frac{\partial T}{\partial t} \tag{5}$$

Where 'α' refers to thermal diffusivity which is given by

$$\alpha = \frac{k}{\rho c} \tag{6}$$

In the two-dimensional case, we analyse the energy balance which will be integrated over a system where the temperature varies in the x-direction and the y-direction. One can imagine a case where we have a thin plate or silicon stacked chip or lateral chip earlier referred. The shape of the chip can change but it is still subject to the heat equation at any point within the chip.

3.1. Finite Volume Formulation

Since the thermal energy is conserved everywhere in the solution domain, the differential thermal Equation 2 can be integrated over the control volume (Ω) and over a time interval from t to $t+\Delta t$ as

$$\int_t^{t+\nabla t}\int_\Omega \frac{\partial}{\partial x}k\left(\frac{\partial T}{\partial x}\right)d\Omega dt + \int_t^{t+\nabla t}\int_\Omega \frac{\partial}{\partial y}k\left(\frac{\partial T}{\partial y}\right)d\Omega dt$$
$$+ \int_t^{t+\nabla t}\int_\Omega Q d\Omega dt = \int_t^{t+\nabla t}\int_\Omega \rho c\left(\frac{\partial T}{\partial t}\right)d\Omega dt \tag{7}$$

The following coordinate system presents CVM formulation:

Referring to the two-dimensional Cartesian mesh in Figure 3, a general nodal point is identified by P and its neighbour in a two-dimensional

geometry, the nodes to the west, east, south and north are identified by W, E, S and N respectively. The four side faces are labelled W, E, S and N which stand for west, east, south and north. The width and height of the control volume are Δx and Δy respectively. If the temperature at a node is assumed to prevail over the whole control volume, the right hand side can be written as

$$\int_t^{t+\nabla t}\int_\Omega \rho c\left(\frac{\partial T}{\partial t}\right)d\Omega dt = \rho c\left(T_p - T_p^0\right)\Delta x\Delta y \tag{8}$$

In Equation 8, superscript '0' refers to temperature at time t; temperature at time level $t+\Delta t$ is not superscripted. After substituting Equation 8 into Equation 7 and applying central differencing to the diffusion terms on the left hand side of Equation 7, we have

$$\int_t^{t+\nabla t}\left(k_e\Delta y\frac{T_E - T_P}{\Delta x} - k_w\Delta y\frac{T_P - T_W}{\Delta x}\right)dt$$
$$+ \int_t^{t+\nabla t}\left(k_n\Delta x\frac{T_N - T_P}{\Delta y} - k_s\Delta x\frac{T_P - T_S}{\Delta y}\right)dt$$
$$+ \int_t^{t+\nabla t}\int_\Omega Q d\Omega dt = \rho c\left(T_p - T_p^0\right)\Delta x\Delta y \tag{9}$$

Figure 3. Two dimensional cartesian mesh

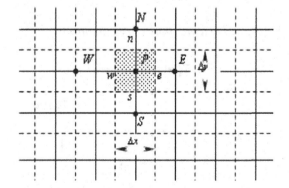

Boundary Conditions

Dirichlet, Neumann, or mixed Boundary conditions are specified for all boundaries. Dirichlet boundary conditions for the rectangle slab/chip are shown in Figure 4. $T_{bc}1 - 4$ is the certain fixed values of temperature at the boundaries mentioned below.

$$T(x=0,y,t)=T_{bc}1, T(x=L,y,t) = T_{bc}2 \qquad (10)$$

$$T(x,y=0,t)=T_{bc}3, T(x,y=L,t)= T_{bc}4 \qquad (11)$$

We have included heat generation and consumption terms in our analysis using MATLAB:

Generalized generation and consumption terms: The heat equation as we have written it in Equation 2 contains a generic generation and consumption term, which can be a function of time and position, and which, in one dimension, appears in the heat equation as f(t,x).

$$\frac{\partial T}{\partial t} = \alpha\left(\frac{\partial^2 T}{\partial x^2}\right) + \alpha\left(\frac{\partial^2 T}{\partial y^2}\right) + \frac{f(t,x)}{\rho c} \qquad (12)$$

No generation or consumption: When there are no generation or consumption terms,

$$f(t,x) = 0 \qquad (13)$$

Non-Uniform heat loss or heat gain: When we have heat loss (or gain) along the rod due to lack of insulation, the energy loss (or gain)[energy/time] has the form

Figure 4. Dirichlet boundary conditions

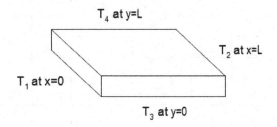

$$qA = hA(T_{surround} - T(x,t)) \qquad (14)$$

Our consumption (or generation) term per unit volume of the material [energy/time/volume] is then

$$f(t,x) = \frac{qA}{V} = \frac{hA\left(T_{surround} - T(x,t)\right)}{V} \qquad (15)$$

Which, substituting Equation 15 into Equation 4 gives us a heat equation of the form:

$$\frac{\partial T}{\partial t} = \alpha\left(\frac{\partial^2 T}{\partial x^2}\right) + \frac{hA\left(T_{surround} - T(x,t)\right)}{\rho c V} \qquad (16)$$

The two-dimensional version would look like:

$$\frac{\partial T}{\partial t} = \alpha\left(\frac{\partial^2 T}{\partial x^2}\right) + \alpha\left(\frac{\partial^2 T}{\partial y^2}\right) + \frac{hA\left(T_{surround} - T(x,t)\right)}{\rho c V} \qquad (17)$$

The pdetool used for solving the two-dimensional heat equation asks for the input in the format:

$$u\varphi - div((grad(u)) + au` = f \qquad (18)$$

We can rearrange equation (12) to give us the same form:

$$\frac{\partial T}{\partial t} = \alpha\left(\frac{\partial^2 T}{\partial x^2}\right) + \alpha\left(\frac{\partial^2 T}{\partial y^2}\right) + \frac{hA\left(T(x,y,t)\right)}{\rho c V}$$
$$= \frac{hA\left(T_{surround}\right)}{\rho c V} \qquad (19)$$

By making the following equalities:

$$u= T, d=1, c=\alpha \qquad (20)$$

$$f = \frac{hA\left(\mathrm{T}_{\mathrm{surround}}\right)}{\rho c\, V} \quad (21)$$

$$\mathrm{u}` = \frac{\partial T}{\partial \mathrm{t}} \quad (22)$$

$$div((grad(\mathrm{u})) = \left(\frac{\partial^2 T}{\partial x^2}\right) + \left(\frac{\partial^2 T}{\partial y^2}\right) \quad (23)$$

$$\mathrm{a} = \frac{hA}{\rho c\, V} \quad (24)$$

For the two-dimensional case, the area is the surface area of the chip, exposed to the surrounding temperature. For a rectangular chip of length L_x and width L_y with one heat loss through one face, the area is given by $A = L_x L_y$. For the same dimensions with heat loss through both faces, $A = 2L_x L_y$. The heat loss through the x and y boundaries is taken care of by the boundary conditions, not the generation term so that surface area does not appear in these equations. The volume would be $V = L_x L_y L_z$, where L_z is the width of the plate.

4. LINKING SIMULATION ANALYSIS THROUGH CLOUD NETWORK

Definition of Cloud Computing

Cloud Computing, as the name suggests is a style of computing where dynamically scalable and often visualized resources are provided as a service over the internet. These services can be consumed by any user over a standard HTTP medium. The user doesn't need to have the knowledge, expertise or control over the technology infrastructure in the "cloud" that supports them. The name cloud computing was inspired by the cloud symbol that's often used to represent the Internet inflow charts and diagrams.

Overview

Figure 5 shows the typical example of the server-client architecture in cloud in which Software as a service (or SaaS) is a way of delivering applications over the Internet-as a service. Instead of installing and maintaining software, you simply access it via the Internet, freeing yourself from complex software and hardware management. SaaS applications are sometimes called Web-based software, on-demand software, or hosted software. Whatever the name, SaaS applications run on a SaaS provider's servers. The provider manages access to the application, including security, availability, and performance. SaaS customers have no hardware or software to buy, install, maintain, or update. Access to applications is easy: you just need an Internet connection. This types of cloud computing delivers a single application through the browser to thousands of customers using a multitenant architecture. Here application is installed in the server operating system. This operating system works as a virtual background for the application over the clients operating background. On the customer side, it means no upfront investment in servers or software licensing; on the provider side, with just one app to maintain, costs are low compared to conventional hosting. Here using MATLAB installed in server machine. The users can access the services offered in this case, refers a thermal simulation and analysis of MCM. Thus any user can visualize the full application given an input through a GUI.

5. RESULTS AND DISCUSSION

Details of Cloud Computing

A cloud service provider offers virtual operating system as per user's requirements. Once the user who wants to host his own server for his applications, softwares and other features accepts the terms and conditions of the cloud service

Figure 5. Server-client architecture

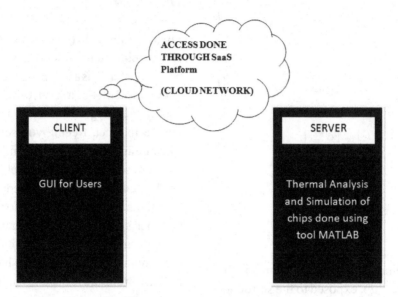

provider and get accessibility to create his own server with his own operating system. After a long study on cloud computing and its service providers we have understood one of the service provider ElasticHosts.com whose specifications match our study.

Figure 6 shows that with the help of API (Tight VNC) the client accesses windows server 2008 R2 over his operating system and uses the application. With the help of this cloud computing the client can operate on both the operating system simultaneously. The client need not install the server operating system.

Elastichosts has many inbuilt operating systems, one among the inbuilt OS we subscribed for Windows server 2008 R2 operating system in which our application is installed and this OS acts as a server. Now the client who wishes to use this particular application has to get an authorization from the server through an API. The API that we

Figure 6. Virtualization through cloud

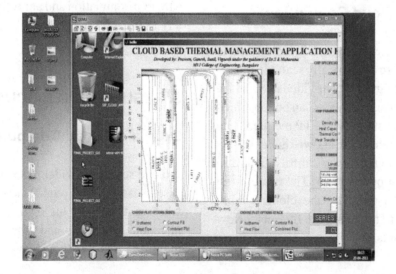

are using for our study is Tight VNC. Through this API the client access the application with the virtual OS as its background.

The application developed for the client is as shown in Figure 7. This user interface is accessed by the client with the background virtual OS. The client first selects the required configuration whether it is stack or series. Then the client inputs the chip parameters such as the density, heat capacity, thermal conductivity and heat transfer coefficient. Then the client should give the module dimensions such as the total length and width of entire sip. Then the width of each chip can be given manually for computation. Finally the contour levels can also be given so that the output graph is modified accordingly. The parameters for silicon was entered and the isotherms, contour fills, heat flow and combined plots can be generated using this application for respective chip configurations (series and stack).

5.1. Silicon Chip

The silicon chip parameters are entered for the values shown in Table 1.

When these parameters are entered by client in this application the isotherm plot is obtained

Table 1. Silicon chip parameters

Density	2.3290gcm^{-3}
Heat capacity	19.789JMol^{-1}K^{-1}
Thermal conductivity	149Wm^{-1}K^{-1}
Heat transfer coefficient	0.85kW/(m^{2}K)

as shown in Figure 8. From the analysis we can observe that the contour lines obtained are parabolic in nature. As seen in the legend (colorbar) the maximum temperature conduction is denoted by colour red and minimum temperature conduction is denoted by colour blue and other intermediate temperature conduction values is shown with varied colours.

The configuration tested here is that three silicon chips are placed in series with heat source given for the entire substrate and the temperature conduction patterns indicate that neighbouring chips generate maximum heat compared to centre chip. And the output graph for heat flows can also be generated for the same parameters and it is been observed that the heat flows are orthogonal to the temperature flow graph.

The same parameters are given to stack accordingly and the plot is generated similarly as shown in Figure 9.

Figure 7. Graphical user interface

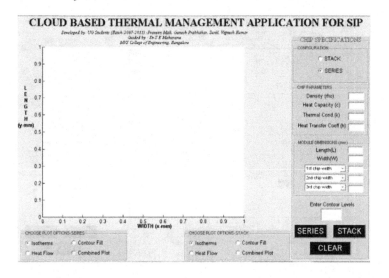

Figure 8. Use of application through cloud: silicon isotherm series plot

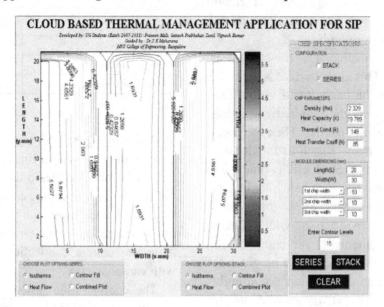

6. CONCLUSION

The MATLAB based simulation technique to analyse thermal characteristics in SIP using cloud has given us an efficient and easier way to analyse the thermal characteristics and give a meaningful conclusion for any type of arrangement of chips. By using of SaaS platform, application delivery is done via a web browser with data encryption, transmission, and access and storage services.

REFERENCES

Guenin, B. (2002). The many flavors of ball grid array packages. *Electronics Cooling*, *8*(1), 32–40.

Figure 9. Use of application through cloud: silicon isotherm stack plot

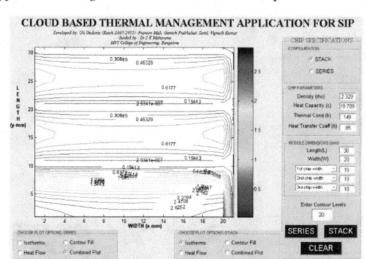

ITRS. (2010). *International technology roadmap for semiconductors.* Retrieved from http://public.itrs.net

Keffler, D. (1999). *Application and solution of heat equation in one and two dimensional systems using numerical methods.* Knoxville, TN: University of Tennessee.

Khaund, B. (2009). *Cloud Architecture-SaaS.* Retrieved from http://www.codeproject.com/KB/webservices/CloudSaaS.aspx

Lall, B., Guenin, B., & Molnar, R. (1995). Methodology for thermal evaluation of multichip modules. *IEEE Transactions on Components Packaging & Manufacturing Technology Part A, 18*(4), 758–764. doi:10.1109/95.477461

Zahn, B. (1998, March). Steady state thermal characterization of multiple output devices using linear superposition theory and a non-linear matrix multiplier. In *Proceedings of the Fourteenth Annual Semiconductor Thermal Measurement and Management Symposium* (pp. 39-46).

This work was previously published in the International Journal of Cloud Applications and Computing (IJCAC), Volume 1, Issue 3, edited by Shadi Aljawarneh & Hong Cai, pp. 12-21, copyright 2011 by IGI Publishing (an imprint of IGI Global).

Chapter 13
Toward Understanding the Challenges and Countermeasures in Computer Anti-Forensics

Kamal Dahbur
New York Institute of Technology, Jordan

Bassil Mohammad
New York Institute of Technology, Jordan

ABSTRACT

The term computer anti-forensics (CAF) generally refers to a set of tactical and technical measures intended to circumvent the efforts and objectives of the field of computer and network forensics (CF). Many scientific techniques, procedures, and technological tools have evolved and effectively applied in the field of CF to assist scientists and investigators in acquiring and analyzing digital evidence for the purpose of solving cases that involve the use or misuse of computer systems. CAF has emerged as a CF counterpart that plants obstacles throughout the path of computer investigations. The purpose of this paper is to highlight the challenges introduced by anti-forensics, explore various CAF mechanisms, tools, and techniques, provide a coherent classification for them, and discuss their effectiveness. Moreover, the authors discuss the challenges in implementing effective countermeasures against these techniques. A set of recommendations are presented with future research opportunities.

INTRODUCTION

The use of technology is increasingly spreading covering various aspects of our daily lives. An equal increase, if not even more, is realized in the methods and techniques created with the intention to misuse the technologies serving varying objectives being political, personal or anything else. This has clearly been reflected in our terminology as well, where new terms like cyber warfare, cyber security, and cyber crime, amongst others, were introduced. It is also noticeable that such attacks

DOI: 10.4018/978-1-4666-1879-4.ch013

are getting increasingly more sophisticated, and are utilizing novel methodologies and techniques. Fortunately, these attacks leave traces on the victim systems that, if successfully recovered and analyzed, might help identify the offenders and consequently resolve the case(s) justly and in accordance with applicable laws. For this purpose, new areas of research emerged addressing Network Forensics and Computer Forensics in order to define the foundation, practices and acceptable frameworks for scientifically acquiring and analyzing digital evidence in to be presented in support of filed cases. In response to Forensics efforts, Anti-Forensics tools and techniques were created with the main objective of frustrating forensics efforts, and taunting its credibility and reliability.

This paper attempts to provide a clear definition for Computer Anti-Forensics and consolidates various aspects of the topic. It also presents a clear listing of seen challenges and possible countermeasures that can be used. The lack of clear and comprehensive classification for existing techniques and technologies is highlighted and a consolidation of all current classifications is presented.

Please note that the scope of this paper is limited to Computer-Forensics. Even though it is a related field, Network-Forensics is not discussed in this paper and can be tackled in future work. Also, this paper is not intended to cover specific Anti-Forensics tools; however, several tools were mentioned to clarify the concepts.

After this brief introduction, the remainder of this paper is organized as follows: we provide a description of the problem space, introduce computer forensics and computer anti-forensics, and provide an overview of the current issues concerning this field; an overview of related work is presented with emphasis on Anti-Forensics goals and classifications; we provide detailed discussion of Anti-Forensics challenges and recommendations; and we provide our conclusion, and suggested future work.

THE PROBLEM SPACE

Rapid changes and advances in technology are impacting every aspect of our lives because of our increased dependence on such systems to perform many of our daily tasks. The achievements in the area of computers technology in terms of increased capabilities of machines, high speeds communication channels, and reduced costs resulted in making it attainable by the public. The popularity of the Internet, and consequently the technology associated with it, has skyrocketed in the last decade (Table 1 and Figure 1). Internet usage statistics for 2010 clearly show the huge increase in Internet users who may not necessary be computer experts or even technology savvy (Thuen, 2007).

Unfortunately, some of the technology users will not use it in a legitimate manner; instead, some users may deliberately misuse it. Such misuse can result in many harmful consequences including, but not limited to, major damage to others systems or prevention of service for legitimate users. Regardless of the objectives that such "bad guys" might be aiming for from such misuse (e.g., personal, financial, political or religious purposes), one common goal for such users is the need to avoid detection (i.e., source determination). Therefore, these offenders will exert thought and effort to cover their tracks to avoid any liabilities or accountability for their damaging actions. Illegal actions (or crimes) that involve a computing system, either as a mean to carry out the attack or as a target, are referred to as Cybercrimes (Internet World Stats, n. d.). Computer crime or Cybercrime are two terms that are being used interchangeably to refer to the same thing. A Distributed Denial of Service attack (DDoS) is a good example for a computer crime where the computing system is used as a mean as well as a target.

Fortunately, cybercrimes leave fingerprints that investigators can collect, correlate and analyze to understand what, why, when and how a

Table 1. World Internet usage – 2010 (adapted from Thuen, 2007)

World Regions	Population (2010 Est.)	Internet Users Dec. 31, 2000	Internet Users Latest Data	Growth 2000-2010
Africa	1,013,779,050	4,514,400	**110,931,700**	2357%
Asia	3,834,792,852	114,304,000	**825,094,396**	622%
Europe	813,319,511	105,096,093	**475,069,448**	352%
Middle East	212,336,924	3,284,800	**63,240,946**	1825%
North America	344,124,450	108,096,800	**266,224,500**	146%
Latin America/ Caribbean	592,556,972	18,068,919	**204,689,836**	1033%
Oceania/Australia	34,700,201	7,620,480	**21,263,990**	179%
WORLD TOTAL	6,845,609,960	360,985,492	1,966,514,816	445%

crime was committed; and consequently, and most importantly, build a good case that can bring the criminals to justice. In this sense, computers can be seen as great source of evidence. For this purpose Computer Forensics (CF) emerged as a major area of interest, research and development driven by the legislative needs of having scientific reliable framework, practices, guidelines, and techniques for forensics activities starting from evidence acquisition, preservation, analysis, and finally presentation.

Computer Forensics can be defined as the process of scientifically obtaining, examining and analyzing digital information so that it can be used

as evidence in civil, criminal or administrative cases (Internet World Stats, n. d.). A more formal definition of Computer Forensics is "the discipline that combines elements of law and computer science to collect and analyse data from computer systems, networks, wireless communications, and storage devices in a way that is admissible as evidence in a court of law" (Nelson, Phillips, & Steuart, 2010). US-CERT defines Computer Forensics – CF – as *"the discipline that combines elements of law and computer science to collect and analyse data from computer systems, networks, wireless communications, and storage devices in a way that is admissible as evidence*

Figure 1. World Internet usage – 2010 (adapted from Thuen, 2007)

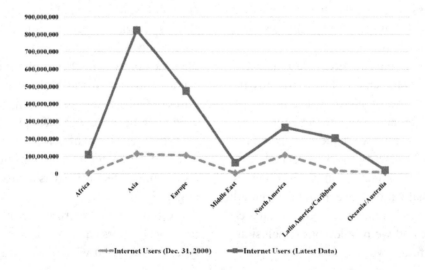

in a court of law" (US-Computer Emergency Readiness Team, 2008).

To hinder the efforts of Computer Forensics, criminals work doggedly to instigate, develop and promote counter techniques and methodologies, or what is commonly referred to as Anti-Forensics. If we adopt the definition of Computer Forensics (CF) as scientifically obtaining, examining, and analysing digital information to be used as evidence in a court of law, then Anti-Forensics can be defined similarly but in the opposite direction. In Computer Anti-Forensics (CAF) scientific methods are used to simply frustrate Forensics efforts at all forensics stages. This includes preventing, impeding, and/or corrupting the acquiring of the needed evidence, its examination, its analysis, or its credibility. In other words, whatever necessary to ensure that computer evidence cannot get to, or will not be admissible in, a court of law.

The use of Computer Anti-Forensics tools and techniques is evident and far away from being an illusion. So, criminals' reliance on technology to cover their tracks is not a claim, as clearly reflected in recent researches conducted on reported and investigated incidents. Based on 2009-2010 Data Breach Investigations Reports (Verizon Business, 2009, 2010), investigators found signs of anti-forensics usage in over one third of cases in 2009 and 2010 with the most common forms being the same for both years. The results show that the overall use of anti-forensics remained relatively flat with slight movement among the techniques themselves. Figure 2 shows the types of anti-Forensic techniques used (data wiping, data hiding and data corruption) by percentage of breaches. As shown in Figure 2, data wiping is still the most common, because it is supported by many commercial off-the-shelf products that are available even as freeware that are easy to install, learn and use; while data hiding and data corruption remain a distant behind.

It is important to note that the lack of understanding on what CAF is and what it is capable of may lead to underestimating or probably over-

Figure 2. Types of anti-forensics – 2010 (adapted from Verizon Business, 2010)

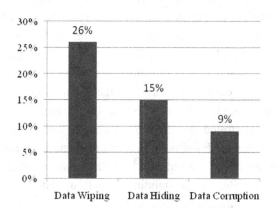

looking CAF impact on the legitimate efforts of CF. Therefore, when dealing with computer forensics, it is important that we address the following questions, among others, that are related to CAF: Do we really have everything? Are the collected evidences really what were left behind or they are only just those intentionally left for us to find? How to know if the CF tool used was not misleading us due to certain weaknesses in the tool itself? Are these CF tools developed according to proper secure software engineering methodologies? Are these CF tools immune against attacks? What are the recent CAF methods and techniques? This paper attempts to provide some answers to such questions that can assist in developing the proper understanding for the issue.

CAF GOALS AND CLASSIFICATIONS

Even though computer forensics and computer ant-forensics are tightly related, as if they are two faces of the same coin, the amount of research they received was not the same. CF received more focus over the past ten years or so because of its relation with other areas like data recovery, incident management and information systems risk assessment. CF is a little bit older, and therefore more mature than CAF. It has consistent definition, well defined systematic approach and complete set of leading best practices and technology.

CAF on the other side, is still a new field, and is expected to get mature overtime and become closer to CF. In this effort, recent research papers attempted to introduce several definitions, various classifications and suggest some solutions and countermeasures. Some researchers have concentrated more on the technical aspects of CF and CAF software in terms of vulnerabilities and coding techniques, while others have focused primarily on understanding file systems, hardware capabilities, and operating systems. A few other researchers chose to address the issue from an ethical or social angle, such as privacy concerns. Despite the criticality of CAF, it is hard to find a comprehensive research that addresses the subject in a holistic manner by providing a consistent definition, structured taxonomies, and an inclusive view of CAF.

Goals of Anti-Forensics

As stated in the previous section, CAF is a collection of tools and techniques that are intended to frustrate CF tools and CF's investigators efforts. This field is growingly receiving more interest and attention as it continues to expose the limitations of currently available computer forensics techniques as well as challenge the presumed reliability of common CF tools. We believe, along with other researchers, that the advancements in the CAF field will eventually put the necessary pressure on CF developers and vendors to be more proactive in identifying possible vulnerabilities or weaknesses in their products, which consequently should lead to enhanced and more reliable tools.

CAF can have a broad range of goals including: avoiding detection of event(s), disrupting the collection of information, increasing the time an examiner needs to spend on a case, casting doubt on a forensic report or testimony. In addition, these goals may also include: forcing the forensic tool to reveal its presence, using the forensic tool to

attack the organization in which it is running, and leaving no evidence that an anti-forensic tool has been run (Verizon Business, 2009, 2010).

Classifications of Anti-Forensics

Several classifications for CAF have been introduced in the literature. These various taxonomies differ in the criteria used to do the classification. The following are the most common approaches used:

1. Categories Based on the Attacked Target
    ◦ *Attacking Data*: The acquisition of evidentiary data in the forensics process is a primary goal. In this category CAFs seek to complicate this step by wiping, hiding or corrupting evidentiary data.
    ◦ *Attacking CF Tools*: The major focus of this category is the examination step of the forensics process. The objective of this category is to make the examination results questionable, not trustworthy, and/or misleading by manipulating essential information like hashes and timestamps.
    ◦ Attacking the Investigator: This category is aimed at exhausting the investigator's time and resources, leading eventually to the termination of the investigation.
2. CAF Techniques vs. Tactics

This categorization makes a clear distinction between the terms anti-forensics and counter-forensics (Hartley, 2007), even though the two terms have been used interchangeably by many others as the emphasis is usually on technology rather than on tactics.

    ◦ *Counter-Forensics*: This category includes all techniques that target the forensics tools directly to cause them to crash, erase collected evidence, and/or break completely (thus disallowing the investigator from using it). Compression bombs are good example on this category.

- *Anti-Forensics*: This category includes all technology related techniques including encryption, steganography, and alternate data streams (ADMs).

3. Traditional vs. Non-Traditional

- *Traditional Techniques*: This category includes techniques involving overwriting data, Cryptography, Steganography, and other data hiding approaches beside generic data hiding techniques.

- *Non-Traditional Techniques*: As opposed to traditional techniques, these techniques are more creative and impose more risk as they are harder to detect. These include:

  - *Memory injections*: where all malicious activities are done on the volatile memory area.

  - *Anonymous storage*: utilizes available web-based storage to hide data to avoid being found on local machines.

  - *Exploitation of CF software bugs*: including Denial of Service (DoS) and Crashers, amongst others.

4. Categories Based on Functionality

This categorization includes data hiding, data wiping and obfuscation. Attacks against CF processes and tools is considered a separate category based on this scheme

CHALLENGES OF ANTI-FORENSICS

Because Computer Anti-Forensics (CAF) is a relatively new discipline, the field faces many challenges that need considered and addressed. In this section, we have attempted to identify the most pressing challenges surrounding this area, highlight the research needed to address such challenges, and attempt to provide perceptive answers to some the concerns.

Ambiguity

Aside from having no industry-accepted definition for CAF, studies in this area view anti-forensics differently; this leads to not having a clear set of standards or frameworks for this critical area. Consequently, misunderstanding may be an unavoidable end result that could lead to improperly addressing the associated concerns. The current classification schemes, stated above, which mostly reflect the author's viewpoint and probably background, confirm as well as contribute to the ambiguity in this field. A classification can only be beneficial if it has clear criteria that can assist not only in categorizing the current known techniques and methodologies but will also enable proper understanding and categorization of new ones. The attempt to distinguish between the two terms, anti-forensics and counter-forensics based on technology and tactics is a good initiative but yet requires more elaboration to avoid any unnecessary confusions.

To address the definition issue, we suggest to adopt a definition for CAF that is built from our clear understanding of CF. The classification issue can be addressed by narrowing the gaps amongst the different viewpoints in the current classifications and excluding the odd ones.

Investigation Constraints

A CF investigation has three main constraints/challenges, namely: time, cost and resources. Every CF investigation case should be approached as separate project that requires proper planning, scoping, budgeting and resources. If these elements are not properly accounted for, the investigation will eventually fail, with most efforts up to the point of failure being wasted. In this regard, CAF techniques and methodologies attempt to attack the time, cost and resources constraints of an investigation project. An investigator may not able to afford the additional costs or allocate the additional necessary resources. Most importantly, the time factor might play a critical role in the investigation as evidentiary data might lose value with time, and/or allow the suspect(s) the opportunity to cover their tracks or escape. Most, if not all, CAF techniques and methodologies (including data wiping, data hiding, and data corruption) attempt to exploit this weakness. Therefore, it proper project management is imperative before and during every CF investigation.

Integration of Anti-Forensics into Other Attacks

Recent researches show an increased adoption of CAF techniques into other typical attacks. The primary purposes of integrating CAF into other attacks are undetectability and deletion of evidence. Two major areas for this threatening integration are Malware and Botnets (Brand, 2007).

Malware, malicious software, being a Worm, Virus or a Trojan-horse is using anti-forensics techniques to avoid detection and analysis of its code by using encryption, destruction, data hiding, debugger detection and data contraception amongst others. While encryption makes the code unreadable, destruction deletes related generated files, and data hiding hides the data into unexpected locations. With debugger detection the malware can detect if it is being debugged or not, upon which a different code path is triggered for execution. Data contraception ensures that data is never stored on the disk and writes are kept minimum and on dynamic memory where possible.

Botnets on the other side utilise the same techniques and yet are more dangerous. The term "Botnet" refers to a collection of compromised machines that are under the control of the "Botmaster" (Secureworks.com, 2008). Botnets are constructed from a collection of different types of malware, incorporating techniques developed for viruses, worms and Trojan horses. Botnets are mostly utilized in Distributed Denial of Service Attacks (DDoS).

Botnets make forensic analysis harder by using encryption to hide the traffic between the Botmaster and the controlled Zombies. Provided the time constrain that the investigator has, these Malwares and Botnets armed with these techniques will make his or her efforts labour and time intensive which might eventually lead to overlooking critical evidences. Botnets in relation to Anti-Forensics is typically discussed in the context of Network Forensics as it has to do more with network logs and traffic.

Breaking the Forensics Software

CF tools are, of course, created by humans, just like other software systems. Rushing to release their products to the market before their competition, companies tend to, unintentionally, introduce vulnerabilities into their products. In such cases, software development best practices, which are intended to ensure the quality of the product, might be overlooked leading to the end product being exposed to many known vulnerabilities, such as buffer overflow and code injection. Because CF software is ultimately used to present evidence in courts, the existence of such weaknesses is not tolerable. Hence, all CF software, before being used, must be subjected to thorough security testing that focuses on robustness against data hiding and accurate reproduction of evidence.

The Common Vulnerabilities and Exposures (CVE) database is a great source for getting updates on vulnerabilities in existing products (Mitre.org, 2011). Some studies have reported several weaknesses that may result in crashes during runtime leaving no chance for interpreting the evidence (Guidance Software, 2011). Regardless of the fact that some of these weaknesses are still being disputed (Newsham, Palmer, & Stamos, 2007), it is important to be aware that these CF tools are not immune to vulnerabilities, and that CAF tools would most likely take advantage of such weaknesses. A good example of a common technique that can cause a CF to fail or crash is the "Compression Bomb"; where files are compressed hundreds of times such that when a CF tool tries to decompress, it will use up so many resources causing the computer or the tool to hang or crash.

Privacy Concerns

Increasingly, users are becoming more aware of the fact that just deleting a file does not make it really disappear from the computer and that it can be retrieved by several means. This awareness is driving the market for software solutions that provide safe and secure means for files deletion. Such tools are marketed as "privacy protection" software and claim to have the ability to completely remove all traces of information concerning user's activity on a system, websites, images and downloaded files. Some of these tools do not only provide protection through secure deletion; but also offer encryption and compression. Moreover, these tools are easy use, and some can even be downloaded for free. WinZip is a popular tool that offers encryption, password protection, and compression. Such tools will most definitely complicate the search for and acquiring of evidence in any CF investigation because they make the whole process more time and resources consuming.

Privacy issues in relation to CF have been the subject of detailed research in an attempt to define appropriate policies and procedures that would maintain users' privacy when excessive data is acquired for forensics purposes (Srinivasan, 2007).

Nature of Digital Evidence

CF investigations rely on two main assumptions to be successful: (1) the data can be acquired and used as evidence, and (2) the results of the CF tools are authentic, reliable, and believable. The first assumption highlights the importance of digital evidence as the basis for any CF investigation; while the second assumption highlights the critical role of the trustworthiness of the CF tools in order for the results to stand solid in courts.

Digital evidence is more challenging than physical evidence because of its more susceptible to being altered, hidden, removed, or simply made unreadable. Several techniques can be utilized to achieve such undesirable objectives that can complicate the acquisition process of evidentiary digital data, and thus compromise the first assumption.

CF tools rely on many techniques that can attest to their trustworthiness, including but limited to: hashing; timestamps; and signatures during examination, analyses and inspection of source files. CAF tools can in turn utilize new advances in technology to break such authentication measures, and thus comprise the second assumption.

The following is a brief explanation of some of the techniques that are used to compromise these two assumptions:

- *Encryption* is used to make the data unreadable. This is one of the most challenging techniques, as advances in encryption algorithms and tools empowered it to be applied on entire hard drive, selected partitions, or specific directories and files. In all cases, an encryption key is usually needed to reverse the process and decrypt the desired data, which is usually unknown to an investigator, in most cases. To complicate matters, decryption using brute-force techniques becomes infeasible when long keys are used. More success in this regard might be achieved with keyloggers or volatile memory content acquisition.

- *Steganography* aims at hiding the data, by embedding it into another digital form, such as images or videos. Commercial Steganalysis tools, that can detect hidden data, exist and can be utilized to counter Steganography. Encryption and Steganography can be combined to obscure data and make it also unreadable, which can extremely complicate a CF investigation.

- *Secure-Deletion* removes the target data completely from the source system, by overwriting it with random data, and thus rendering the target data unrecoverable. Fortunately, most of the available commercial secure-deletion tools tend to underperform and thus miss some data (Geiger, 2005). More research is needed in this area to understand the weaknesses and identify the signatures of such tools. Such information is needed to detect the operations and minimize the impact of these tools.

- *Hashing* is used by CF tools to validate the integrity of data. A hashing algorithm accepts a variable-size input, such as a file, and generates a unique fixed-size value that corresponds to the given input. The generated output is unique and can be used as a fingerprint for the input file. Any change in the original file, no matter how minor, will result in considerable change in the hash value produced by the hashing algorithm. A key feature in hashing algorithms is "Irreversibility" where having the hash value in hand will not allow the recovery of the original input. Another key feature is "Uniqueness" which basically means that the hash values of two files will be equal if and only if the files are absolutely identical.

Other algorithms like MD5, MD6, Secure Hashing Algorithms (SHA, SHA-1, SHA-2) amongst others are harder to break, however, are all subject for analyses and research in this direction. Researchers showed that MD5 can be broken within reasonable time frames ((Wang & Yu, 2005). *"We used this attack to find collisions of MD5 in about 15 minutes up to an hour computation time. The attack is a differential attack, which unlike most differential attacks, does not use the exclusive-or as a measure of difference, but instead uses modular integer subtraction as the measure. We call this kind of differential a modular differential. An application of this attack to MD4 can find a collision in less than a fraction of a second."*

Forensics tools rely on hashes computation to identify original files based on known good files libraries. This process relies on the assertion that a file cannot be modified in such a way that its MD5 hash does not change. New advances in hash attacks focusing on finding hash collisions will require that the investigator do deep bit-by-bit inspection for each file to assure its integrity which is eventually time consuming and requires more resources. Guidance Software appears to enhance EnCase's performance in this area by teaming up with other companies like Bit9 (Guidance Software, 2011; Hartley, 2007).

- *Timestamps* are associated with files and are critical for the task of establishing the chain of events during a CF investigation. The time line for the events is contingent on the accuracy of timestamps. CAF tools have provided the capability to modify timestamps of files or logs, which can mislead an investigation and consequently coerce the conclusion. Many tools currently exist on the market, some are even freely available, that make it easy to manipulate the timestamps, such as Timestamp Modifier and SKTimeStamp (Hartley, 2007).

- *File Signatures*, also known as Magic Numbers, are constant known values that exist at the beginning of each file to identify the file type (e.g., image file, word document, etc.). Hexadecimal editors, such as WinHex, can be used to view and inspect these values. Forensics investigators rely on these values to search for evidence of certain type. When a file extension is changed, the actual type file is not changed, and thus the file signature remains unchanged. CAF tools intentionally change the file signatures in their attempt to mislead the investigations as some evidence files are overlooked or dismissed. Complete listing of file signatures or magic numbers can be found on the web (Kessler, 2011).

- *CF Detection* is simply the capability of CAF tools to detect the presences of CF software and their activities or functionalities. Self-Monitoring, Analysis and Reporting Technology (SMART) built into most hard drives reports the total number of power cycles (Power_Cycle_Count), the total time that a hard drive has been in use (Power_On_Hours or Power_On_Minutes), a log of high temperatures that the drive has reached, and other manufacturer-determined attributes. These counters can be reliably read by user programs and cannot be reset. Although the SMART specification implements a DISABLE command (SMART 96), experimentation indicates that the few drives that actually implement the DISABLE command continue to keep track of the time-in-use and power cycle count and make this information available after the next power cycle. CAF tools can read SMART counters to detect attempts at forensic analysis and alter their behavior accordingly. For example, a dramatic increase in Power_On_Minutes might in-

dicate that the computer's hard drive has been imaged (McLeod, 2005).

- *Business Needs: Cloud Computing* (CC) is yet another emerging area that is driven by business needs and does not have a unified definition. For the purpose of this paper the simplest definition will be used where CC is seen as a huge group of interconnected computers. These computers can be personal computers, network servers, private or public devices.

Cloud Computing (CC) as a business model is typically suited to SMEs (Small and Medium Enterprises) who do not have enough resources to invest in building their own IT infrastructure. Hence, they tend to outsource this to third parties who will in turn lease their infrastructure and probably applications as services. SMEs who choose to outsource their IT services, would seek service level agreements (SLAs) from the service provides in order to define responsibilities and services limitations.

The companies at the forefront of this new concept include Amazon, Google, Yahoo, and Microsoft. CC can be considered as the new generation of traditional outsourcing where services are being actually leased over the internet while concealing the structure behind it.

This new concept introduces more challenges to the computer forensics investigator due to mainly the fact that the data is on the cloud hosted somewhere that we do not necessary know exactly where, being transferred across countries with different legislations and most importantly might reside on a machine that hosts other instances of other enterprises. In some instances, the data for the same enterprise might be stored across multiple data centres. These issues complicate the data acquisition, examination, analyses, and would mostly make the building of case terribly hard.

Cloud computing cannot be discussed without Virtualization. Virtualization is the key feature and enabler of the current cloud computing architecture. It allows sharing of software and hardware for multiple users. VMWare from Microsoft is one of the most commonly used virtualization engine. In virtualization, multiple virtual machines with their associated data are hosted on physically one powerful server that is capable of running all these instances with acceptable performance. Typically these servers have TB storage space to be able to host data for multiple clients.

The impact of Anti-Forensics tools can be devastating in the virtual environment because of the amount of data that can be targeted and the difficulty that the investigator would have when attempting to acquire evidence. Privacy concerns would probably get major attention as well especially when the data of more than one client is hosted on a server and this server needs to be inspected as part of a case related to a company but not the other. Avoiding business interruption besides legal and compliance matters must be carefully analysed and managed in such environments.

Whenever virtualization is used, both the hosting server and the guest instances must be hardened since a breach in any side being the host or the guest will lead to massive breach all across the physical server. Also, working towards laws and regulations unity across various countries will assist the computer forensics investigator chasing criminals regardless of their attacks origins.

Table 2 provides a list of some commercial CAF software packages available on the market. The tools listed in the table are intended as examples; none of these tools were purchased or tested as part of this paper work.

RECOMMENDATIONS

Based on our findings, we see room for improvement in the field of CAF that can address some of the issues surrounding this field. We believe that such recommendations, when adopted and/or implemented properly, can add value and consoli-

date the efforts for advancing this field. Below is a list and brief explanation of the recommendations:

a)   Spend More Efforts to Understand CAF

More efforts should be spent in order to reach an agreed upon comprehensive definition for CAF that would assist in getting better understanding of the concepts in the field. These efforts should also extend to develop acceptable best practices, procedures and processes that constitute the proper framework, or standard, that professionals can use and build onto. CAF classifications also need to be integrated, clarified, and formulated on well-defined criteria. Such fundamental foundational efforts would eventually assist researchers and experts in addressing the issues and mitigating the associated risks.

Awareness of CAF techniques and their capabilities will prevent, or at least reduce, their success and consequently their impact on CF investigations. Knowledge in this area should encompass both techniques and tactics. Continued education and research are necessary to stay atop of latest developments in the field, and be ready with appropriate countermeasures when and as necessary.

Table 2. Examples of anti-forensic tools

Category	Tool Name
Privacy and Secure Deletion	• Privacy Expert • SecureClean • PrivacyProtection • Evidence Eliminator • Internet Cleaner
File and Disk Encryption	• TruCrypt • PointSec • Winzip 14
Time stamp Modifiers	• SKTimeStamp • Timestamp Modifier • Timestomp
Others	• The Defiler's Toolkit – Necrofile and Klimafile • Metasploit Anti-Forensic Investigation Arsenal (known affectionately as MAFIA)

b) Define Laws that Prohibit Unjustified Use of CAF

Existence of strict and clear laws that detail the obligations and consequences of violations can play a key deterrent role for the use of these tools in a destructive manner. When someone knows in advance that having certain CAF tools on one's machine might be questioned and possibly pose some liabilities, one would probably have second thoughts about installing such tools.

Commercial non-specialized CAF tools, which are more commonly used, always leave easily detectable fingerprints and signatures. They sometimes also fail to fulfil their developers' promises of deleting all traces of data. This can later be used as evidence against a suspected criminal and can lead to an indictment. The proven unjustified use of CAF tools can be used as supporting incriminatory evidence in courts in some countries. *"In June 2005, Robert M. Johnson, a former publisher of Newsday and New York state education official, was charged with destruction of evidence for using counter-forensic software after learning he might be the target of a child pornography investigation (U.S. v Robert Johnson)"* (Findlaw.com).

To address the privacy concerns, such as users needs to protect personal data like family pictures or videos, an approved list of authorized software can be compiled with known fingerprints, signatures and special recovery keys. Such information, especially recovery keys, would then be safeguarded in possession of the proper authorities. It would strictly be used to reverse the process of CAF tools, through the appropriate judicial processes.

c) Utilize Weaknesses of CAF Software

In some cases, digital evidence can still be recovered if a data wiping tool is poorly used or is functioning improperly. Hence, each CAF software must be carefully examined and continuously analyzed in order to fully understand its exact behaviour and determine its weaknesses and vulnerabilities (Geiger, 2005) (Justice.gov, 2003). This can help to develop the appropriate course of actions given the different possible scenarios and circumstances. This could prove to be valuable in saving time and resources during an investigation.

d) Hardening of CF Software

CAF and CF thrive on the weaknesses of each other. To ensure justice CF must always strive to be more advanced than its counterpart. This can be achieved by conducting security and penetration tests to verify the software is immune to external attacks. Also, it is imperative not to submit to market pressure and demand for tools by rapidly releasing products without proper validation. The best practices of software development must not be overlooked at any rate. When vulnerabilities are identified, proper fixes and patches must be tested, verified and deployed promptly in order to avoid zero-day attacks.

CONCLUSION

Computer Anti-Forensics (CAF) is an important developing area of technology. Because CAF success means that digital evidence will not be admissible in courts, Computer Forensics (CF) must evaluate its techniques and tactics very carefully. Also, CF efforts must be integrated and expedited to narrow the current exiting gap with CAF. It is important to agree on an acceptable definition and classification for CAF which will assist in implementing proper countermeasures. Current definitions and classifications all seem to concentrate on specific aspects of CAF without truly providing the needed holistic view.

It is very important to realize that CAF is not only about tools that are used to delete, corrupt, or hide evidence. CAF is a blend of techniques and tactics that utilize technological advancements in areas like encryption and data overwriting amongst other techniques to obstruct investigators' efforts.

Many challenges exist and need to be carefully analyzed and addressed. In this paper we attempted to identify some of these challenges and suggested some recommendations that might, if applied properly, mitigate the risks.

FUTURE WORK

This paper provides solid foundation for future work that can further elaborate on the various highlighted areas. It suggests a definition for CAF that is closely aligned with CF and presents several classifications that we deem acceptable. It also discusses several challenges that can be further addressed in future research. CAF technologies, techniques, and tactics need to receive more attention in research, especially in the areas that present debates on hashes, timestamps, and file signatures.

Research opportunities in Computer Forensics, Network Forensics, and Anti-Forensics can use the work presented in this paper as a base. Privacy concerns and other issues related to the forensics field introduce a raw domain that requires serious consideration and analysis. Cloud computing, virtualization, and related laws and regulations concerns are topics that can be considered in future research.

REFERENCES

Biggs, S., & Vidalis, S. (2010). Cloud computing storms. *International Journal of Intelligent Computing Research, 1*(1).

Brand, M. (2007). *Forensics analysis avoidance techniques of malware*. Perth, Australia: Edith Cowan University.

Findlaw.com. (n. d.). *U.S. vs. Robert Johnson - Child pornography indictment*. Retrieved from http://news.findlaw.com/hdocs/docs/chldprn/usjhnsn62805ind.pdf

Garfinkel, S. (2007). Anti-forensics: Techniques, detection and countermeasures. In *Proceedings of the 2nd International Conference in i-Warefare and Security* (p. 77).

Geiger, M. (2005). *Evaluating commercial counter-forensic tools*. Paper presented at the Digital Forensic Research Workshop, Pittsburgh, PA.

Guidance Software. (n. d.). *Computer forensics solutions and digital investigations*. Retrieved from http://www.guidancesoftware.com/

Gurav, U., & Shaikh, R. (2010). Virtualization – A key feature of cloud computing. In *Proceedings of the International Conference and Workshop on Emerging Trends in Technology* (pp. 227-229).

Harris, R. (2006). Arriving at an anti-forensics consensus: Examining how to define and control the anti-forensics problem. *Digital Investigation, 3*, 44–49. doi:10.1016/j.diin.2006.06.005

Hartley, M. W. (2007). *Current and future threats to digital forensics*. ISSA Journal.

Internet World Stats. (n. d.). *The Internet big picture, world Internet users and population stats*. Retrieved from http://www.internetworldstats.com/stats.htm

Justice.gov. (2003). *United States of America vs. H. Marc Watzman*. Retrieved from http://www.justice.gov/usao/iln/indict/2003/watzman.pdf

Kessler, G. (2011). *File signature table*. Retrieved from http://www.garykessler.net/library/file_sigs.html

Kessler, G. C. (2007). *Anti-forensics and the digital investigator*. Burlington, VT: Champlain College.

McLeod, S. (2005). *SMART anti-forensics.* Retrieved from http://www.forensicfocus.com/smart-anti-forensics

Mitre.org. (2011). *Common vulnerabilities and exposures (CVE) database.* Retrieved from http://cve.mitre.org/

Nelson, B., Phillips, A., & Steuart, C. (2010). *Guide to computer forensics and investigations* (4th ed.). Cambridge, MA: Course Technology.

Newsham, T., Palmer, C., & Stamos, A. (2007). *Breaking forensics software: Weaknesses in critical evidence collection.* Retrieved from http://www.isecpartners.com

Secureworks.com. (2008). *Security 101: Botnets.* Retrieved from http://www.secureworks.com/research/newsletter/2008/05/

Srinivasan, S. (2007). *Security and privacy vs. computer forensics capabilities.* ISACA Online Journal.

Thuen, C. (2007). *Understanding counter-forensics to ensure a successful investigation.* Retrieved from http://citeseerx.ist.psu.edu/viewdoc/summary?doi=10.1.1.138.2196

Trickyways.com. (2009). *How to change timestamp of a file in Windows.* Retrieved from http://www.trickyways.com/2009/08/how-to-change-timestamp-of-a-file-in-windows-file-created-modified-and-accessed/

US-Computer Emergency Readiness Team. C. (2008). *Computer F=Forensics.* Retrieved from http://www.us-cert.gov/reading_room/forensics.pdf

Verizon Business. (2009). *2009 data breach investigations report.* Retrieved from http://www.verizonbusiness.com/about/news/podcasts/1008a1a3-111=129947--Verizon+Business+2009+Data+Breach+Investigations+Report.xml

Verizon Business. (2010). *2010 data breach investigations report.* Retrieved from http://www.verizonbusiness.com/resources/reports/rp_2010-data-breach-report_en_xg.pdf?&src=/worldwide/resources/index.xml&id=

Wang, X., & Yu, H. (2005). How to break MD5 and other hash functions. In *Proceedings of the Annual International Conference on the Theory and Applications of Cryptographic Techniques* (pp. 19-35).

Whitteker, M. (2008, 11). Anti-forensics: Breaking the forensics process. *ISSA Journal.*

This work was previously published in the International Journal of Cloud Applications and Computing (IJCAC), Volume 1, Issue 3, edited by Shadi Aljawarneh & Hong Cai, pp. 22-35, copyright 2011 by IGI Publishing (an imprint of IGI Global).

Chapter 14
Power Aware Meta Scheduler for Adaptive VM Provisioning in IaaS Cloud

R. Jeyarani
Coimbatore Institute of Technology, India

N. Nagaveni
Coimbatore Institute of Technology, India

Satish Kumar Sadasivam
IBM Systems and Technology Group, India

Vasanth Ram Rajarathinam
PSG College of Technology, India

ABSTRACT

Cloud Computing provides on-demand access to a shared pool of configurable computing resources. The major issue lies in managing extremely large agile data centers which are generally over provisioned to handle unexpected workload surges. This paper focuses on green computing by introducing Power-Aware Meta Scheduler, which provides right fit infrastructure for launching virtual machines onto host. The major challenge of the scheduler is to make a wise decision in transitioning state of the processor cores by exploiting various power saving states inherent in the recent microprocessor technology. This is done by dynamically predicting the utilization of the cloud data center. The authors have extended existing cloudsim toolkit to model power aware resource provisioning, which includes generation of dynamic workload patterns, workload prediction and adaptive provisioning, dynamic lifecycle management of random workload, and implementation of power aware allocation policies and chip aware VM scheduler. The experimental results show that the appropriate usage of different power saving states guarantees significant energy conservation in handling stochastic nature of workload without compromising the performance, both when the data center is in low as well as moderate utilization.

DOI: 10.4018/978-1-4666-1879-4.ch014

INTRODUCTION

With the advent of cloud computing, large scale data centers are becoming common in the computing industry. However, these data centers equipped with high performance infrastructures consume huge power causing global warming by emitting CO_2 footprint, giving a serious environmental threat to today's world. One of the major causes for energy inefficiency in the data center is the idle power wasted when servers run at low average utilization. Even at 10% of CPU utilization, the power consumed is over 50% of the peak power (Neugebaur & McAuley, 2001; Ragavendra et al., 2008). This results in more power consumption per workload during off-peak load. Pinheiro and Rajamony quoted that 22% of energy consumption of a single server is needed to cool it. A study on data center issues also shows that energy consumption of data centers worldwide doubled between 2000 and 2006 (Pinheiro et al., 2001; Elnozahy, 2003). Incremental US demand for data center energy between 2008 and 2010 is the equivalent of 10 nuclear power plants (Kaplan et al., 2008). To handle this issue, data centers perform consolidation of various workloads onto a set of common servers with the help of live state migration facility which is enabled through virtualization technology (James & Ravi, 2005).

In a cloud environment, the power and energy management strategies need to consider the characteristics of servers and incoming workloads. Modern microprocessor technology supports the processing elements (PE) such as core, chip and host to be set in different sleep states depending on the demand (Lee et al., 2007). The sleep states are also referred to as power saving states. A PE conserves different amount of power at different sleep states. The power drawn during wake up time is comparatively insignificant. The shallow sleep states realize lower power conservation with lower wake up latency and the deep sleep states realize higher power conservation with higher wake up latency. IBM's Power family machines support nap and sleep modes (Sinharoy et al., 2005; Kim et al., 2011; Cardosa et al., 2009). The nap is a low-power state designed for short processor idle periods (Malcolm et al., 2010). It provides modest power reduction over a software idle loop, and the wake-up latency of the nap state is less than $5\mu s$. Instruction execution begins immediately upon wakeup. The second idle mode is sleep state. It is a lower-power, higher latency standby state intended for processing cores that will be unused for an extended period of time.

To conserve power significantly, we have to provision the resources in such a way that the required computing resources are well utilized and the idle resources are kept in appropriate power saving states. In the proposed work as we consider IaaS cloud wherein the term workload refers to custom configuration of Virtual Machine (VM) to be launched and so the terms workload and VM request are used interchangeably. The study on workload characteristics in a typical data center reveals that there are wide variations in the number of workloads arriving to a data center as well as the amount of resources required by each workload in a particular instance of time and it concludes that incoming VM requests are highly dynamic in nature. Hence, by dynamically predicting the arrival pattern of the workloads based on recent utilization history of data center, the required number of processing cores for the incoming VM requests can be provisioned for immediate allocation. The remaining resources that are not provisioned can be transitioned to suitable power saving states, and thus idle powers of processor cores are not wasted.

Hence, our objective is to build Power Aware Meta scheduler (PAMS) which finds right fit infrastructure for launching VMs and conserves power by realizing the internal power saving without compromising the performance. The major challenge lies in efficiently handling the stochastic nature of incoming workloads by adaptive resource provisioning, while realizing energy conservation. The PAMS saves power efficiently at the core level as well as at the chip level by using various static and dynamic consolidation policies during VM placement and VM migration. As the cloud environment is very large, even 1% of energy saving in each node contributes to significant amount of power conservation. Hence, the power aware provisioning policy is of strong relevance in green computing.

The Clouds exhibit variation in supply and demand patterns, making the system highly transient. The varying demand represents dynamic incoming workload pattern, characterized by heterogeneous and competing QoS requirements along with dynamic application scaling requirements, typical in the cloud environment. The varying supply pattern represents the resource dynamics due to availability of hosts and change in the current load that affects the workload assignment (Sudha Sadhasivam et al., 2009; Dyachuk & Mazzucco, 2010; Jeyarani et al., 2011). These changing supply and demand patterns make the process of evaluating the performance of various provisioning policies in real cloud environment such as Amazon EC2, Microsoft Azure for different workload mixes very difficult. Hence for modeling heterogeneous dynamic cloud infrastructure, we extended an event driven simulator, CloudSim and simulated various scenarios and evaluated the performance of proposed power aware provisioning policies in fulfilling the QoS requirements of cloud consumers extensively (Calheiros et al., 2011).

The remaining part of the paper has been organized as follows. The next section details the literature survey on the concepts involved in power aware scheduling and the recent work on power conservation in data centers. The system architecture is then described, as well as the proposed Power Aware Meta Scheduler for adaptive provisioning. Several simulated experiments are also analyzed. The final section concludes the present work and suggests the scope for future work in resource provisioning and power conservation in cloud environment.

LITERATURE SURVEY

As datacenters run large scale distributed applications using more and more computing nodes and storage resources, the energy consumption raises quickly. To make efficient use of computing systems and reduce their environmental and social impact, both computer research community and industry are focusing on green computing. There are hardware techniques to solve the energy problems, but in many cases it requires software intervention to achieve the best results.

POWER MANAGEMENT TECHNIQUES

The following subsection discusses various power saving techniques, suitable for cloud environment.

Dynamic Voltage Frequency Scaling

One of latest techniques used in modern processors for power conservation is Dynamic Voltage Frequency Scaling (DVFS), in which the power consumption of the server is optimized by dynamically changing the CPU voltage/frequency, without affecting the performance adversely. DVFS is a kind of internal power saving method which exploits the various operating states of a processor core based on the load. For example, let us consider the system has just one task to perform and the workload requires only 10 cycles to execute, but the deadline associated with the task is 100 seconds. In this case, the processor can slow down to 1/10 cycles/sec, saving power and meeting the deadline (Venkatachalam & Franz, 2005). This dynamic slow down process is more efficient than running the task at full speed and idling for the remainder of the period. One of the software strategy for reducing the energy consumption using DVFS is given by Lorch and Smith (2001) and their work focused on making the operating system determine how slowly the processor can be run without disturbing the user, so that the maximal energy reduction can be obtained.

In DVFS the inappropriate increase in voltage / frequency may raise the system power demand. Similarly, the reduction in voltage / frequency reduces the number of instructions a processor can execute in a given amount of time. This increases the run time of the program segments which are CPU bound, resulting in performance degradation. Moreover when the program is I/O bound, the change in CPU speed will not have an effect on the execution time of the program, thus DVFS may result in non linearity on the execution time. Thus changing the frequency of the server may lead to changes in the scheduling order of tasks (Lorch & Smith, 2001). Most of the existing methods of power conservation deal with DVFS, in spite of the challenges in providing optimal energy/performance control.

The success of DVFS lies in the accurate prediction of execution time of the specific workload. Even if the DVFS algorithm predicts correctly what the processor's workload will be, determining how fast to run the processor is nontrivial. The non determinism in real systems removes any strict relationships between clock frequency, execution time and power consumption. The theoretical studies on dynamic voltage scaling may sound reasonable, but are not guaranteed to hold in real time systems. The dynamic voltage scaling may be combined with resource hibernation to power down other energy hungry parts of the chip, to realize the significant power saving.

Dynamic Component Deactivation

Certain policies use strategy of dynamically switching off specific number of servers, based on simple workload prediction model (Beloglazov et al., 2011). The major issue in frequent switching the servers on/off is component failures. Dynamic Component Deactivation (DCD) method is used in certain components that are idle and do not support

performance scaling. It helps in significant power conservation by leveraging workload variability.

Power Saving States

The dynamic power management techniques such as DVFS and DCD can be implemented in the hardware as part of an electronic circuit. But the implementation and reconfiguration of dynamic power management algorithms and policies in hardware are difficult. Therefore there is a shift in the implementation in software level. To address this issue, the organizations such as Intel, Toshiba and Microsoft have published Advanced Configuration and Power Interface (ACPI) specification to provide standardized nomenclature for various power saving states and also define software interfaces for managing them (Kant, 2009). It is an open standard and it defines unified operating system centric device configuration and power management interface. The interface can be used by software developers to leverage flexibility in adjusting system power states.

ACPI defines five system states. The working state is denoted by S0. The state S3 denotes standby or inactive with state saved into DRAM. And S5 denotes the hibernating or inactive with states saved into secondary storage. In a cloud environment, the power and energy management strategies need to consider the characteristics of servers, while provisioning the incoming workloads. Nowadays the contemporary servers used in cloud data centers are modeled as multi chip modules. (MCM) Modern microprocessor technology supports the processing elements (PE) such as core, chip and host to be set in different sleep states depending on the demand. The sleep states are also referred to as power saving states. A PE conserves different amount of power at different sleep states. The time taken to come to live state from any sleep state is called wake up latency. And the power drawn during wake up time is comparatively insignificant. The power conservation and wake up latency are inversely

Table 1. Power Conservation and Wake Up Latency for different sleep modes

Sleep Modes	% of Power Conservation	WakeUp Latency
Nap	20	5us
Sleep	40	1ms
Winkle	80	few 100ms
Idle	100	Few seconds

related in a particular sleep state as shown in Table 1. The power conservation for various power saving states has been depicted in Figure1.

Many modern power-aware processors and microcontrollers have built-in support for active, idle and sleep operating modes. In sleep mode, substantially more energy savings can be obtained, but it requires a significant amount of time to switch into and out of that mode. Hence, a significant amount of energy is lost due to idle gaps between executing tasks that are shorter than the required time for the processor to enter the sleep mode.

Dynamic Consolidation

To handle power conservation, datacenters perform the dynamic consolidation of various workloads onto a set of common servers with the help of live state migration facility which is enabled through virtualization technology. At the same time, the idle servers are transitioned to different power saving modes, based on the dynamic workload prediction. We have considered

dynamic server farm, where the servers are at different power saving modes for making power aware provisioning. Even though workloads are combined, the energy usage does not add linearly due to significant percentage of idle energy. So consolidation can be done optimally during allocation to realize energy conservation without giving performance degradation. When cloud consumers outsource their computational needs to the cloud, they specify QoS in terms of strict service level agreements (SLAs). Sometimes due to aggressive consolidation and variability in workload, there arise performance losses in terms of increased response time, time outs or failure in worst case. Therefore cloud providers have to take measures to adhere SLAs, while minimizing the power consumption.

Existing work on power conservation have not considered the presence of servers in different power saving states which are inherent in modern processor technology. We have introduced a novel approach in which the meta scheduler makes decision regarding the mapping of incoming workload (job) with the server by considering the power saving state of idle servers, apart from the server capability in fulfilling the requirement of the job. When all the active servers are heavily loaded say more than 80 percent of their capacity, and there are two servers S1 and S2 both can fulfill requirements of an incoming job, with S1 being in a deep sleep state conserving more power and S2 being in a shallow sleep state conserving comparatively lesser power, with quick wake up latency. Then our heuristic meta scheduler schedules the job

Figure 1. Percentage of power conservation in various power saving modes

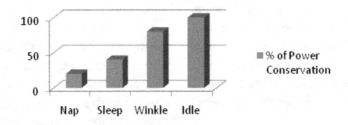

on to S2, resulting in a lesser increment in power consumption, without making the incoming job to wait. As the complexity of servers and their real time management increases, performance and power aware scheduling in real time become very much challenging, in large heterogeneous data centers. Our previous work on power aware meta scheduler, considered the load on the server, in such a way that the lightly loaded servers are preferred for allocation compared to heavily loaded servers, resulting lesser increment in the power drawn by server farm (Jeyarani et al., 2011).

Related Work

Pinheiro et al. have proposed a technique called load concentration to save power while providing the QoS requirements. The author uses a static estimation method to predict the performance by considering the demand for various resources (Pinheiro et al., 2001). This method is inefficient if the total demand exceeds the available resource capacity leading to throughput degradation. Our work predicts the demand of resources for the incoming jobs dynamically based on the recent history and makes adaptive provisioning to meet SLA.

The authors Lefevre and Orgerie (2010) have focused on switching off the unused nodes in the data center. They have stated that the increase in power consumption due to VM migration is insignificant, whereas the migration latency is quite significant (Lefevre & Orgerie, 2010). They have incorporated electrical sensors in the cloud infrastructure to log power consumption of each node and making the resource manager access the log to take power related decisions. But the policies are not stated clearly. Also they collect input such as whether the submitted job is to be run immediately or later, its permission to aggregate with other jobs, its restrictions on migration, etc. which will make the cloud scheduler very complex. Also simple On/Off model affects the reliability of the hardware.

Srikantaiah et al. (2009) highlighted the importance of key observable characteristics such as resource utilization, performance and energy consumption in designing an effective consolidation strategy. Their work considered CPU and disk resource combination for consolidation and determined optimal utilization level at which maximum power can be saved (Srikantaiah et al., 2009). However, the performance degradation due to consolidation has not been discussed. In our work the allocation algorithm uses worst fit policy for static consolidation of workloads and also dynamic consolidation is done with limited migration. This result in negligible transition overhead compared to the amount of power saved and it is suitable for a generic cloud environment.

Gandhi et al. (2009) aimed at optimal power allocation by determining the optimal frequency of the servers' CPU at which the mean response time is minimal and it is done with the help of power to frequency relationship. They studied the server power consumption with respect to CPU frequency scaling techniques at different states, namely T-states and P-states and combination of both, for CPU intensive workloads (Gandhi et al., 2009). The results showed that the power to frequency relationship is linear at T-states and P-states, and it is a cubic square relationship for the combination of the two states. The authors stated that it is not always optimal to run the servers at the maximum speed to get the best performance.

Cardosa et al. (2009) presented intelligent power-performance trade-off mechanisms for power conservation in virtualized server environments. Given the minimum and maximum resource requirements of each VM and their corresponding charges, they have analyzed the best placement of VMs onto servers which maximizes profit for the cloud (Cardosa et al., 2009). The problem has been formulated as multi-knapsack problem in which items placed into the knapsack can be elastic in size between minimum and maximum resource requirements. As the cloud data center is highly dynamic in terms of workload, the static

VM placement may not be optimal always. So we have included in our proposed work dynamic consolidation of VMs through live VM migrations. In order to minimize the overhead due to migration, our system restricts the number of migrations as well as remaining execution time of each specific VM.

Lee and Zomaya (2010) have shown energy efficient task scheduling in large scale distributed systems. The devised algorithm uses a novel objective function called Relative Superiority which takes into account minimum makespan and energy consumption (Lee & Zomaya, 2010). For a given ready task, its RS value on each processor is computed using the current best combination of the processor and voltage supply level for that task. The processor with maximum RS value is selected for allocation. As the cloud is large scale having millions of processors, finding the best RS value for each single task may be time consuming.

Milenkovic et al. (2009) describe power aware resource management through VM migration along with a scheme called platooning, in which a large pool of servers is divided into sub-pools supporting each other for handling varying workloads (Milenkovic et al., 2009). H. Abdelsalam et al. have described a pro-active energy-aware change management scheme for cloud computing environments that handles interactive web applications (Abdelsalam et al.,2009). The optimal number of servers and the optimal running frequencies needed to satisfy the client's SLAs are found by the cubic relationship between the power consumption and frequency at which a server is running.

Rusu et al. (2006) have developed QoS based power management scheme for the heterogeneous clusters. The system encompasses two important components namely front end manager and a local manager. While the former finds the servers that should be turned on or off for a given system load, the local manager exploits DVFS to conserve power (Rusu et al., 2006). The main drawback is the on/off policy relies on the table

of values, computed offline. Also without server consolidation through VM migration, the on/off policy may not be much effective.

Kim et al. (2011) have investigated power-aware provisioning of Virtual machines for real-time cloud services. The users submit their real-time applications to the broker who analyses and decides on the configuration of needed VM. They have proposed adaptive DVFS algorithms for power aware provisioning meeting the deadline requirements of the submitted real time applications (Kim et al., 2011). Here the provider has to bare the penalty, when deadlines are not met and it may be due to improper configuration decided by the broker. Verma et al. (2008) have used bin packing approaches such as First Fit Decreasing (FFD) and incremental First Fit Decreasing (iFFD) for power aware VM provisioning (Verma et al., 2008). Borja et al. have contributed work on Virtual Infrastructure Management in Private and Hybrid Clouds, focusing on advanced reservation and resource preemption (Sotomayor et al., 2008, 2009)

Considerable amount of work have been done in the area of power-efficient computing. Most of the existing work on power conservation either uses DVFS or host On/Off policy or server consolidation through VM migration. Some of the research work proves the effectiveness of combination of two techniques. But none of them have exploited the use of different power saving states of various components present in multi chip modules, the major constituent of cloud data centers. In this paper, we have proposed a novel holistic approach which integrates extensive exploitation of power saving states of various components with short term dynamic prediction, while preserving the performance through chip-aware dynamic consolidation with constrained VM migration. Our proposed system conserves power significantly without any performance degradation, vital requirement of cloud environment.

POWER CONSERVATION IN IaaS CLOUD

System Architecture

Figure 2 depicts the proposed system architecture. It shows the interaction of the Power Aware Meta-scheduler with other components for adaptive VM provisioning and VM scheduling. The system consists of the following important entities.

Datacenter

It represents a pool of hosts and each host is a Multi Chip Module (MCM) containing more than one chip. Each chip contains eight processing cores. Every datacenter is associated with a provider, who configures the datacenter broker for job dispatching. The datacenter has a module called VM Allocation Policy, which is used to identify the optimal host for the incoming VM request. Similarly, each host is associated with VM Scheduler and is used to select the most optimal chip for launching the incoming VM request. The VM allocation policy as well as the VM scheduler is performance aware as they try to create a VM within a chip. Also they are power aware as the former tries to pack VM requests within minimum number of hosts and the later packs the VM requests within a few chips in a selected host.

Datacenter Broker

According to the configurations set by the provider, the datacenter broker receives the VM requests from the cloud users, batches them at a specific interval of time and sends the batch to the meta scheduler.

Figure 2. System architecture - interaction of meta scheduler with other components in cloud environment

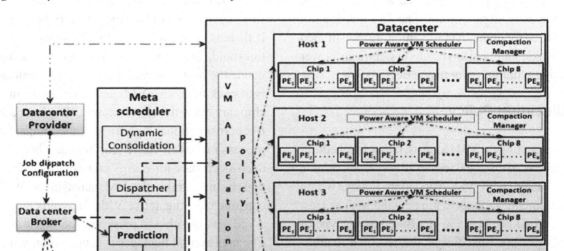

Meta Scheduler

It lies in between datacenter broker and datacenter. It plays a vital role in saving energy. It performs six important tasks viz. prediction, provisioning, dispatching, monitoring, VM Life Cycle Management and Dynamic Consolidation which have been explained in the following sections.

Prediction and Provisioning

The VM requests from the clients arrive at Broker randomly. The broker does request batching at regular intervals and transfers them to meta scheduler. After monitoring the arrival pattern for a specific period of time, the meta scheduler starts predicting the resource requirements dynamically and makes provisioning of adequate resources for satisfying the future VM requests. In the present work, it predicts the number of PEs required for the next immediate batch and reserves them in nap state. While making reservation, it takes into account the number of PEs already in nap state as well as the number of PEs that will become free at the next dispatch time. The PEs that are not required for immediate allocation are transitioned to winkle state, and this results in significant power saving without performance degradation.

Allocation Policy

On accepting the VM requests, the meta scheduler transfers them to datacenter. Every datacenter is associated with VM Allocation Policy, which implements variants of bin packing such as Best Fit, Worst Fit, First Fit, etc. Based on the type of policy, a host that fulfills the resource requirement is allocated for VM creation. If none of the hosts has sufficient PEs in Nap, then a host with the required number of PEs in winkle state is considered for launching VM creation. Thus the allocation policy identifies the most suitable host, in which the VM could be created within a chip and at the same time, the policy confirms that

every host is fully utilized. On getting the VM request, the host delegates it to VM Scheduler, which then selects a chip that has sufficient PEs in nap state for launching VM.

Monitoring and VM Life Cycle Management

The meta scheduler also monitors the expiry of VMs and it notifies the datacenter. Similar to resource allocation, resource reclamation is also jointly done by VM Allocation policy and VM scheduler to release the host and the PEs respectively. Thus in the proposed extended cloudsim framework, VM life cycle management is done dynamically facilitating better resource utilization.

Dynamic Consolidation and Compaction Manager

The meta scheduler checks for the possibility of resource aggregation by frequent analysis of the utilization factor of individual host. To bring dynamic consolidation, the meta scheduler first finds the number of free PEs in every busy host. If there is more number of free PEs than a given threshold, then it triggers compaction manager inside the host. The role of compaction manager is to pack the active VMs such that one or more chips are made free. For example, if eight PEs are free across various chips in a host, the compaction process makes an entire chip free and puts it in deep sleep state for energy conservation. This is done by migrating active VMs across chips. While consolidating the active VMs, it avoids placing a VM across a chip to preserve QoS constraints. The compaction manager is triggered only for a limited number of times to avoid overhead due to consolidation.

POWER AWARE META SCHEDULER FOR ADAPTIVE VM PROVISIONING

Generation of Dynamic Workload Patterns

To depict the realistic cloud scenario, the dynamic workload pattern is generated. In this, there are three attributes that are randomized viz., batch size representing the number of VMs that arrives for a given batch, required PEs representing the processing cores requested by each VM based on the user's need in executing their parallel workload, duration representing the run time period during which the resources are needed for a given VM. For example, a VM might need 4 PEs for a period of three time units from the time of dispatch.

Table 2 shows the sample values used for randomizing the number of required PEs for each VM. The first column denotes minimum and maximum number of PEs requested by each VM. Consider there are 100 VM requests in a batch. 5% of VM requests need PEs in the subrange 1 to 2. Another 5% of VM requests need PEs in the subrange 3 to 6. The 90% of the VM requests need PEs in the subrange 7 to 8.

In the same manner, the size of VM request batch and duration in which a VM needs the resources are also randomized. The algorithm for dynamic workload pattern generation is shown in Algorithm 1.

Table 2. Randomizing Number of Required PEs for VMs in a batch

Number of PEs Requested in Range	Percentage of VM Requests	Number of PEs Requested in SubRange
	5	1-2
1-8	5	3-6
	90	7-8

Algorithm Generation of Dynamic Workload Patterns (n)

Input: The number of batches required, n
Output: Random workload pattern
Method: This algorithm generates a series of VM request batches, each batch containing varying number of VM requests. In each VM request the number of PEs requested as well as the duration for which the VM needs the resources are randomized.

Algorithm 1. Generation of Dynamic Workload Patterns

Step 1: Set batchSizePerc[3], batchSizeSubRange[3] to generate Batch size

Step 2: Set reqPEPerc[3],reqPESubRange[3] to generate required PEs in each VM

Step 3: Set durationPerc[3],durationSubRange[3] to generate Duration for each VM

Step 4: for i←1 to n do

Step 5: batchSize[i] ← Randomize(batchSizePerc, batchSizeSubRange)

Step 6: for j←1 to batchSize[i] do

Step 7: reqPE[i][j] ← Randomize(reqPEPerc, reqPESubRange)

Step 8: duration[i][j] ← Randomize(durationPerc, durationSubRange)

Step 9: for end

Step 10: for end

Procedure Randomize(perc[],subRange[],m,n)

Begin //Generate a random number in the range 1-100

Rnum1←RANDOM(100)

ifRnum1 isbetween perc[1] and perc[2] then //Generate a random number in the subRange[1]

Rnum2←RANDOM(subRange[1])

else ifRnum1 is between perc[2] and perc[3] then // Generate a random number in the subRange2

Rnum2←RANDOM(subRange[2])

else Rnum2←RANDOM(subRange[3]) //Generate a random number in subRange3

Return Rnum2

End

Dynamic Run Time Prediction

The system predicts dynamic arrival pattern of VM requests based on recent history of arrivals, and uses it for provisioning resources. In the dynamic cloud environment, we cannot switch off a PE or Chip or Host, as the wake up latency involved is significant; instead, idle resources are kept in suitable power saving mode for energy conservation. We use simple average method to predict the number of PEs that will be requested by the forthcoming VM request batch.

Adaptive Provisioning Policy

Based on the dynamic prediction, the meta scheduler performs adaptive provisioning to fulfill the expected resources. Let the scheduling process start at time t and the request batching duration be t_o. The steps for adaptive provisioning are shown in Algorithm 2.

Algorithm 2. Adaptive VM Provisioning Policy

Step 1: Get the predictedRequiredPEs for the next dispatch time $t+t_o$

Step 2: Set PEsToBeProvisioned ← predictedRequiredPEs + X% of Shallow Sleep Power Saving State more provisioning

Step 3: Find the PEs that will become free at $t+t_o$; Let it be denoted as futureFreePEs;

Step 4: If futureFreePEs are more than PEsToBeProvisioned then excessPEs are transitioned to Deep Sleep Power Saving State

Step 5: else

Find the number of PEs insufficient for allocation;

This number is either satisfied from PEs in Shallow Power Saving State or from PEs in Deep Power Saving State

Step 6: The excessPEs found after provisioning are transitioned to Deep Sleep Power Saving State

Step 7: End.

Chip Aware VM Scheduler

This component focuses on creating each VM within a chip. A VM created with n PEs from the same chip outperforms a VM that is created by n PEs across several chips in a host, because the latter suffers from increased internal latency. The VMs so launched ensure better performance and thus chip aware VM scheduler facilitates to improve QoS, while conserving power. When there is more than one chip available for VM creation, then the VM Scheduler applies best fit policy in choosing the chip. This policy tries to pack many VMs within a chip, if possible. Thus the usage of all the chips in a host is avoided, while launching VMs, contributing to power saving.

EXPERIMENTAL SET UP AND RESULTS OF SIMULATION

The simulated environment was set up using an event driven simulator named cloudsim toolkit. The simulator has been extended to include the entities such as Broker, MCM, Data Center Manager (DC Manager), power aware allocation policies, chip aware VM scheduler and compaction manager. A major extension is the dynamic life cycle management of VMs in the datacenter. A resource pool containing 15 hosts is created as shown in Figure 2. The processing cores within the node are homogeneous. The power consumption is modeled as 20 watts per processor core.

The input to the cloud system is a batch of workload with a highly dynamic arrival pattern. The workload represents configuration of resources such as the number of cores, memory size, storage, bandwidth for a Virtual Machine (VM) and an associated execution time. The submitted workload requires immediate resource allocation. In this work, the dynamism in the input refers to the variation in the number of Processing Elements (PE) required, the duration of time needed by each VM request as well as variation in the arrival rate of VM requests. The workloads expect the cloud to be responsive immediately. In the context of IaaS model, the custom VM needs to be created and handed over for immediate usage by the client. The experimental results are averaged over five runs.

Experiment 1: Behavior of Provisioning Policy for Random Workload Pattern

Using dynamic workload pattern generation algorithm, three kinds of patterns namely smooth variation, drastic variation and mixed variation in the number of required PEs are generated. The patterns are subjected to Best Fit allocation policy along with chip aware VM scheduler. The percentage of power saved with three kinds of workload patterns has been depicted in Figure 3. We conclude that our proposed provisioning policy results in considerable power conservation for any random workload pattern compared to host On/Off policy.

Experiment 2: Exploitation of Single Power Saving State

This was conducted to show that there is significant power saving when we consider the power saving modes of microprocessors. For a given random workload pattern, initially power conservation is computed using a provisioning policy with

host On/Off scheme. To handle the unexpected workload surges, an additional pool of hosts are reserved and kept on, other than the number of hosts needed that is found through prediction. It does not consider any power saving modes. The same experiment is repeated with a provisioning policy which exploits a single power saving state called nap. Here the number of cores needed for executing next batch of workload and the additional pool of cores, reserved for handling workload surges are kept in the nap state. The results show that there is improvement in power saving when the power saving modes are exploited and it is depicted in Figure 4.

Experiment 3: Exploitation of Multiple Power Saving States

The prediction module finds the number of processor cores to be provisioned for executing the next batch of workloads, through simple moving average method. In addition, we make 5% of needed cores as extra provisioning to handle the prediction error. In this experiment, the predicted number of cores are kept in the nap state and the additional number of cores are kept in winkle state which conserves comparitively more power than a nap state host. The simulation results of exploiting multiple power saving modes show

Figure 3. Power saving in various patterns

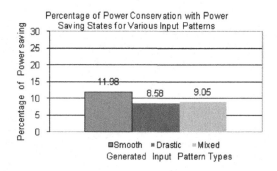

Figure 4. Comparison of Power Conservation in single PSS with On/Off Policy

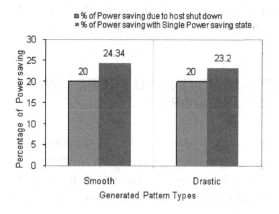

that slightly better power conservation is realised and it is depicted in Figure 5.

Experiment 4: Performance of Various Allocation Policies with Increase in Provisioning Percentage

In this experiment, a number of VMs allocated within a chip for different policies are found by varying additional resource provisioning percentage. The more the number of VMs within a chip, the more improvements in the performance. The result shows that worst fit policy outperforms by launching more VMs within chip, resulting in better performance. It is graphically depicted in Figure 6.

Experiment 4: Trade Off Between Performance and Power Conservation

In this experiment, the additional provisioning percentage was gradually increased and its impact on power conservation and performance in terms of Number of VMs scheduled within a chip is observed. As overprovisioning is given, power conservation decreases drastically. But there is a significant improvement in the performance through the increase in percentage of VMs assigned within a chip. The cloud administrator can select an optimal point at which both curves meet. From the graph shown in Figure 7, it is inferred that with 18% of overprovisioning, there is 48000 watt power saving with 85% of VMs placed within a chip ensuring better performance.

CONCLUSION AND SCOPE FOR FUTURE WORK

As the main challenge in cloud computing is to facilitate the cloud provider to manage unexpected demand for resources in an energy efficient manner, an efficient power aware meta

Figure 5. Power conservation under multiple power saving modes

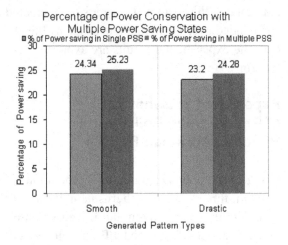

Figure 6. Performance of various allocation policies

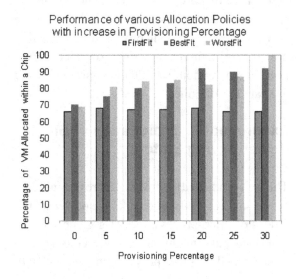

Figure 7. Trade off between performance and power conservation

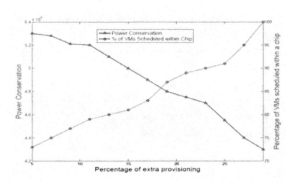

scheduler has been proposed in this paper. The roles and responsibilities of the meta scheduler and its interaction with the other components have been clearly architected. The power conservation without performance degradation is achieved through adaptive provisioning policy along with power aware allocation policy and chip aware VM scheduler. Our future work will aim at implementing machine learning algorithm for predicting the workload arrival pattern.

REFERENCES

AbdelSalam, H., Maly, K., Mukkamala, R., Zubair, M., & Kaminsky, D. (2009). Towards energy efficient change management in a cloud computing environment. In R. Sadre & A. Pras (Eds.), *Proceedings of the 3rd International Conference on Autonomous Infrastructure, Management and Security: Scalability of Networks and Services* (LNCS 5637, pp. 161-166).

Beloglazov, A., Buyya, R., Lee, Y. C., & Zomaya, A. (2011). *A taxonomy and survey of energy-efficient data centers and cloud computing systems.* Advances in Computers.

Calheiros, R. N., Ranjan, R., Beloglazov, A., De Rose, C. A. F., & Buyya, R. (2011). ClouSim: A toolkit for modeling and simulation of cloud computing environments and evaluation of resource provisioning algorithms. *Software, Practice & Experience, 41*, 23–50. doi:10.1002/spe.995

Cardosa, M., Korupolu, M. R., & Singh, A. (2009). Shares and utilities based power consolidation in virtualized server environments. In *Proceedings of the IFIP/IEEE International Symposium on Integrated Network Management* (pp. 327-334).

Dyachuk, D., & Mazzucco, M. (2010). On allocation policies for power and performance. In *Proceedings of the 11th IEEE/ACM International Conference on Grid Computing*, Belgium.

Elnozahy, E., Kistler, M., & Rajamony, R. (2003). Energy-efficient server clusters. In B. Falsafi & T. N. Vijaykumar (Eds.), *Proceedings of the Second International Workshop on Power-Aware Computer Systems* (LNCS 2325, pp. 179-197).

Gandhi, A., Harchol-Balter, M., Das, R., & Lefurgy, C. (2009). Optimal power allocation in server farms. In *Proceedings of the 11th International Joint Conference on Measurement and Modeling of Computer Systems* (pp. 157-168).

Jeyarani, R., Nagaveni, N., & Vasanth Ram, R. (2011). Self adaptive particle swarm optimization for efficient virtual machine provisioning in cloud. *International Journal of Intelligent Information Technologies, 7*(2), 25–44.

Kant, K. (2009). Data center evolution- A tutorial on state of the art, issues, and challenges. *Computer Networks, 53*(17), 2939–2965. doi:10.1016/j.comnet.2009.10.004

Kaplan, J. M., Forest, W., & Kindler, N. (2008). *Revolutionizing data centre energy efficiency.* Chicago, IL: McKinsey & Company.

Kim, K. H., Beloglazov, A., & Buyya, R. (2011). Power-aware provisioning of virtual machines for real-time cloud services. *Concurrency and Computation: Practice and Experience.*

Le, H. Q., Starke, W. J., Fields, J. S., O'Connell, F. P., Nguyen, D. Q., & Ronchetti, B. J. (2007). IBM POWER6 microarchitecture. *IBM Journal of Research and Development, 51.*

Lee, Y. C., & Zomaya, A. Y. (2010). Resource allocation for energy efficient large-scale distributed systems. *Information Systems. Technology and Management, 54*(1), 16–19.

Lefevre, L., & Orgerie, A.-C. (2010). Designing and evaluating an energy efficient cloud. *The Journal of Supercomputing, 51*, 352–373. doi:10.1007/s11227-010-0414-2

Lorch, J. R., & Smith, A. J. (2001). Improving dynamic voltage scaling algorithms with PACE. In *Proceedings of the ACM SIGMETRICS Conference on Measurement and Modeling of Computer Systems* (pp. 50-61).

Milenkovic, M., Castro-Leon, E., & Blakley, J. R. (2009). Power-aware management in cloud data centers. In M. G. Jaatun, G. Zhao, & C. Rong (Eds.), *Proceedings of the First International Conference on Cloud Computing* (LNCS 5931, pp. 668-673).

Neugebauer, R., & McAuley, D. (2001) Energy is just another resource: Energy accounting and energy pricing in the nemesis OS. In *Proceedings of the 8th IEEE Workshop on Hot Topics in Operating Systems*, Schloss Elmau, Germany (pp. 59-64).

Pinheiro, E., Bianchini, R., Carrera, E. V., & Heath, T. (2001). Load balancing and unbalancing for power and performance in cluster-based systems. In *Proceedings of the Workshop on Compilers and Operating Systems for Low Power* (pp. 182-195).

Raghavendra, R., Ranganathan, P., Talwar, V., Wang, Z., & Zhu, X. (2008). No power struggles: Coordinated multi-level power management for the data center. In *Proceedings of the 13th International Conference on Architectural Support for Programming Languages and Operating Systems* (pp. 48-59).

Rusu, C., Ferreira, A., Scordino, C., Watson, A., Melhem, R., & Moss'e, D. (2006). Energy-efficient real-time heterogeneous server clusters. In *Proceedings of the 12th IEEE Real-Time and Embedded Technology and Applications Symposium*, San Jose, CA (pp. 418-428).

Sinharoy, B., Kaal, R. N., Tendler, J. M., Eickerneyer, R. J., & Joyner, J. B. (2005). POWER 5 system micro architecture. *IBM Journal of Research and Development*, *49*(4-5), 505–521. doi:10.1147/rd.494.0505

Smith, J. E., & Nair, R. (2005). *Virtual machines-Versatile platforms for systems and processes*. San Francisco, CA: Morgan Kaufmann.

Sotomayor, B., Montero, R. S., Liorente, I. M., & Foster, I. (2008). Capacity leasing in cloud systems using OpenNebula engine. In *Proceedings of the Workshop on Cloud Computing and its Applications*, Chicago, IL.

Sotomayor, B., Montero, R. S., Liorente, I. M., & Foster, I. (2009). Virtual infrastructure management in private and hybrid clouds. *IEEE Internet Computing*, *13*(5), 14–22. doi:10.1109/MIC.2009.119

Srikantaiah, S., Kansal, A., & Zhao, F. (2009). Energy aware consolidation for cloud computing. *Cluster Computing*, *12*, 1–15.

Sudha Sadhasivam, G., Jeyarani, R., Vasanth Ram, R., & Nagaveni, N. (2009). Design and implementation of an efficient two level scheduler for cloud computing environment. In *Proceedings of the International Conference on Advances in Recent Technologies in Communication and Computing*, Kottayam, India.

Venkatachalam, V., & Franz, M. (2005). Power reduction techniques for microprocessor systems. *ACM Computing Surveys*, *37*(3), 195–237. doi:10.1145/1108956.1108957

Verma, A., Ahuja, A., & Neogi, A. (2008). pMapper: Power and migration cost aware application placement in virtualized systems. In *Proceedings of the ACM/IFIP/USENIX 9th International Middleware Conference* (pp. 243-264).

Ware, M., Rajamani, K., Floyd, M., Brock, B., Rubio, J. C., Rawson, F., & Carter, J. B. (2010). Architecting for power management: The IBM POWER 7 approach. In *Proceedings of the IEEE 16th International Symposium High Performance Computer Architecture* (pp. 1-11).

This work was previously published in the International Journal of Cloud Applications and Computing (IJCAC), Volume 1, Issue 3, edited by Shadi Aljawarneh & Hong Cai, pp. 36-51, copyright 2011 by IGI Publishing (an imprint of IGI Global).

Chapter 15
Custom–Made Cloud Enterprise Architecture for Small Medium and Micro Enterprises

Promise Mvelase
CSIR Meraka Institute, South Africa

Nomusa Dlodlo
CSIR Meraka Institute, South Africa

Quentin Williams
CSIR Meraka Institute, South Africa

Matthew Adigun
University of Zululand, South Africa

ABSTRACT

Small, Medium, and Micro enterprises (SMMEs) usually do not have adequate funds to acquire ICT infrastructure and often use cloud computing. In this paper, the authors discuss the implementation of virtual enterprises (VE) to enable SMMEs to respond quickly to customers' demands and market opportunities. The virtual enterprise model is based on the ability to create temporary co-operations and realize the value of a short term business opportunity that the partners cannot fully capture on their own. The model of virtual enterprise is made possible through virtualisation technology, which is a building block of cloud computing. To achieve a common goal, enterprises integrate resources, organisational models, and process models. Through the virtual business operating environment offered by cloud computing, the SMMEs are able to increase productivity and gain competitive advantage due to the cost benefit incurred. In this paper, the authors propose a virtual enterprise enabled cloud enterprise architecture based on the concept of virtual enterprise at both business and technology levels. The business level comprises of organisational models, process models, skills, and competences whereas the technology level comprises of IT resources.

DOI: 10.4018/978-1-4666-1879-4.ch015

1. INTRODUCTION

In today's rapidly changing world, businesses are seeking greater agility in their decision-making and operational processes. The advent of cloud computing provides an extraordinary opportunity to deliver the IT agility that organizations require. Cloud computing has changed and will keep on changing the role of IT in organisationsal structure. There are two major factors at play – the technology aspect and the organizational aspect (Schooff, 2009).

On the technology side, cloud computing represents a shift from highly specialised, isolated technologies to simpler, clustered on-demand technologies. This shift requires a corresponding change in the kinds of expertise that IT personnel possess, with less emphasis on technical knowhow and more emphasis on real-time service management, end-user communication and a basic understanding of the economics of cloud usage. On the organisational side of things, one can expect a shift from internal IT towards managed service provider (MSP), as well as an overall reduction in the need for personnel for the day-to-day running of IT.

SMMEs have been on the cloud for some time now, and are the main drivers of the largest and fastest growth curve surrounding cloud technology. Adopting SaaS early on, many SMMEs have contributed to the rapid expansion of SaaS providers like Salesforce.com. Now small and medium businesses are looking to put much of their IT into the cloud, if only for the sole reason of saving money. SMMEs also typically have less existing infrastructure, more flexibility, and smaller capital budgets for purchasing in-house technology. Similarly, SMMEs in emerging markets are typically unburdened by established legacy infrastructures, thus reducing the complexity of deploying cloud solutions.

Unfortunately, many cloud computing providers and cloud application vendors are overlooking the emerging SMME market and just concentrating on the big players.

In the current e-business environments, individual enterprises, including SMMEs cannot survive on their own. It is crucial that SMMEs can engage effectively with their partners and customers. Some larger enterprises require a certain way of e-business interaction with their partners. We believe that virtual enterprise is one of the business models that can facilitate cloud computing for SMMEs.

It is clear that emerging technologies including cloud have the potential of transforming and automating the business processes of SMMEs and enable them to engage with trading partners and customers in global networks (Dai, 2009).

Identified characteristics of cloud computing are massive scale, homogeneity, resilient computing, low cost software, virtualization, geographic distribution, service orientation, advanced security technologies, resource pooling (Mell & Grance, 2009). In cloud computing, dynamically scalable and often virtualized resources are offered as a service (Behnia, 2009). Therefore, SMMEs do not need to have knowledge of, be experts in, nor have control over the technology infrastructure in the cloud that supports them (Menken & Blokdijk, 2009). In addition, cloud computing applies a model to enable available, convenient and on-demand network access to a shared pool of configurable computing resources that can be provisioned and released on a fly with minimal management effort or service provider interaction (F5 Networks, 2009). The development and success of cloud computing is due to the maturity reached by virtualisation, web 2.0, grid computing, service oriented architecture technologies and some other technologies.

Cloud computing offers IT infrastructure as an Internet service. Ongoing advances in ICT infrastructure and far more sophisticated applications provide individuals and organizations with the ability to connect to data anywhere and anytime. Cloud computing is powerful in that it is based on modularity system. The use of virtualization and cloud platform allows organizations to break down services and systems into smaller components, which can function separately or across a widely distributed network. The cloud services are possible through any devices that have Internet access (Greengard, 2010). Virtualization is a vital enabling technology for many cloud computing environments. Much of the power of virtualisation comes from the platform independence that virtual machines afford (Matthews, Garfinkel, Hoff, & Wheeler, 2009). IT computing becomes a utility, similar to power, and organizations are billed either based on usage of resources or through subscription, this is achieved through virtualization (Menken & Blokdijk, 2009).

Cloud computing as a service delivered over the network adopts a service driven operating model, therefore it places a strong emphasis on service management. A business has a different language and vocabulary from IT. The separation between business and IT in terms of skills and goals may be linked by cloud computing supported through service-oriented architecture, by outsourcing IT services to various service providers under service level agreements (SLA) contracts. SLA assurance is therefore a critical objective for every provider.

Cloud computing infrastructure enables operational services; and also carries along the idea of an application programming interface (API). Although there is a lot of publicity on cloud computing, there is yet no one official definition of what it is (no agreed standards). The most useful definition to date is that from the American National Institute of standards and Technology (NIST), "Cloud computing is a model for enabling convenient, on-demand network access to a shared pool of configurable computing resources (e.g., networks, servers, storage, applications, and services) that can be rapidly provisioned and released with minimal management effort or service provider interaction."

1.1. Cloud Service Delivery Models

The cloud applications are usually in the form of software as a service, platform as a service and infrastructure as a service. Their primary target is end–users-individuals, SMMEs or enterprises.

1.1.1. Software as a Service (SaaS)

This layer is what most people recognise since it represents the customer's interface. Services must be in a large scale to qualify for cloud computing. SaaS customers tap into computing resources provided by a third party company on a pay per use basis, or as a subscription and users do not have to worry about the underlying cloud datacenter infrastructure (e.g., Salesforce.com, Google apps) (Schulz, 2009). The primary target is end users, individuals, and SMMEs. The examples of cloud software as a service deliverables include hosted application software such as word processor, spreadsheet and email applications, as well as social networking and photo and video sharing solutions.

1.1.2. Platform as a Service (PaaS)

PaaS targets developers, supplying them with a programming-language-level environment and a set of well defined application protocol interfaces (API) supported by the provider. Development platforms let developers write their applications and upload their code into the cloud, where the application is accessible and can be run in a web-based manner (e.g., Facebook, Windows azure). Developers have no need to worry about issues of scalability when application usage grows (Weinhardt, Anandasivam, Blau, & Stober, 2009).

1.1.3. Infrastructure as a Service (IaaS)

IaaS consist of two categories: those providing storage capabilities and those providing computing power supplied on demand. The consumer does not manage or control the underlying infrastructure but has control over operating systems, deployed applications, and possibly select the networking components such as host firewalls and load balancers. Computing resources offered as a service take the form of virtual machines (VM) (e.g., Amazon Web Services, Rackspace, and GoGrid) (Kushida, Breznitz, & Zysman, 2010).

1.2. Cloud Deployment Models

Private cloud: The cloud infrastructure is operated or leased by a single organization and is operated solely for that organization. There is increased use of dedicated resources in this deployment model. It may be managed by the organization or a third party and may exist on premise or off premise.

Community cloud: The cloud infrastructure is shared by several organizations and supports a specific community that has shared concerns (e.g., mission, security requirements, policy, and compliance considerations). This implies greater resource sharing and increased agility. It may be managed by the organizations or a third party and may exist on premise or off premise.

Public cloud: The cloud infrastructure is made available to the general public or a large industry group and is owned by an organization selling cloud services.

Hybrid cloud: The cloud infrastructure is a composition of two or more clouds (private, community, or public) that remain unique entities but are bound together by standardized or proprietary technology that enables data and application portability (e.g., cloud bursting).

Virtual private cloud: A VPC is a combination of cloud computing resources with a virtual private network (VPN) infrastructure to give users the abstraction of a private set of cloud resources that are transparently and securely connected to their own infrastructure. To create a VPC, dynamically configurable pools of cloud resources are connected to enterprise sites with VPNs (private

cloud Amazon EC2) (Wood, Shenoy, Gerber, Ramakrishnan, & Van der Merwe, 2009).

SMMEs may not have enough resources to respond to a business opportunity or customer demand. To overcome this, SMMEs can engage in virtual enterprises collaboration to enable them to respond on varying market opportunities and act on customer demand to remain competitive. One of the defining characteristics of a VE is the collaboration of resources. To cope with temporal resource unavailability, a VE will include many members with identical capabilities. In this manner some resources may be idle. In addition, the use of traditional outsource does not offer the VE agility and control it requires. To alleviate these SMMEs should get resources from an external source. One of such sources is cloud computing platforms. Access to these resources is not always guaranteed to be available or is too expensive to access.

We envision a way to overcome this problem by creating a VE enabled cloud computing platform to enable SMMEs. Due to the opportunistic nature of the VE business and the corresponding reservation of the participating enterprises to make long term investments, we argue that SMMEs participating in a VE collaboration face challenges of achieving the level of coordination of an extended enterprise without having prior long term relationship within a supply chain and without an expectation for future economies of scale. ICT support for the co-ordination of services is missing in the VE that offers services (Service Provider). Therefore, this can be achieved by shifting the ICTs to the cloud; hence adaptability, flexibility and agility can be realised. SOA as an underlying architecture of services in the cloud provide evidence of this issue. In the SOA approach, applications are developed by coordinating the behavior of autonomous components distributed over an overlay network.

The fact that cloud computing lowers participation and entrance barriers will help SMMEs to gain access to technologies and facilitate direct participation in SMME VE alliances. Due to the cloud computing technologies, ICT solutions and e-business models can be crafted to allow SMMEs to gain access to ICT resources without purchasing the infrastructure. Cloud computing architecture and VE architecture both utilises SOA and virtualisation as enabling technologies. It is against this background that we propose a VE enabled cloud enterprise architecture for enabling SMMEs, based on the cloud technologies aimed at providing SMMEs with an opportunity to quickly respond to market opportunities and customer demands and give them the flexibility, adaptability and agility they need to remain competitive. Research has been done for proposing various VE architectures and cloud computing architectures independently, but there have been no attempts of using them together.

2. RELATED WORK

a) Virtual Enterprise Architectures

Virtual enterprise architecture is presented in several previous works. Ahonen, Alvarengaa, Provedela, and Paradab (2010), suggested a client-broker-server architecture. The author implemented system architecture with the functionalities of a virtual enterprise that follows the broker architectural pattern defined in Buschmann, Meunier, Rohnert, Sommerlad, and Stal (1996). The pattern describes how the components representing clients and servers may communicate with each other with the help of a special broker component. This intermediary component helps the client to locate the services they need without burdening them with unnecessary details of communication. The architecture makes services offered by separate enterprises look in the client perspective as if they were offered only by one enterprise. The selected architectural pattern makes it possible to extend the broker component from a simple message gateway to an intelligent agent, which, for example, negotiates with several service components in order to find the most appropriate service for a requesting client.

In Jagdev and Thoben (2001), the authors discuss the key types of collaborations (i.e., supply chain, extended enterprise and virtual enterprise). In the supply chain participating enterprises are termed node, they agree to contribute their expertise towards the completion and supply of a common end-product. Each node in the supply chain, act as both a customer and a supplier. A supply chain needs to be a sequential set of nodes, taking the form of enterprise network.

Extended enterprises types of collaboration are evolutionary in nature. For example when organizations have known each other and conducted business as a supply chain for some time. During this time sufficient level of trust has developed to automate the sharing of day-to-day operational data. Each organisation will be prepared to invest in the modern ICT tools for the effortless sharing of information; this implies their willingness to be committed to long term relationship. The seamless exchange of relevant operational information on top of existing long term, successful relationship distinguishes the extended enterprise from other long term collaboration.

In virtual enterprise type of collaboration, organisation responds to the dynamic and globalised markets of today. Virtual enterprise priority is to realize customer needs. These needs can be wide and unique (e.g., a large project based contract) or small but with numerous variations. The authors give an example of a number of complementary companies specializing in the repair and maintenance of household items which may form a virtual enterprise to give a comprehensive service to its potential customers.

Aerts, Szirbik, and Goossenaerts (2002) suggested that the strength of VEs enforce strong requirements on their ICT support. The ICT infrastructure must be highly flexible for the VE to be agile. He proposes a mobile agent-based ICT architecture to provide the required flexibility.

Mobile agents play a role of manager for rush orders to supervise their fulfillment from where a customer places an order. This architecture uses a scheduler system and real time application request to ensure the fulfillment of orders. At the VE level there is scheduling and trading of services which determine whether it is possible to fill the order and allocate the proper enterprises to the order.

The enterprise architecture (EA) deals with the structure of an enterprise, relationships and interactions of its elements. EA presents a holistic approach to reconcile IT and business concerns in an enterprise. The virtual enterprises architecture is built upon the principles of EA. Service-oriented architecture (SOA) to implement EA shows to be an enabler of VE at Business and Technology Levels. The authors argue that enterprise architecture alone is not sufficient to overcome the three major challenges of VE which are flexibility, adaptability, and agility. SOA provides agile, reconfigurable and loosely-coupled infrastructure for enterprise integration of diverse processes and platforms of different enterprises. At business level, SOA services composition and reuse helps VE infrastructure to quickly assemble and react to changes as they occur. SOA's service components can be reused and composed across multiple VE instances hence making it easier for business to flexibly participate in different VE at the same time (Goel, Schmidt, & Gilbert, 2009).

b) Cloud Computing

Huerta-Canepa and Lee (2010) propose a virtual cloud computing platform using mobile phones where a mobile device can be a virtual cloud computing provider. He argues that a mobile device is resource constrained, to solve that mobile device should get resources from an external source, in this case cloud computing the proposed system uses a resource sharing mechanism. If a user wants to execute tasks which need more resources than available at the device, the system listens for nodes in the surrounding area. If resources are available the system interrupts the application loading and modifies the application in order to use the virtual cloud.

Hata, Kamizuru, Honda, Shimizu, and Yao propose a dynamic IP-VPN architecture that runs on virtual private cloud deployment model. It is connected to the users via IP-VPN on the Internet. There are different kinds of IP-VPN such as IPSec, L2TP or SSL-VPN. They assume that multiple protocol programmability on a single platform is one of external requirements to provide private cloud to enterprise networks at low cost. Agility is also required which can be achieved through cloud computing environments as another external requirement. Cloud computing environments provide service by using virtual machine on one platform. This network structure will eliminate the use of corporate networks for integration process to the virtual private network to further reduce cost.

It is extremely desirable that cloud resources be seamlessly integrated into an enterprise's current infrastructure without having to deal with substantial configuration, address management, or security concerns. Existing commercial solutions present cloud servers as isolated entities with their own IP addresses space that is outside of the customer's control. To achieve seamless integration, Wood et al. (2009) propose cloudNet, a cloud platform architecture which utilises virtual private networks (VPNs) that securely and seamlessly link cloud and enterprise sites. CloudNet uses VPNs to provide secure communication channels and to allow customer's greater control over network provisioning and configuration. The idea of virtual private clouds is used to create a flexible, secure resource pool transparently connected to enterprises via VPNs.

Grigoriu (2009) views a company that is developed according to business process utility (BPU), virtual enterprise, cloud computing, enterprise architecture, service oriented architecture to look like a cloud enterprise architecture (i.e., a virtual enterprise with SOA-like architecture with its' business functions, processes and their

IT resource supplied over the Web by a cloud of business and IT service providers). Cloud computing covers both the IT applications and technology layers of enterprise architecture. SOA offers transparent distribution, loose coupling, technology transparency, and interfaces easing the integration process to the cloud. SOA is an enabler of virtual enterprise, cloud computing, and business process utility.

3. BENEFITS OF CLOUD COMPUTING FOR SMMEs

Whilst this subscription or rent based model appeals to enterprises of all sizes, the SMMEs stand to benefit the most from this paradigm shift. With no start-up capital required, the SMMEs can quickly develop their computing capacity. Cloud computing makes things simpler for SMMEs to manage by placing the responsibility of maintaining and upgrading the service on the service provider. Since they do not require all the services on a daily basis, the issue of underutilised IT resources is eliminated.

There are two major benefits of cloud computing for SMMEs. The first is the removal of main IT infrastructure investment. Secondly the users of the system can access their data from anywhere in the world. Any computer with internet access can do. These benefits can result in a major change in the way that businesses of all sizes exploit computers. Primarily it will be the SMMEs that benefit the most.

For an SMME, cloud computing can provide an ideal platform to host the business system requirements. The system is maintained in a secure environment where data is backed up on a daily basis. The business does not need to invest on expensive IT infrastructure. Neither is it required to employ staff or a company to run the computer system. All that is necessary is a simple rental agreement and fast

Internet access (Marcus, 2009). Without the opportunities available in the cloud, most SMMEs would have to go without core enterprise systems such as CRM, ERP, calendar sharing, e-mail, and even business intelligence. But what they have come to realise early on, is that all of these can be enjoyed out of the cloud for at lower cost.

Cloud provides the SMME a better opportunity to obtain a competitive advantage through technology, and potentially makes these benefits more in reach. The reason being services residing on the cloud reduces or eliminates many of the overhead expenses associated with software solutions. The benefit is these overhead costs can be avoided to let the provider handle those aspects of the applications. As a result, the SMME can focus completely on the benefits without having to deal with the problem of maintenance. This also allows small companies to compete more effectively with some of the larger businesses, effectively balancing the playing field.

Cloud meets organisational needs, for the SMME on the cloud computing can be the ideal solution to best meet organizational needs for efficiency and accuracy of data because fewer resources are needed to run on each individual computer (Marcus, 2009).

As a result, small businesses and start-ups in particular are able to deliver speed and agility, update operations and reduce time to market, as well as improve customer engagement and success. Therefore, there is improved focus on business rather than technology. While some critics question the perception of storing significant business data on outside servers, the approach does have strong appeal to SMMEs who do not have resources, both financial and human, to store data on in-house servers and secure it properly.

4. VIRTUAL ENTERPRISE AND CLOUD COMPUTING

The virtual enterprise (VE) concept is based on distributed business functions and utilities, outsourced to partners that work together with the firm to deliver the product to end customers. A VE collects data on markets and customer needs and combines it with the newest design methods and computer-integrated network that includes suppliers, distributors, retailers and consumers. Enterprise architecture layers uses virtualisation as an interface for providing business functions; this is called business layer virtulisation (Wood, Shenoy, Gerber, Ramakrishnan, & Van der Merwe, 2009).

A VE "is an ad-hoc coalition of independent enterprises and organisations, collaborating to achieve an explicit and specific goal of responding to a specific situation, by leveraging resources, skills and competences of the members of the coalition". A VE has no dominant partner, legal existence or physical ownership of resource inventories. Members can join or leave the coalition at any time, but within contractual limits. A Virtual Enterprise is dissolved as soon as its explicit goal is achieved" (Amit, Heinz, & David, 2010).

One of the most striking features of the VE is its opportunistic nature. SMMEs may use the VE strategy to meet unexpected change and unforeseen events, in this way become agile. One of the beneficial results is that unused capabilities or planned overcapacity can be made productive. To cope with the temporary unavailability of a particular type of capability, a VE will include several members with similar capabilities (redundancy). This will help the VE to achieve agility. These characteristics distinguish the VE from a more long-term inter-organisational structure, such as the supply chain or the extended enterprise (Aerts, Szirbik, & Goossenaerts, 2002).

Supply chains and extended enterprises might evolve towards virtual enterprises in response to competitive pressures and in order to better deliver the differentiation to satisfy customers, they are then called supply chain VE or extended enterprise VE, but then the opposite may also happen. When a VE business venture has the potential of a stable market, partners in a VE may develop longer-term relationships and evolve into a supply chain or an extended enterprise, in case one particular partner takes the lead (Aerts, Szirbik, & Goossenaerts, 2002). In reality a fraction of a business that links in the value chain is outsourced to partners. Hence cloud computing, i.e. outsourcing IT to third parties, becomes part of the Virtual Enterprise (Camarinha-Matos & Afsarmanesh, 1998).

A VE aims to use Information technology such as computer networks, business process/workflow management systems and service oriented architectures, to achieve a dynamic business partnership that can easily respond to business opportunities as they occur. The establishment of cooperation agreements between enterprises is not something new, but the use of IT to support the information networking is one of the characteristics of the VE concept. The general purpose for enterprises to be involved in a VE alliance is to leverage core competencies, resources and to each provide benefits to the other for a specific set of business opportunities. Only a small headquarters staff to deal with administrative and a management detail is needed, the actual work is performed by the geographically dispersed partners joined through computer hardware and software (Barnett, Presley, Johnson, & Liles, 1994).

The coalition may gain benefits which they would not get if acting independently due to resource limitation. Above all, the key point to the formation of a VE is the rapid integration of business processes of the participating actors to enable them to respond quickly to customer demand and provide a new solution for an unpredicted opportunity. Communication between distributed sites is necessary for capturing a new opportunity quickly. Participating organizations establish a relationship through a suitable service provider. A VE may consist of a number of business process utilities or functions outsourced to various service providers.

Service oriented architecture (SOA) is an important technology in enabling the virtual enterprise (VE). SOA also underlies the architecture of services in the cloud. SOA is essentially a collection of services; hence true service-orientation allows cloud services to be componentized, pluggable, compassable and loosely coupled. These services communicate with each other. The communication can involve simple data passing or it could involve two or more services coordinating some activity. A way of connecting services to each other is needed both in VE and cloud computing architectures (SOA). The provision of business functions is made possible by means of orchestration. Orchestration allows you to change the way your business functions, as needed, to define or redefine any business process immediately. This mechanism provides the business with the flexibility, adaptability and agility needed to compete in today's changing market environment. To meet the changing needs of the domain, orchestration must provide a dynamic, flexible and adaptable mechanism.

5. THE CONCEPT OF VIRTUALISATION TECHNOLOGY

Virtualisation is a technology that abstracts away the details of physical hardware and provides virtualised resources for high-level applications. A virtualised server is commonly called a virtual machine (VM). Virtualisation technologies have facilitated the realisation of a new model called cloud computing.

Virtualisation forms the foundation of cloud computing, by providing the capability of pooling computing resources from clusters of servers and dynamically assigning or reassigning virtual resources to applications on-demand. The basic idea is to have one resource pretend to be another; it allows one physical computer to become several different computers. This concept is essential to cloud computing and to storage as a service (Menken & Blokdijk, 2009).

Basically, virtualisation reduces complexity. In that way, a virtualised network is easier to manage, hence less cost to administrating a complex solution. Enhanced scalability, flexibility, and utilisation provide additional cost savings. Automation is an additional functionality available through virtualisation. Without it, resolving problems would require an understanding of the specific characteristics of each component in the solution. The biggest benefits for enterprises to take advantage of virtualisation are:

- Standardised configuration of storage devices- A virtual environment requires a minimum level of standardisation on the storage components. With a consistent configuration throughout the drives, it is easier to drive availability uptime in the environment. The virtualisation software allows all operating systems to map to the same file system.
- Easier migration- Data migration can result in lengthy downtime; it can be reduced through virtualisation.

Cloud computing leverages virtualisation technology to achieve the goal of providing computing resources as a utility. Therefore, virtualized technology offers many benefits for SMMEs that locate their ICT part of the business in the cloud. The benefits include easier manageability, elimination of compatibility issues, fault isolation, increased security, efficient use of resources, portability, problem-free testing, rapid deployment, reduced costs, the ability to separate applications, Easier Manageability and Improved Uptime all of which are critical to SMME growth.

6. MOTIVATION FOR CLOUD COMPUTING

Cost savings are still the primary motivation behind the adoption of cloud-based services. Nonetheless, there is evidence that the other business benefits of cloud computing are gaining ground. One of the main benefits of the cloud is its capability to make an enterprise better prepared to react and respond to unpredicted changes, or to easily add-on new services as needed (F5 Networks, 2009).

Cost savings are the primary motivation behind the adoption of cloud-based services. The improvement in storage, processing power, or technology offered by cloud computing enables innovations that were not possible before, therefore allowing the SMMEs competitive advantage and productivity. Cloud computing will provide the agility and control that traditional outsource cannot offer. If an enterprise is not satisfied with the services of the cloud service provider, it is allowed to switch to another far easier than when changing IT outsourcers. The proposed cloud enterprise architecture will be the vehicle that will enable SMMEs to get digitally connected with their partners and customers, therefore act on business opportunities as well as competition advantage (Hinchcliffe, 2009).

7. PROPOSED ARCHITECTURE

A VE enabled cloud enterprise structure and operation consist of the SMMEs in VE alliance and all its' business process utility and IT cloud computing services and their providers.

In Figure 1 we define the cloud enterprise architecture for SMMEs participating in a VE setting as business context, business services, business processes and IT services. Business context layer is responsible for the definition of business goals, strategies, structure, policies and performance metrics and indicators. The main users of services at this level are business owners and executives who are hardly ever IT experts. The main functions of a business such as human resources, payroll, accounting, etc. are defined as coarse-grained services, called "business services" in the business services layers. Users such as business or IT architects may define or select the required business services from out-of-box business services blueprints. The IT services layer represents the services that are obtainable in the cloud. Finally, the business processes layer is the illustration of selection, design, integration and composition of IT services in the form of workflows that fulfill the needs of outlined business services. Figure 1 also shows the users of various layers.

In this architecture, the SMMEs in the virtual setting share business context, business services, and business processes to improve competitive advantage, and quickly respond to market opportunities. Hence the VE enabled cloud enterprise architecture comes down to a value system, which involves a number of companies' value chain that is collaborating to deliver the end product to the customer. The aspect of value chain is not covered in this paper.

Rather than relying on established organizations (e.g., Google cloud, Amazon EC), SMMEs in the VE setting form their own private cloud, where they collaborate their existing IT infrastructure, skills, processes, organizational models and core-competencies. There should be in place, strategies on sharing competencies. We should remember that the SMMEs also compete with one another; hence they cannot expose all their competencies. The reason that they do not host their IT services on a third party Cloud SP, is because of the advantages that come from the collaboration of resources within the VE. Hosting their entire IT infrastructure to the cloud SP could cost them even more, and the resources underutilized. Assuming that the SMMEs in the alliance understand each other better, there can be no one SMME dominating.

Figure 1. A VE enabled cloud enterprise architecture

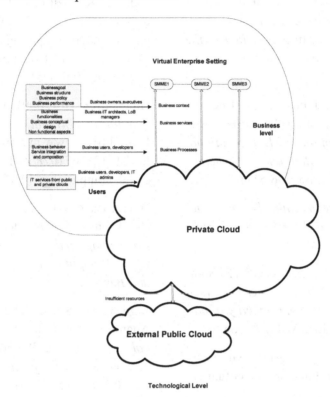

In case of insufficient resources the SMMEs can then tap into external public cloud. This comes down to a hybrid deployment model. Therefore, cloud computing capabilities provide the VE (SMMEs) alliance with agility, flexibility, and adaptability required due to its' highly flexible ICT infrastructure.

8. CONCLUSION

Most of the big organisation are into cloud, this include Dell, Amazon, HP, Intel and more. Cloud computing is a new enterprise model and the current big wave in computing. It has many benefits and some downside as well. Many big companies started by outsourcing some of its IT to the cloud until they can afford their own cloud infrastructure. Although cloud computing offers huge opportunities to the IT industry, the maturity

of cloud computing is currently at its infancy with many issues still to be addressed.

REFERENCES

F5 Networks. (2009). *Cloud computing: Survey results*. Seattle, WA: F5 Networks.

Aerts, A. T. M., Szirbik, N. B., & Goossenaerts, J. B. M. (2002). A flexible agent-based ICT for virtual enterprises. *Journal of Computers in Industry, 49*(3). doi:10.1016/S0166-3615(02)00096-9

Ahonena, H., Alvarengaa, A. G., Provedela, A., & Paradab, V. (2010). A client-broker-server architecture of a virtual enterprise for cutting stock. *International Journal of Computer Integrated Manufacturing, 14*(2), 194–205. doi:10.1080/09511920150216314

Amit, G., Heinz, S., & David, G. (2010). Formal models of virtual enterprise architecture: Motivations and approaches. In *Proceeding of the Pacific Climate Information System Workshop* (pp. 1207-1217).

Barnett, W., Presley, A., Johnson, J., & Liles, D. H. (1994). An architecture for the virtual enterprise. In *Proceedings of the IEEE International Conference on Systems, Man, and Cybernetics*, San Antonio, TX (Vol. 1, pp. 506-511).

Behnia, K. (2009). *Time to rethink IT service delivery & bring the clouds down to earth*. Virtual Strategy Magazine.

Buscmann, F., Meunier, R., Rohnert, H., Sommerlad, P., & Stal, M. (1996). *Pattern-oriented software architecture- A system software of patterns*. Chichester, UK: John Wiley & Sons.

Camarinha-Matos, L. M., & Afsarmanesh, H. (1998). Towards and architecture for virtual enterprises. *Journal of Intelligent Manufacturing*, *9*(2), 189–199. doi:10.1023/A:1008880215595

Dai, W. (2009). The impact of emerging technologies on small and medium enterprises (SMEs). *Journal of Business Systems. Governance and Ethics*, *4*(4), 53–60.

Goel, A., Schmidt, H., & Gilbert, D. (2009). Towards formalizing virtual enterprise architecture. In *Proceedings of the Distributed Object Computing Conference Workshops* (pp. 238-242).

Greengard, S. (2010). Cloud computing and developing nations. *Communications of the ACM*, *53*(5), 18–20. doi:10.1145/1735223.1735232

Grigoriu, A. (2009). *The cloud enterprise*. Retrieved from http://www.bptrends.com/publicationfiles/TWO_04-09-ART-The_Cloud_Enterprise-Grigoriu_v1-final.pdf

Hata, H., Kamizuru, Y., Honda, A., Shimizu, K., & Yao, H. (2010). Dynamic IP-VPN architecture for cloud computing. In *Proceedings of the Information and Communication Technologies Conference* (pp. 1-5).

Hinchcliffe, D. (2009). *Eight ways that cloud computing will change business*. Retrieved from http://www.majorcities.org/pics/download/1_1247671028/ZDNET_Eight_ways_that_cloud_computing_will_change_business.pdf

Huerta-Canepa, G., & Lee, D. (2010). A virtual cloud computing provider for mobile devices. In *Proceedings of the 1st ACM Workshop on Mobile Cloud Computing & Services: Social Networks and Beyond*.

Jagdev, H. S., & Thoben, K. D. (2001). Anatomy of enterprise collaborations. *Journal of Production Planning and Control: The Management of Operations*, *12*(5), 437–451. doi:10.1080/09537280110042675

Kushida, K. E., Breznitz, D., & Zysman, J. (2010). *Cutting through the fog: Understanding the competitive dynamics in cloud computing*. Berkeley, CA: The Berkely Round Table on the International Economy (BRIE).

Marcus, B. (2009). *How on the cloud computing can benefit small and medium enterprises (SMMEs)*. Retrieved from http://www.stumbleupon.com/url/www.helium.com/items/1624860-how-on-the-cloud-computing-can-benefit-small-and-medium-enterprises-smes

Matthews, J., Garfinkel, T., Hoff, C., & Wheeler, J. (2009). Virtual machine contracts for datacenter and cloud computing environments. In *Proceedings of the 6th International Conference on Autonomic Computing and Communications* (pp. 25-30).

Mell, P., & Grance, T. (2009). *Effectively using the cloud computing paradigm*. Gaithersburg, MD: Nist Information Technology Laboratory.

Menken, I., & Blokdijk, G. (2009). *Cloud computing specialist certification kit: Virtualization.* Brisbane, Australia: Emereo Publishing.

Schooff, P. (2009). *The insider's guide to business and IT agility.* Retrieved from http://www.ebizq.net/soaregistercloudfutures.html

Schulz, W. (2009). *What is SaaS, Cloud Computing, PaaS and IaaS.* Retrieved from http://www.s-consult.com/2009/08/04/what-is-saas-cloud-computing-paas-and-iaas/

SOA. (n. d.). *Service-oriented architecture definition.* Retrieved from http://www.service-architecture.com/web-services/articles/service-oriented_architecture_soa_definition.html

Weinhardt, C., Anandasivam, A., Blau, B., & Stober, J. (2009). Business models in the service world. *IEEE Computer, 11*(2), 28–33.

Wood, T., Shenoy, P., Gerber, A., Ramakrishnan, K. K., & Van der Merwe, J. (2009). *The case for enterprise-ready virtual private clouds.* Paper presented at the Workshop on Hot Topics in Cloud Computing.

This work was previously published in the International Journal of Cloud Applications and Computing (IJCAC), Volume 1, Issue 3, edited by Shadi Aljawarneh & Hong Cai, pp. 52-63, copyright 2011 by IGI Publishing (an imprint of IGI Global).

Chapter 16
Architectural Strategies for Green Cloud Computing:
Environments, Infrastructure and Resources

P. Sasikala
Makhanlal Chaturvedi National University of Journalism and Communication, India

ABSTRACT

Opportunities for improving IT efficiency and performance through centralization of resources have increased dramatically in the past few years with the maturation of technologies, such as service oriented architecture, virtualization, grid computing, and management automation. A natural outcome of this is what has become increasingly referred to as cloud computing, where a consumer of computational capabilities sets up or makes use of computing in the cloud network in a self service manner. Cloud computing is evolving, and enterprises are setting up cloud-like, centralized shared infrastructures with automated capacity adjustment that internal departmental customers utilize in a self service manner. Cloud computing promises to speed application deployment, increase innovation, and lower costs all while increasing business agility. This paper discusses the various architectural strategies for clean and green cloud computing. It suggests a variety of ways to take advantage of cloud applications and help identify key issues to figure out the best approach for research and business.

INTRODUCTION

Today is the age of information technology. The facets of work and personal life are moving towards the concept of availability of everything online. Significant technological advances are often made during periods of crisis and change. Thus it is unsurprising that today's CIO's and IT professionals, confronted with extraordinary challenges of spiking energy bills, underutilized data centers, accelerated data growth, during a time of restricted capital and economic uncertainty

DOI: 10.4018/978-1-4666-1879-4.ch016

are gravitating towards innovative efficiency enhancing technological models (http://www.cioforum.com/; http://www.cio.in/). Cloud computing is one such model. Cloud computing is the latest evolution of Internet-based computing (Marks & Lozano, 2010). The Internet provided a common infrastructure for applications. It deploys as a complete platform for supporting scalable applications in a way that improves the efficiency of both IT management and operations. The architectures used by true cloud computing platforms - rapid scalability, flexibility, resource pooling and usage-based pricing - are significantly different from what are now deemed "classic" IT computing models. With these differences comes the opportunity for significant gains in asset efficiency, capital utilization and business responsiveness (Sasikala, in press). The potential benefits of cloud computing are overwhelming. However, attaining these benefits requires that each aspect of the cloud platform support the key design principles of the cloud model. One of the core design principles is dynamic scalability, or the ability to provision and decommission servers on demand (http://cloudcomputing.qrimp.com/portal.aspx) Unfortunately, the majority of today's database servers are incapable of satisfying this requirement. Cloud computing is not a fad it is driven by some tangible and very powerful benefits. Whether the cloud is provided as an internal corporate resource, as a service hosted by a third-party, or as a hybrid of these two models, there are some very real advantages to this model (http://www.gartner.com/). These advantages derive from specialization and economies of scale. The combination of all the benefits is driving cloud computing from mere buzzword to disruptive and transformational tsunami. The report from IDC says that due to the emergence of cloud computing, IT marketplace is undergoing a change and it expects that investment on cloud services will reach to $42 billion by 2012 (Wagner, 2010).

Goldman Sachs, Wells Fargo Securities, Gartner and other prominent observers of the technology industry predict that cloud computing is the most significant IT shift of this decade (http://cloudcomputing.qrimp.com/portal.aspx; http://www.gartner.com). SYS-CON's Cloud Computing Journal lists the top 150 most active players in the cloud ecosystem (Geelan, 2009). A robust ecosystem of solutions providers is emerging around cloud computing. In this paper we discuss the various feasible applications of the latest Java Enterprise Edition platform and the potential roles of web profiles in the present context along with multi-cloud framework approach. The paper also elucidates cloud databases and suggests on preparing data for cloud within the limits of CAP theorem. The paper highlights the open source private cloud computing benefits mainly of the Eucalyptus. Cloud computing scenario in India in general and for the small and medium enterprises are also discussed. Finally, cloud applications in education and the resources available are explained in detail.

JAVA™ PLATFORM APPLICATIONS IN CLOUD COMPUTING

The evolution in Sun's enterprise Java™ platform, Java EE (Enterprise Edition), over the years lead to a remarkable transformation for a mature, widely deployed, well supported server side development platform (http://in.sun.com/java/). The focus of Java EE 5 was squarely on reducing complexity by embracing the concepts of annotations, POJO programming, zero configuration systems and freedom from XML hell. One of the major criticisms of Java EE has been that it is simply too large. They have grown in complexity and size, leading to very large downloads and very large runtimes. In majority of cases, the full EE environment just isn't needed. Indeed, a majority of small to medium range Java web applications do not utilize the full Java EE stack. One can imagine the same to be true of SOA applications that would use features like messaging, transactions, persistence, and web services but have no need for presentation tier technologies like JSP or JSF. Profiles are designed to address this issue. In recognition of this, most of the well known application servers have released "lite" versions of themselves. Java EE 6 is a big step in the journey towards the ideal of a simple, streamlined and well integrated platform (http://www.oracle.com/technetwork/java/javaee/tech/index.html). Java EE 6 is the industry standard platform for building enterprise-class computing applications coded in the Java programming language. Based on the solid foundation of Java Platform, Java EE 6 adds libraries and system services that support the scalability, accessibility, security, integrity, and other requirements of enterprise-class applications. Java EE 6 also includes a rich set of innovations best reflected in the technologies that comprise the platform including brand new APIs like WebBeans 1.0 and JAX-RS 1.1 or even mature APIs like Servlet 3.0.

JAVA EE 6 WEB PROFILE

Profiles are a way of slimming down the EE environment. A profile defines a minimum feature set for a compliant product to implement, although a product could also choose to offer additional features, apparently with the provision that a TCK must exist for those features. The idea of Profiles has been used successfully in the standards world for a relatively long time. Java ME supports the idea of Profiles geared towards particular device runtime environments. Profiles are used to better organize the increasingly complex world of web services standards such as the WS -I Basic Profile, the WS-I Basic Security Profile and the like. Essentially, a profile is a subset of the enormous number of APIs/specs etc that a JavaEE platform supports. For example, a proposed Java EE Web Profile may only include APIs that are likely to be used in most Java web applications.

The first profile we will be tackling is the Web Profile. Utilising the new, lightweight Java EE 6 Web Profile to create next-generation web applications, and the full power of the Java EE 6 platform for enterprise applications is the thrust area (Chinnici, 2008). It is easy-to-use and quite extensible and hence embraces open-source frameworks using web fragments. Developers will benefit from productivity improvements with more annotations, more POJOs, simplified packaging, and less XML configuration. GlassFish Server Open Source Edition 3.0 and Oracle GlassFish Server 3.0 provide feature-rich implementation (Goncalves, 2010). Oracle GlassFish Server is the first commercial Java EE 6 product available.

If you look at any new web profile in more detail, you see that it is a specified minimal configuration targeted for small footprint servers that should support something called "typical" web applications. It is thought of as a minimal specification, so a vendor is free to add additional services in their concrete implementation. Clear from a technology point for view. For me this was simply a new way of pruning; getting rid of some older specs and finding a way to have more Java EE certified servers again. But what I really do not get is the actual hype (and that's what I am feeling) around this. Although Java EE 6 defines the rules for creating new Profiles through separate JSRs, only one Profile, the Web Profile is included in platform this time. Should WebBeans be included in Java EE 6 in its current form? If not, what changes could be made to the JSR? Note also that EJB Lite, and not the full version of the EJB specification is included in the Web Profile.

WebBeans is perhaps the most groundbreaking API developed in the Java EE 6 time frame. WebBeans fills a number of gaps in Java EE. Although WebBeans is inspired by Seam, Google Guice as well as Spring, it does not directly mirror any of them. Indeed, WebBeans adds a number of unique innovations of its own. WebBeans unifies the JSF, JPA and EJB 3 programming models to truly feel like a single, well-integrated platform. WebBeans implicitly manages the life-cycle of all registered components in appropriate contexts. WebBeans brings a robust set of dependency injection features to the platform in a completely type -safe and Java-centric way. WebBeans enhances the Java EE Interceptor model by adding the ability to bind interceptors to annotations instead of having to bind interceptors to target object classes themselves. As impressive as this is, it is just the tip of the iceberg. WebBeans adds a lot of other very cool features that constitutes the next generation of integration features for Java EE.

I should point out that the cloud computing in recent times will set trend towards more modularization in application servers to work very well with extensibility, by delivering a simpler, more effective way to manage the versioning and availability of libraries to applications. As we expand the freedom to combine technologies in a cloud product, the issue of compatibility requirements that span multiple technologies comes to the fore. One concern we heard around profiles is that they might weaken compatibility. The Web Profile itself may evolve in the future by adding more technologies (but not by subtracting them, except via pruning), based on experience acquired. It's also the case that the full Java EE 6 platform will always be there on the cloud, offering an ample choice of certified Java EE 6 products from different vendors, to address needs that go well beyond what the Web Profile offers. Seriously, we need to look beyond just JSPs and Servlets to equate a profile to Java EE 6. Web Developers need frameworks to help perform development (JPA, Web Beans and EJB3.1 Lite) fit the bill. I think a set of layered profiles would suit everyone best. I think there should be more fine grained approach. The most basic version should be the bare bones and ready for servlet or RMI deployment only (not even jsp) as it can be used for rich client applications, or GWT. Also, why not provide a mechanism to add-remove things easily (console and gui) so you don't need

this profiles stuff at all? Hence cloud computing will be the future answer for all these issues and it promise a great potential. Java EE 6 applications can be easily deployed on Amazon, RightScale, Elastra, Joyent, and other clouds.

Amazon.com, Inc. (http://www.amazon.com/) is a US-based multinational electronic commerce company, headquartered in Seattle, Washington. It is America's largest online retailer, with nearly three times the Internet sales revenue of the runner up, Staples, Inc., as of January 2010.

Short for Amazon Elastic Computer Cloud, Amazon EC2 is a commercial Web service that lets customers "rent" computing resources from the EC2 (http://aws.amazon.com/ec2/).

EC2 provides storage, processing, and Web services to customers. EC2 is a virtual computing environment, that enables customers to use Web service interfaces to launch instances with a variety of operating systems, load them with your custom applications, manage your network's access permissions, and run your image using as many or few systems as you need. Amazon EC2 reduces the time required to obtain and boot new server instances to minutes, allowing you to quickly scale capacity, both up and down, as your computing requirements change. Amazon EC2 changes the economics of computing by allowing you to pay only for capacity that you actually use. Amazon EC2 provides developers the tools to build failure resilient applications and isolate themselves from common failure scenarios.

The Amazon Simple Storage Service (Amazon S3) is a scalable, high-speed, low-cost Web-based service designed for online backup and archiving of data and application programs (http://aws.amazon.com/s3/). According to the Amazon Web services pages, the S3 was intentionally designed with a minimal feature set and was created to make Web-scale computing easier for developers (Murty, 2010). The service gives subscribers access to the same systems that Amazon uses to run its own Web sites. The S3 allows uploading, storage, and downloading of practically any file or object up to five gigabytes (5 GB) in size. Amazon.com imposes no limit on the number of items that a subscriber can store. Subscriber data is stored on redundant servers in multiple data centers. The S3 employs a simple Web-based interface and uses encryption for the purpose of user authentication. Subscribers can choose to keep their data private or make it publicly accessible. Users can also, if they so desire, encrypt data prior to storage. Rights may be specified for individual users. When a subscriber stores data on the S3, Amazon.com tracks usage for billing purposes but does not otherwise access the data unless required to do so by law.

Reduced Redundancy Storage (RRS) is a new storage option within Amazon S3 that enables customers to reduce their costs by storing non-critical, reproducible data at lower levels of redundancy than Amazon S3's standard storage (http://aws. amazon.com/about-aws/whats-new/2010/05/19/ announcing-amazon-s3-reduced-redundancy-storage/http://aws.amazon.com/rds/). RRS provides a lower cost, less durable, highly available storage option that is designed to sustain the loss of data in a single facility. RRS is ideal for non-critical or reproducible data.

Amazon Relational Database Service (Amazon RDS) is a web service that makes it easy to set up, operate, and scale a relational database in the cloud (http://aws.amazon.com/fws/). It provides cost-efficient and resizable capacity while managing time-consuming database administration tasks, freeing you up to focus on your applications and business. Amazon RDS gives you access to the full capabilities of a familiar MySQL database.

Amazon Fulfillment Web Service (Amazon FWS) allows merchants to access Amazon.com's world-class fulfillment capabilities through a simple web services interface (http://aws.amazon. com/sqs/). Merchants can programmatically send order information to Amazon with instructions to physically fulfill customer orders on their behalf.

Amazon Simple Queue Service (Amazon SQS) offers a reliable, highly scalable, hosted queue for storing messages as they travel between computers (http://aws.amazon.com/sns/). By using Amazon SQS, developers can simply move data between distributed components of their applications that perform different tasks, without losing messages or requiring each component to be always available. Amazon SQS makes it easy to build an automated workflow, working in close conjunction with the Amazon Elastic Compute Cloud (Amazon EC2) and the other AWS infrastructure web services. Amazon SQS works by exposing Amazon's web-scale messaging infrastructure as a web service. Any computer on the Internet can add or read messages without any installed software or special firewall configurations. Components of applications using Amazon SQS can run independently, and do not need to be on the same network, developed with the same technologies, or running at the same time.

Amazon Simple Notification Service (Amazon SNS) is a web service that makes it easy to set up, operate, and send notifications from the cloud (http://aws.amazon.com/cloudwatch/). It provides developers with a highly scalable, flexible, and cost-effective capability to publish messages from an application and immediately deliver them to subscribers or other applications. It is designed to make web-scale computing easier for developers.

Amazon CloudWatch is a web service that provides monitoring for AWS cloud resources, starting with Amazon EC2 (http://aws.amazon. com/fps/). It provides customers with visibility into resource utilization, operational performance, and overall demand patterns—including metrics such as CPU utilization, disk reads and writes, and network traffic. To use Amazon CloudWatch, simply select the Amazon EC2 instances that you'd like to monitor; within minutes, Amazon CloudWatch will begin aggregating and storing monitoring data that can be accessed using the AWS Management Console, web service APIs or Command Line Tools.

Amazon Flexible Payments ServiceTM (Amazon FPS) is the first payments service designed from the ground up for developers (http://docs.amazonwebservices.com/AmazonVPC/2010-08-31/NetworkAdminGuide/http://aws.amazon.com/ebs/). It is built on top of Amazon's reliable and scalable payments infrastructure and provides developers with a convenient way to charge Amazon's tens of millions of customers (with their permission, of course!). Amazon customers can pay using the same login credentials, shipping address and payment information they already have on file with Amazon. With Amazon FPS, developers can accept payments on their website for selling goods or services, raise donations, execute recurring payments, and send payments.

Amazon Virtual Private Cloud (VPC) is a commercial cloud computing service that provides a virtual private cloud, allowing enterprise customers to access the Amazon Elastic Compute Cloud over an IPsec based virtual private network (http://aws.amazon.com/amis/Sun%20Microsystems/2794). For Amazon it is an endorsement of the hybrid approach, but it's also meant to combat the growing interest in private clouds. Unlike traditional EC2 instances which are allocated internal and external IP numbers by Amazon, the customer can assign IP numbers of their choosing from one or more subnets.

Amazon Elastic Block Store (EBS) provides block level storage volumes for use with Amazon EC2 instances (http://www.oracle.com/technetwork/middleware/glassfish/overview/index.html). Amazon EBS volumes are off-instance storage that persists independently from the life of an instance. Amazon Elastic Block Store provides highly available, highly reliable storage volumes that can be attached to a running Amazon EC2 instance and exposed as a device within the instance. Amazon EBS is particularly suited for applications that require a database, file system, or access to raw block level storage.

Sun Microsystems Inc. recently announced the release of Hardened OpenSolaris 2009.06 on Amazon EC2's cloud computing service (Shachor, 2001). This 32-bit AMI gives the power and secu-

rity of OpenSolaris combined with the flexibility of Amazon's cloud computing service, and is optimized for Amazon EC2's cloud computing environment.

Oracle announced recently the release of Oracle GlassFish Server 3.0.1 and GlassFish Server Open Source Edition 3.0.1 (http://blogs.sun.com/ec2/entry/mysql_ami_on_opensolaris_2008). GlassFish Server 3.0.1 is the first product release under Oracle. The Oracle-branded release includes numerous critical fixes and updated platform support. The Oracle GlassFish Enterprise Server 3 is the industry's first application server to support the Java Platform, Enterprise Edition 6 (the Java EE 6 platform) standard. Its greatly improved flexibility and ease of use reduces cost by providing improved developer productivity and simplified application architecture.

mod_jk (layered approach, support both Apache1.3.x and Apache2.xx, Better support for SSL) is a replacement to the elderly mod_jserv (was too complex, supported only Apache, couldn't reliably identify whether a request was made via HTTP or HTTPS) (Wikipedia, n. d.). It is a completely new Tomcat-Apache plug-in that handles the communication between Tomcat and Apache. There are 2 ways to have your applications automatically start at boot time on Solaris: SMF and Legacy Init Scripts. Legacy Init Scripts are just what the name suggests, the good ol' RC scripts you've used forever on System V based UNIX. Solaris's Service Management Facility (SMF) is dependency aware, human friendly, and very smart. With SMF we start to look at our applications and daemons as services. Using the svcs command we can view running services, if you add the "-a" option it'll show you all services running or not.

MySQL Database 5.1 AMI is 32bit AMI is based on OpenSolaris 2008.11 AMI (http://aws.amazon.com/ec2/pricing/). It contains MySQL 5.1 binaries with DTrace probes, and SysBench. SysBench is a modular, cross-platform and multi-threaded benchmark tool for evaluating

OS parameters that are important for a system running a database under intensive load. There are also sample Dtrace scripts which utilize the probes in MySQL to collect information about SQL calls. Users should check /export/home/mysql README files for directions. The MySQL and SysBench binaries are located in the /usr/local/mysql and /sysbench directories. An existing user 'mysql' is authorized to start the MySQL server and run tests.

Service Management Facility (SMF) is a feature of the Solaris operating system that creates a supported, unified model for services and service management on each Solaris system and replaces init.d (http://www.rightscale.com/) SMF introduces: Dependency order, Configurable boot verbosity, Delegation of tasks to non-root users, Parallel starting of services, and Automatic service restart after failure. All these capabilities are made possible by treating Services as "first class objects". That is, they are more than just user-executed software to the OS. The SMF concept is - Instances: a service is a collection of configurations. An instance is a running execution of a defined service executable. One can have many instances of a defined service, such as multiple webservers listening on different ports referencing different WWW root directories (Figure 1).

```
GlassFish: svcadm restart/enable/
disable svc:/application/GlassFish/
domain1:default
```

Figure 1. Java EE 6 on Amazon

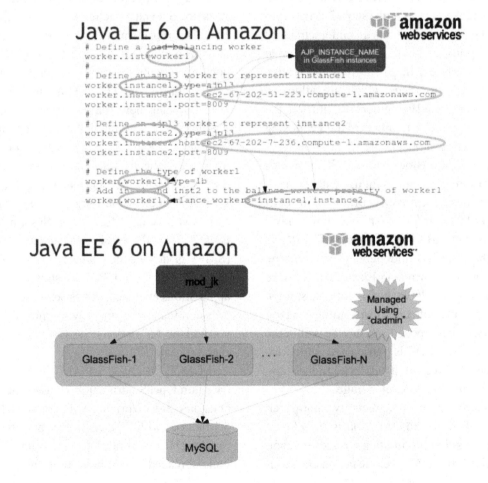

```
MySQL SMF: svcadm enable mysql
mod_jk: svcadm restart/refresh/
enable/disable
svc:/network/http:apache22
```

The following are the ways to deploy them:

- Launch MySQL AMI, create database, user, privileges, etc;
- Launch 1 or more GlassFish AMI (Set AJP_INSTANCE_NAME in each GlassFish);
- Administer multiple instances using cl-admin (--target instance-list OR set AS_TARGET="..."; cladmin create-jdbc-connection-pool ...; cladmin deploy ~/samples/hello.war);
- Launch mod_jk AMI (Configure "worker.properties").

Amazon EC2 Pricing (http://www.elastra.com): Pay only for what you use. There is no minimum fee. Estimate your monthly bill using AWS Simple Monthly Calculator. The prices listed are based on the Region in which your instance is running. For a detailed comparison between On-Demand Instances, Reserved Instances and Spot Instances, see Amazon EC2 Instance Purchasing Options.

RightScale: a web based cloud computing management platform for managing cloud infrastructure from multiple providers (http://www.joyent.com/). RightScale (Figure 2) provides a cloud computing platform and consulting services that enable companies to create scalable web solutions running on Amazon Web Services (AWS). SaaS is used to manage servers in multiple IaaS. It automates everything that keeps operations busy. It includes, Providing a library of pre-configured assets, Design: Cloud-Ready ServerTemplates, Deploy: Group of Servers, Macros, Full Automation: Autoscaling, Active monitoring based on real-time triggers, Configuration, Macros, and Best practices. It offers Professional Services in practice.

It can be deployed in the following way (Figure 3): Launches a new virtual server with clean install of Ubuntu; Install GlassFish Server Open Source Edition 3.0; Detects database in the deployment (Installs MySQL Connector/J Driver; Creates a JDBC Connection Pool and Resource); Install samples (Archives (WAR/EAR/...) stored in S3).

Figure 2. RightScale cloud applications

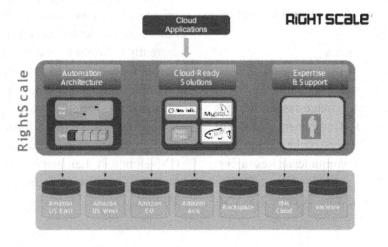

Figure 3. High availability deployment

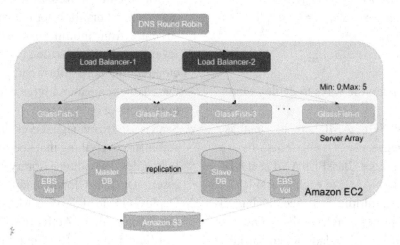

WHAT IS ELASTRA?

Elastra software enables enterprises to automate modeling, deployment and policy enforcement of their application (http://www.simplecloud. org). Elastra products work with provisioning and virtualization tools to deliver the IT infrastructure required running complex applications. Elastra users enjoy faster application deployment, increased resource utilization and improved IT governance. The Key Benefits include:

- Model and provision application infrastructure
- Automate changes to application infrastructure
- On-demand internal, external & virtualized resources
- Enforce IT policy rules

Cloud Computing is a Sustainable Business Model: Design, deploy, manage system designs on private/public clouds (Component: A piece of software such as GlassFish or Apache; Connectors: Enables components to communicate; Resources: Network storage); Manage a hybrid cloud (Design → Deployment(s)) (VMWare vCenter 2.5, VM-

Ware vSphere4, AWS); Enterprise Cloud Server (ECS) or AWS Edition.

WHAT IS JOYENT?

Joyent is a San Francisco, CA based company that provides Infrastructure as a Service and Platform as a Service for thousands of customers including popular web sites such as LinkedIn, Gilt Groupe, and Kabam. Joyent's customers use our integrated stack of enterprise-class Smart Technologies for cloud computing to improve application performance, solve capacity problems and reduce costs for business-critical (http://www. elastra.com) offers high performance public and private cloud computing services using Joyent SmartDataCenter for service providers. Joyent is currently believed to operate the largest installation of OpenSolaris in the world. In addition to providing IaaS and PaaS, Joyent is a supporter of many open source projects including Node. js and Illumos. High-performance and reliable public, private, and hybrid cloud; Environment (Development Language: Java, PHP, Ruby… Server: GlassFish, Apache, Nginx… Database: MySQL, Oracle, etc.).

MULTI-CLOUD VENDORS / FRAMEWORK

Deploying on multiple clouds provides the choice to select the best provider and reduces vendor lock. This is a big win for our community.

Simple Cloud: The Simple Cloud API brings cloud technologies to PHP and the PHPilosophy to the cloud, starting with common interfaces for three cloud application services: File Storage Services; Document Storage Services; and Simple Queue (http://incubator.apache.org/libcloud/index.html). You can start writing scalable, highly available, and resilient cloud applications that are portable across all major cloud vendors today. The Simple Cloud API will enable developers to invoke cloud services in a common way across cloud providers. The Simple Cloud API combines the benefits of open source community processes with the active participation of the cloud vendors themselves. The Simple Cloud API is an example of Microsoft's continued investment in the openness and interoperability of its platform. The Simple Cloud API tackles the main challenges of cloud computing adoption, portability and interoperability. Zend has invited the open source community and software vendors of all sizes to participate. IBM, Microsoft, Rackspace, Nirvanix, and GoGrid have already joined the project as contributors.

Elastra: Elastra software enables enterprises to automate modeling, deployment and policy enforcement of their application infrastructure (http://www.jclourds.org). Elastra products work with provisioning and virtualization tools to deliver the IT infrastructure required running complex applications. Elastra users enjoy faster application deployment, increased resource utilization and improved IT governance. Key Benefits include: Model and provision application infrastructure; Automate changes to application infrastructure; On-demand internal, external & virtualized resources and Enforce IT policy rules.

Libcloud: a unified interface to the cloud, is a standard client library for many popular cloud providers, written in python and java (http://www.deltacloud.org). Libcloud is a client library for interacting with many of the popular cloud server providers. It was created to make it easy for developers to build products that work between any of the services that it supports. Libcloud was originally created by the folks over at Cloudkick, but has since grown into an independent free software project licensed under the Apache License (2.0). Libcloud represents a fundamental change in the way clouds are managed, breaking the barriers of proprietary, closed clouds. Libcloud will make life easier for our customers. We appreciate and support this standardization tool. I'm excited to see the development of projects, like libcloud, that help make the lives of the cloud computing community easier by offering a standardized way to communicate with their provider of choice. We believe in an open cloud and are thrilled to see libcloud push the movement forward.

Right Scale: Thousands of companies depend on RightScale every day to run business-critical applications in the cloud (http://www.joyent.com/). The RightScale Cloud Management Platform offers significant advantages over alternative cloud management approaches. Whether you are just getting started and need a simple on-ramp to the cloud or require support for complex deployments spanning multiple clouds, RightScale provides complete automation yet gives you the flexibility, control, and portability you need. The RightScale Cloud Management Platform works with cloud infrastructures to deliver a more productive environment for managing your cloud deployments. If you are just getting started and need a fast on-ramp to deploying in the cloud, RightScale provides cloud-ready ServerTemplates and best practice deployment architectures. If you require support for complex deployments, RightScale delivers complete automation that reduces complexity and administrative burden, yet gives you the flexibility, control, and portabil-

Figure 4. RightScale cloud management platform

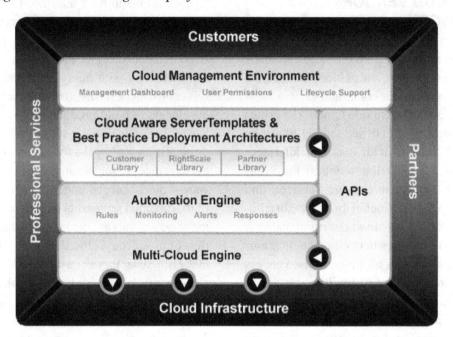

ity you need. If you plan to use multiple clouds, RightScale enables you to manage deployments across multiple clouds – private, public, or hybrid. The RightScale Cloud Management Platform is shown in Figure 4.

Jclouds: is an open source library that helps you get started in the cloud and reuse your java development skills (https://www.cloudkick.com/). It is an OSS JAVA Framework for cloud enablement and inter cloud communication. The api allows you to freedom to use portable abstractions or cloud-specific features. It supports many clouds including Amazon, VMWare, Azure, and Rackspace. Key value storage include: global name space; key, value with metadata; http accessible; sites on demand; and unlimited scaling.

δ•CLOUD: Start an instance on an internal cloud, then with the same code start another on EC2 or Rackspace (http://www.dasein.org/). Deltacloud protects your apps from cloud API changes and incompatibilities, so you can concentrate on managing cloud instances the way you want. Deltacloud API and Aggregator are free and open source (LGPL, GPL). Deltacloud Core gives you: REST

API (simple, any-platform access); Support for all major cloud service providers; and Backward compatibility across versions, providing long-term stability for scripts, tools and applications. Deltacloud Aggregator provides a web UI in front of the Deltacloud API. With Deltacloud Aggregator, users can view image status and stats across clouds, all in one place; migrate instances from one cloud to another; and manage images locally and provision them on any cloud. The Deltacloud Core makes it easy for cloud providers to add their cloud to the Deltacloud common API.

Cloudkick: is a best way to manage infrastructure on the cloud, or in the data center (ACM Symposium on Cloud Computing, 2010). Fast set-up, robust tools, free updates, and predictable pricing are a few reasons why companies are using Cloudkick to manage their servers. The Cloudkick agent gives you the ability to execute simple monitoring scripts. So, in addition to a multitude of built-in checks, you can monitor anything you want, including application-specific metrics like the number of users logged into your site. Cloudkick's REST API lets you query information about your servers and monitors quickly and easily, giving you the ability to integrate Cloudkick into your existing tools and applications. Cloudkick instantly consolidates all your infrastructure management into one dashboard, eliminating the time and hassle involved in maintaining a host of different tools. See a human-friendly overview of all your servers that includes: name; status; IP address; performance guages for CPU, Memory, and Disk space; and controls to perform simple actions like opening a terminal or rebooting a machine. Adding servers from any of our 8 supported clouds EC2, Rackspace, Slicehost, Linode, GoGrid, SoftLayer, VPS.NET, Rimuhosting takes seconds – just enter your API keys and you're good-to-go. Adding physical servers is also a snap; just install the Cloudkick Agent.

Dasein.org: Dasein is a suite of Open Source software (Third International Conference on Cloud Computing, 2010). Dasein began in 1996 as an effort by Valtira's founder, George Reese, to create utilities for the developer of enterprise Java systems. Dasein has grown into a place where you can find all kinds of developer tools and other software under an Open Source license. The Dasein Persistence Library is a suite of Java classes that enable you to create EJB-like components without the complexity of EJBs and without the inflexibility and performance issues of automated persistence systems. The Dasein Java utilities address common needs among creators of enterprise Java applications. They include internationalization tools, enhancements to Java's calendar API, type-safe meta-data management, and high-performance cache management.

We can summarise that, Java EE 6 is lightweight, flexible, easy-to-use; GlassFish Server Open Source Edition 3.0 and Oracle GlassFish Server 3.0 provides feature-rich implementation and Java EE 6 applications can be easily deployed on Amazon, RightScale, Elastra, Joyent, and other clouds

CLOUD DATABASES: PREPARING YOUR DATA

Cloud database usage patterns are evolving, and business adoption of these technologies accelerates that evolution (International Conference on Cloud Computing, 2010). Consumer centric cloud database applications have been evolving with the adoption of Web 2.0 technologies. User generated content, particularly in the form of social networking, have placed somewhat more emphasis on updates. Dynamic scalability one of the core principles of cloud computing has proven to be a particularly vexing problem for databases (Barry, n. d.). The reason is simple; most databases use a shared nothing architecture. The shared nothing architecture relies on splitting / partitioning the data into separate silos of data, one per server. Since data partitioning and cloud databases are inherently incompatible, Amazon, Facebook and Google have taken another approach to solve the cloud database challenge. They have created a persistence engine technically not a database that abandons typical ACID compliance in favor replicated tables of data that store and retrieve information while supporting dynamic or elastic scalability. The database architecture called shared disk, which eliminates the need to partition data, is ideal for cloud databases (http://en.wikipedia.org/wiki/Relational_database). Whether you are assembling, managing or developing on a cloud computing platform, you need a cloud compatible database. As with every tectonic shift in technology, there is a Darwinian ripple effect as we realize which technologies support these changes and which are relegated to legacy systems. Because of their compatibility, cloud computing will usher in an ascendance of the shared disk database.

RELATIONAL DATABASES

A relational database matches data by using common characteristics found within the data set. The resulting groups of data are organized and are much easier for many people to understand. Relational databases are facilitated through Relational Database Management Systems (RDBMS). Almost all database systems we use today are RDBMS, including those of Oracle, SQL Server, MySQL, Sybase, DB2, TeraData, and so on (Oppel, 2004). Relational databases are currently the predominant choice in storing financial records, medical records, manufacturing and logistical information, personnel data and much more (Churcher, 2007). Some of the Pros and Cons of Relational Databases are discussed here. The relational model for databases provides the basic DBMS characteristics. The relational model for databases described by Dr. Codd contains 12 rules. DBMSs vary in the way in which they comply with these rules; however, commercial relational databases generally conform to these rules (Varley, 2009). The strengths of RDBMS are: Flexible and well-established; Sound theoretical foundation and use over many years has resulted in stable, standardized products available; Standard data access language through SQL; Costs and risks associated with large development efforts and with large databases are well understood; and The fundamental structure, i.e., a table, is easily understood and the design and normalization process is well defined. Thus, Data Integrity, ACID Capabilities, High Level Query Model and Data Normalization are the plus points. Data Independence The weaknesses of RDBMS are: Performance problems associated with re-assembling simple data structures into their more complicated real-world representations; Lack of support for complex base types, e.g., drawings; SQL is limited when accessing complex data; Knowledge of the database structure is required to create ad hoc queries; and Locking mechanisms defined by RDBMSs do not allow design transactions to be supported, e.g., the "check in" and "check out" type of feature that would allow an engineer to modify a drawing over the course of several working days Thus, Scaling Issues (By Duplication-Master-Slave; By Sharding/Division

-Not transparent), Fixed Schema, Mostly disk-oriented (Performance), and May fair poorly with large data are the minus points.

The non-relational DBMS are a new kind of Databases for handling Web Scale and differ a lot regarding their data model and therefore the conceptual data design will also differ a lot (Lewis, 2008). Scalability, Replication and Availability, Performance, Deployment Flexibility, Modelling Flexibility and Faster Development are the added advantages in favour of this (Hogan, 2008). At the same time they are not friendly as for as the following points are concerned. Lack of Transactional Support, Data Integrity is Application's responsibility, Data Duplication/Application Dependent, Eventually Consistent (mostly), No Standardization and being New Technology.

Today, the database remains the central application in the entire, Worldwide Realm of Enterprise IT. The Worldwide Web is the next truly transformational development within the realm. In this context be sure that Cloud Computing will improve enterprises and lives. The first distinction I make is between the Enterprise Cloud and the Consumer Cloud. I see the distinction between Public Cloud and Private Cloud as being "a simple matter of execution," as we say. These are both subsets of the Enterprise Cloud phylum. I do understand that there is a substantial Social Networking dimension to enterprises today, and Social Networking was created to screw, or, serve consumers. I also realize that several major businesses today leverage Enterprise Cloud technology to deliver consumer services-think Apple. As with RDBMS a generation ago, everybody needs a Cloud Computing strategy today. Cloud Capable RDBMS are, MySQL, Oracle DB, PostgreSQL, IBM DB2 etc. and almost every RDBMS can run in an IAAS Cloud Platform. However Microsoft SQL services and Amazon web services are basically Cloud Native RDBMS.

Types of Non-Relational DBMS include Key Value Stores, Document Stores, Column Stores and Graph Stores. The Key Value Data-Bases consist of Object that is completely Opaque to DB, supported by mostly GET, PUT & DELETE operations, and may have limits on size of Objects. Recently, a lot of new non-relational databases have cropped up both inside and outside the cloud. One key message this sends is, "if you want vast, on-demand scalability, you need a non-relational database. The inspiration is from the Amazon Dynamo. Reliability at massive scale is one of the biggest challenges Amazon.com face. It is one of the largest e-commerce operations in the world; even the slightest outage has significant financial consequences and impacts customer trust. The Amazon.com platform, which provides services for many web sites worldwide, is implemented on top of an infrastructure of tens of thousands of servers and network components located in many data centers around the world. At this scale, small and large components fail continuously and the way persistent state is managed in the face of these failures drives the reliability and scalability of the software systems. Amazon Dynamo, the next generation of virtual distributed storage, is a highly available key-value storage system that some of Amazon's core services use to provide an "always-on" experience. To achieve this level of availability, Dynamo sacrifices consistency under certain failure scenarios. It makes extensive use of object versioning and application assisted conflict resolution in a manner that provides a novel interface for developers to use.

Key Value DataBases on Cloud can be seen in, Amazon S3, Project Voldemort, Redis, Scalaris, MemCachedDB and Tokyo Tyrant. In the case of Document Data-Bases, Object is not completely opaque to DB, Every Object has its own schema (FirstName="Bob", Address="5 Oak St.", Hobby="sailing". FirstName="Jonathan", Children=("Michael,10", "Jennifer,8")), Can perform queries based on Object's attributes, Possible to describe relationships between Objects, Joins and Transactions are not supported, and Good for XML or JSON objects. Some of the cloud examples with this are Amazon SimpleDBTM

BETA, CouchDB relax and riak. Column-Store Data-Bases are Richer than Document Stores, has Multi-Dimensional Map (Tables-Row, Column, Time-Stamp), Supports Multiple Data Types and usually use an Underlying DFS (Browne, 2009). This is inspired by Google Big Table. Bigtable is a distributed storage system for managing structured data that is designed to scale to a very large size: petabytes of data across thousands of commodity servers. Many projects at Google store data in Bigtable, including web indexing, Google Earth, and Google Finance. These applications place very different demands on Bigtable, both in terms of data size (from URLs to web pages to satellite imagery) and latency requirements (from backend bulk processing to real-time data serving). Despite these varied demands, Bigtable has successfully provided a flexible, high-performance solution for all of these Google products. Bigtable gives clients dynamic control over data layout and format (http://www.eucalyptus.com/). Column-Store Data-Bases on Cloud include Google App Engine, Cassandra, H·BASE, and HYPERTABLE.

While Making a Choice the fallowing Key Factors should be considered. Application Architecture Requirements, Platform choices and Non-Functional Requirements- Consistency, Availability, Partition, Security and Data Redemption. At the same time we must look into Different Solutions for Different Requirements. In the case of Feature First, Corporate Data, Consistency Requirements, Business Intelligence and Legacy Application must be looked in. Such RDBMS can be seen on Amazon Cloud, RackSpace (IaaS) or Microsoft Azure/Amazon RDS (PaaS). Second in the case of Consumer Facing Application, Big Files (Images, BLOB's, Files), Geographical Distribution, Mostly writes and Not heavy requirement on Rich Queries should be addressed. This can be seen in Key-Value Data Stores (Amazon S3, Project Voldemort, Redis). Third in the case of Hundreds of Government Documents with different schemas, we need to serve on Web with Data Mining as seen in Document Data-Stores (Amazon

Figure 5. Mixing and matching of earlier scenarios

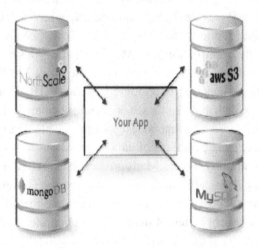

SimpleDB, Apache Couche DB, MongoDB). If we lok at Scale First Scenario, then Huge Data-Set, Analytical Requirements, Consumer Facing and High Availability over Consistency must be addressed as in Column Data-Stores (Google App Engine, Hbase, Cassandra). The other way around is Mix and Match of Earlier Scenarios as given in Figure 5.

Polyglot Persistence cover, RDBMS for low-volume and high value, Key-Value DB for large files with little queries, Memcached DB for short-lived Data and Column DB for Analytics. Finally we need to look at the CAP theorem (http://open.eucalyptus.com). Also known as Brewer's theorem, it states that it is impossible for a distributed computer system to simultaneously provide all three of the following guarantees: Consistency (all nodes see the same data at the same time); Availability (node failures do not prevent survivors from continuing to operate); and Partition Tolerance (the system continues to operate despite arbitrary message loss). According to this theorem, a distributed system can satisfy any two of these guarantees at the same time, but not all three. Hence we need to choose any two based on our requirements as given in Figure 6.

Finally we need to conclude that, One Size does not Fit all, Have many choices, No-SQL DB's providing Alternatives, and RDBMS serve useful purpose. This suggest that lot of work needs to be done in preparing our Data for Cloud applications and emphasis great potential for the

Figure 6. CAP theorem

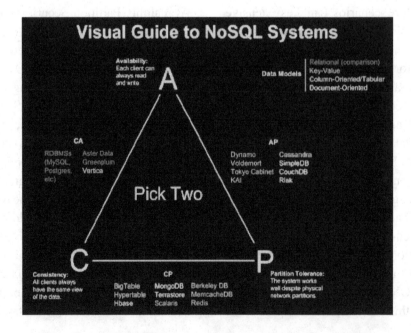

cloud research and development. We can summarise that, one size does not fit all; many choices; No-SQL DB's providing Alternatives; and RDBMS serve useful purpose.

OPEN SOURCE PRIVATE CLOUD COMPUTING

Options for Private Cloud in Open Source: Eucalyptus, ubuntu, enterprise cloud, opennebula.org, nimbus, redhat, etc.

Eucalyptus is Elastic Utility Computing Architecture Linking Your Programs to Useful Systems - is an open source soft ware infrastructure for implementing on-premise clouds on existing Enterprise IT and service provider infrastructure (Springboard Research, 2010a). Eucalyptus Systems delivers private cloud software. This is infrastructure software that enables enterprises and government agencies to establish their own cloud computing environments. With Eucalyptus, customers make more efficient use of their computing capacity, thus increasing productivity and innovation, deploying new applications faster, and protecting sensitive data, while reducing capital expenditure. Eucalyptus is an open source solution that originates from an NSF funded research project at University of California, Santa Barbara. The open source model is generally considered a superior way of creating infrastructure software. Innovation is faster, users and customers have more freedom and flexibility, lock-in is avoided, and secondary benefits accrue from the massive ecosystems that naturally evolve around the most prolific open source products. Users can access to the eucalyptus cloud by HybridFox (Firefox plugin); ElastDream (stand alone client); Euca2ools (Eucalyptus command line tools). These clients have to be configured with user certificates, secret keys. Eucalyptus Benefits: Data center optimization; Automated self service provisioning; Web services based; Elastic resource provisioning. In summary it Provides private cloud Framework

for IT data center; Compatible with EC2 and S3; Open source, transparent. Eucalyptus was based on open-source software components that are used without modification. There are two advantages that it immediately enjoys as a result of this decision. First, because they do not make changes to these packages, they automatically incorporate any improvements or upgrades that the open-source community has contributed. Secondly, by choosing open-source components that are common to most freely available Linux distributions, Eucalyptus can run in pre-existing Linux installations without the addition or modification of the installed software infrastructure. The installation and maintenance of a software platform like Eucalyptus had to be made as easy and transparent as possible. Thus they designed the system to install and behave more as a Linux "tool" (from the system administrator's perspective) rather than a separate platform. Finally, they recognized that the cloud platform itself should be modularized so that it could support multiple APIs and virtualization environments. Thus while the initial versions of the system supported the AWS APIs and Xen hypervisor platform, Eucalyptus is designed to support multiple APIs and virtualization environments simultaneously, in the same cloud.

Eucalyptus enables hybrid cloud and private cloud deployments for enterprise data centers and requires no special purpose hardware or reconfiguration (Springboard Research, 2010b). Leveraging Linux and web service technologies that commonly exist in today's IT infrastructure, Eucalyptus allows customers quickly and easily to create computing clouds "on premise" that are tailored to their specific application needs. At the same time, Eucalyptus supports the popular AWS cloud interface allowing these on-premise clouds to interact with public clouds using a common programming interface. Along with managing virtual machines, the technology supports secure virtualization of the network as well as the storage infrastructure within the cloud environment.

Eucalyptus is compatible with and packaged for multiple distributions of Linux including Ubuntu, RHEL, OpenSuse, Debian, Fedora, and CentOS and will work with a variety of hypervisors and virtualization technologies.

At Eucalyptus Systems, we develop enterprise-grade technology solutions that build upon the open - source Eucalyptus software core. Eucalyptus technology is quickly becoming the standard for on - premise cloud computing, delivering the cost efficiencies and scalability of clouds with the security and control that comes with an organization's own IT infrastructure. Enterprise Eucalyptus can provide capabilities such as end - user customization, self-service provisioning, legacy application support, customized "service level agreements" (SLAs), cloud monitoring, metering, and support for auto-scaling, as a highly available cloud platform. Eucalyptus was designed from the ground up to be easy to install and as non-intrusive as possible. The software framework is a highly modular cooperative set of web services that interoperate using standard communication protocols. Through this framework it implements virtualized machine and storage resources that are interconnected by an isolated layer-2 network. From a client application and/or user perspective, the cloud API is compatible with Amazon's AWS (both SOAP and REST interfaces are supported) although other interfaces are available as customizations.

Eucalyptus Components

Each Eucalyptus service component exposes a well defined language agnostic API in the form of a WSDL document containing both the operations that the service can perform and the input/output data structures (Springboard Research, 2010b). Inter-service authentication is handled via standard WS-Security mechanisms. There are five high level components, each with its own Web-service interface, that comprise a Eucalyptus installation (Figure 1). The components within the

Eucalyptus system are: Cloud Controller (CLC is the entry-point into the cloud for administrators, developers, project managers, and end users); Cluster Controller (CC generally executes on a cluster front end machine); Node Controller (NC is executed on every node that is designated for hosting VM instances); Storage Controller (SC implements block accessed network storage); Walrus (put/get storage that allows users to store persistent data); and Management Platform (provides an interface to various Eucalyptus services and modules). With these components, Eucalyptus (Figure 7) can be configured to support a wide variety of infrastructure features and topologies.

CLC is the Cloud Controller which virtualizes the underlying resources (servers, storage, and network). The Cluster Controllers (CCs) form the front-end for each cluster defined in the cloud. NCs are the machines on which virtual machine instances run. The Storage Controller (SC) provides block storage service (similar to Amazon EBS) while the Walrus storage system spans the entire cloud and is similar to the Amazon S3 in functionality. A Management Platform provides a one-stop console for the cloud administrator to configure and manage the cloud. The Management Platform also exports various interfaces for the administrator, project manager, developer, and other users, with customizable levels of access and privileges.

Benefits of Eucalyptus Cloud

The features so essential to improving the efficiency of an IT infrastructure include the following: Data center optimization; Automated self service; Web services based; Scalable data center infrastructure; Elastic resource provisioning; Open source innovation; and Hybrid cloud capability.

- It enables virtualization of servers, network, and storage in a secure manner, thereby reducing the cost, increasing the

Figure 7. Eucalyptus

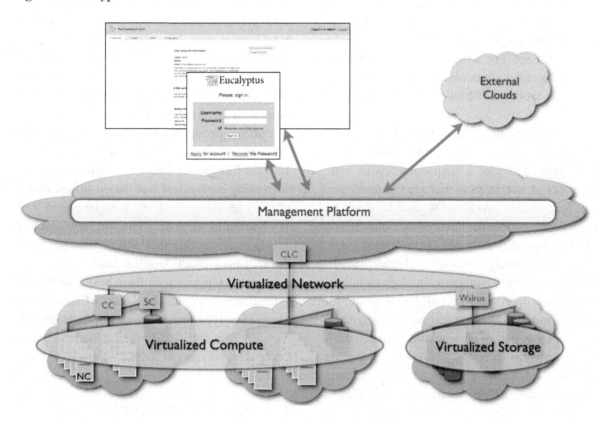

ease of maintenance, and providing user self-service.

• The modular design of Eucalyptus enables a variety of user interfaces, bringing the benefits of virtualization technology to a broad range of users (admins, developers, managers, hosting customers) and provides a platform for service providers to devise profitable consumption-based pricing models.

• VM and Cloud snapshot features provide an exhaustive set of opportunities to improve cluster reliability, template manipulation, and automation. This makes the cloud easy to use, reduces learning time for the average user, and reduces the turnaround time for projects.

• Leverages existing virtualization technology, supports Linux-based operating sys-

tems, and supports multiple hypervisors. Ease of cluster/availability-zone management provides the administrator/user with a host of opportunities to form logical groups of servers, storage, and networks on a per-project/per-user/per-customer basis.

• The core Eucalyptus framework will continue to remain open-source. This provides the users access to the source code as well as an opportunity to leverage the contributions from a world-wide community of developers.

• The ongoing effort to develop public cloud compatible interfaces is a unique advantage for users who intend to cloudburst to other public clouds (also known as hybrid clouds - a private on-premise cloud, in this case, a Eucalyptus cloud, working seamlessly with a public cloud).

- A vibrant ecosystem built around the Amazon AWS can be leveraged. For example, RightScale, CohesiveFT, Zmanda, rPath are just a few of the partners that deliver solutions for Amazon AWS that in turn work seamlessly with Eucalyptus.

Thus, focusing on the cost advantage for IT infrastructure and data center management, Eucalyptus provides a unique framework with a variety of interfaces to manage the resources. The hardware, network, and storage can easily be consolidated under the Eucalyptus cloud, hiding the heterogeneity in hardware, software stack, policies, and configurations. Eucalyptus Systems develops enterprise grade technology solutions built on the open source Eucalyptus software for private and hybrid cloud computing. Originally developed as part of an academic research project, Eucalyptus technology is quickly becoming the standard for on-premise cloud computing, delivering the cost efficiencies and scalability of cloud architecture with the security and control of deploying on an organization's own IT infrastructure. Eucalyptus Systems' mission is to support the open source Eucalyptus platform and to deliver private and hybrid cloud computing solutions for large-scale enterprise deployments.

CLOUD COMPUTING IN INDIA

India is growing at faster pace in information technology sector thereby showing a great potential for the cloud computing services. According to Springboard Research report (June 2010) SAAS India i.e., software as a service in India will register a compounded annual growth rate of 76% in the time period of 2007-2011 (Popli & Rao, 2009). Cloud computing services has huge opportunity in Indian market due to the large number of Small and Medium businesses (SMBs) which is at around 35 million and they want easy to use, reliable and scalable application that helps them to grow and

expand their business (Rittinghouse & Ransome, 2009). This makes India as the fastest growing SAAS market in Asia Pacific region. According to Jeremy Cooper, VP-Marketing (APAC), Salesforce.com 'software as a service' provider started its services in India in September 2005 and since then the adoption rate of cloud computing is increasing. Seeing SAAS success on September 2008 IBM launched cloud computing center in India at Bangalore. This center will cater to the increasing demand of web based infrastructure sharing services. IBM India collaborated with IIT Kanpur to come up with some new developments in computing that will help in academic advancement. Bharti Airtel has launched the cloud computing services with their NetPc model and other giant companies like Reliance Communications, TCS, HCL technologies, Wipro, Netmagic, Verizon, Novatium etc. have also launched cloud computing services in India.

CLOUD COMPUTING FOR THE SMALL AND MEDIUM ENTERPRISES IN INDIA

SMEs are said to be the lifeblood of any vibrant economy. They are known to be the silent drivers of a nation's economy. SMEs are leading the way for entering new global markets and for innovations in the emerging economic order. In India 95% of the industrial units are SMEs which give over 50% of the industrial output (Rao, 2010). Thus SMEs form the backbone of the Indian economy. In the cloud computing environment the SMEs will not have to own the infrastructure so they can abstain from any capital expenditure and instead they can utilize the resources as a service and pay as per their usage of the resources provided by the cloud (Aggarwal & Barnes, 2010). SaaS will provide an opportunity for the SMBs to automate their business by reducing their investment in IT infrastructure (D'Monte, 2010). Cloud based services helps the industries to reduce their cost that are involved in on-premise ERP solutions such as hardware, software, upgradation, training and licensing costs. Moreover long implementation cycles with regular maintenance costs adds to the total cost of traditional ERP (Emerge Forum, 2009). According to V Ramaswamy, SMB global head (TCS), SMBs are in need of easy to use technology as reported in Business standard, Jan 2010 (D'Monte, 2010). With the changing needs and increase of customer base there is requirement of CRM and ERP solutions. As technology changes companies requires up gradation in their software this poses obstacles for the SMEs to scale up (Sharma, 2009). In order to operate in limited budget a less complicated and simplified offering is required. At present most of the Indian and Foreign IT companies are focusing on SMEs for their cloud computing offerings (D'Monte, 2010). According to D'Monte (2010), Cloud computing is providing huge opportunities for the Indian IT company that is helping them to develop cost effective business models. Such models help the SMEs to uplift their business in an effective and cost efficient manner. The promoter of 'The India Cloud Initiative' Mukhi said that there is a huge saving of money by using cloud technology as the industries have to pay only for the operating cost (http://www.rightscale.com/products/cloud-computing-uses/cloud-computing-for-education.php). The biggest advantage of a hosted model (cloud computing) is that it eradicates the need to purchase the software licenses and also eliminates the cost associated with developing and operating in-house applications. In a hosted model, the capital investment, security, backup and server maintenance costs are all the provider's responsibilities."

EDUCATION CLOUD PROGRAMS AND APPLICATIONS

Learning institutions can leverage the scalable and robust cloud computing platform to put their computing resource intensive courses on the cloud (http://aws.amazon.com/education/). Education cloud also allows students to study outside the classroom at their own pace. This platform will also provide teachers with more flexibility in designing courses and interacting with students online. The solution is also expected to improve course materials through collaboration. IT administrators can use the new cloud computing application to optimise resources, reduce costs and enhance productivity. Such initiatives will also provide a new training option for local enterprises on top of its existing services. IBM says the institute will now be able to offer more flexibility and efficiency for knowledge sharing. One of IBM's top social priorities is improving education all around the world. IBM has deep market insights and strong industry domain knowledge which help to start building an enabling platform for individualised lifelong learning and they believe this is crucial in enabling people to adapt and excel in a 21st century smarter economy. Following are some of the most powerful education cloud programs that have emerged recently and look forward a big boom in this new sector.

RightScale education program (Sasikala, in press) provides free access to the RightScale Cloud Management Platform to institutions that want to harness the scalable, on-demand power of cloud computing. RightScale enables:

- Fast on-ramp to the cloud with cloud-ready ServerTemplates
- Automation that reduces complexity and administrative burden
- Transparency into underlying architectures, scripts and machine images
- Account management system with separate accounts that can be used by student project or group

RightScale's automated system management, pre-packaged and reusable components and recommended practices are proven as best-of-breed.

The RightScale Cloud Management Platform currently supports Amazon Web Services in the U.S. and in Europe, Eucalyptus, Flexiscale, GoGrid and Rackspace. With the RightScale Platform, any university or educational laboratory can tap the enormous power of cloud computing for a virtually infinite, cost-effective, pay-as-you-go IT infrastructure. We're currently evaluating educational institutions for participation in this program that offers free usage of the RightScale Website or Grid Editions.

RightScale is committed to supporting educational uses of cloud computing (Sasikala, in press). Working with educational institutions will support new applications for cloud computing, and offer the next generation more options for deploying their scientific or business applications. The RightScale Education program provides RightScale's Cloud Management Platform to institutions that want to harness the scalable, on-demand power of Amazon Web Services cloud computing. Sonian provides educational institutions with a secure, scalable and affordable Hosted Email Archiving and eDiscovery service powered by the Amazon Web Services cloud. The Sonian Archive Service supports all major messaging systems and is designed to help educational institutions meet there growing compliance, litigation discovery and storage management needs around email. VMLogix education program for universities and education institutes provides virtual lab management software which enables sophisticated management and automation capabilities to manage a virtual lab (including infrastructure, configuration environments and user management) running on the Amazon Web Services cloud. Capabilities include the ability to self-serve and build, snapshot, share and deploy complex n-tier machine environments on-demand, the ability to manage and orchestrate virtual computer training labs and to drive a dramatic reduction in the manual effort, time and IT resources required to develop and maintain higher quality software applications.

The UCSB Computer Science Department has created the EUCALYPTUS project to foster community research and development of Elastic/Utility/Cloud service implementation technologies, resource allocation strategies, service level agreement (SLA) mechanisms and policies, and usage models. RightScale and Eucalyptus have partnered to help make cloud computing simple and accessible to everyone from universities and students to entrepreneurs and enterprises evaluating large cloud deployments. Harvard Medical School is using RightScale as part of their "Translational Science in the Cloud" seminar. The participants will conduct a series of exercises in biomedical discovery and translational science using cloud computing technology. To accomplish these objectives, teams will create, manage and use a "translational research laboratory" on the cloud based on emerging cloud technology and services. Teams at Harvard will use RightScale to manage their server deployments running on the Amazon Elastic Compute Cloud (Amazon EC2).

Amazon Web Services in Education (http://aws.amazon.com/education/) provides a set of programs that enable the worldwide academic community to easily leverage the benefits of Amazon Web Services for teaching and research. With AWS in Education, educators, academic researchers, and students can apply to obtain free usage credits to tap into the on-demand infrastructure of Amazon Web Services to teach advanced courses, tackle research endeavors and explore new projects – tasks that previously would have required expensive up-front and ongoing investments in infrastructure. With AWS you can requisition compute power, storage, database functionality, content delivery, and other services — gaining access to a suite of elastic IT infrastructure services as you demand them. AWS enables the academic community to inexpensively and rapidly build on global computing infrastructure to pursue course projects and accelerate their productivity and research results, while enjoying the same benefits of reliability, elasticity, and cost-effectiveness used by industry. The AWS in Education program offers:

- Teaching Grants for educators using AWS in courses (plus access to selected course content resources)
- Research Grants for academic researchers using AWS in their work
- Project Grants for student organizations pursuing entrepreneurial endeavors; Tutorials for students that want to use AWS for self-directed learning
- Solutions for university administrators looking to use cloud computing to be more efficient and cost-effective in the university's IT Infrastructure

Educators: AWS provides a cost-effective way to teach courses in distributed computing, artificial intelligence, data structures, and other compute and storage-intensive subject matter. In the past, such courses would have required extensive hardware and network infrastructure. Now, it's merely a matter of providing each student with access to the global computing infrastructure and storage capacity of the AWS cloud.

Researchers: AWS in Education will review and support selected research projects with grants that offer free access to most AWS infrastructure services. Often, large research projects require extensive compute power and storage infrastructure to complete. Now, researchers around the world have access to the global computing infrastructure and storage capacity of the AWS cloud. Instead of purchasing a large amount of hardware, researchers can get started by simply opening an AWS account. And, with services like Amazon Elastic MapReduce, much of the heavy lifting of provisioning and configuring Hadoop clusters for data-intensive processing is eliminated.

Students: AWS in Education is proud to support student organizations around the world and compelling entrepreneurial student initiatives including Project Olympus at Carnegie Mellon,

Teams in Engineering Service at the University of California, San Diego, and the "3 Day Start Up" event at the University of Texas, Austin. AWS provides Project Grants supporting free usage of AWS to student organizations and student entrepreneurial projects. For students wanting to use AWS for self-directed learning, getting started is easy with our tutorials on asynchronous messaging, consensus algorithms with Amazon EC2, priority queues with Amazon SQS, and Representational State Transfer with Amazon S3.

Education IT: AWS in Education offers solutions for university administrators looking to use cloud computing to be more efficient and cost-effective in the university IT infrastructure. Organizations in the AWS ecosystem, including Independent Software Vendors (ISVs) and Systems Integrators (SIs), are building applications and services on AWS for education customers. AWS in Education works with many ISVs and SIs to bring solutions for common education infrastructure challenges like storage, disaster recovery, archiving and content delivery.

HyperStratus is a Silicon Valley-based consulting firm offering cloud computing education, strategy, application architecture, and project management and implementation services. HyperStratus clients include Sun Microsystems, Applied Biosystems, and the Silicon Valley Education Foundation (SVEF). Moonwalk is a privately owned company specializing in large-scale data management solutions. Moonwalk is used by major organizations across the Banking, Healthcare, Government, Research and Development, Education and Aerospace industries.

GREEN COMPUTING AND CLOUD COMPUTING

Nowadays, green computing and cloud computing are the hot topic everywhere. However, what is called Green Computing and Cloud Computing? Green Computing is the study and practice of using computing resources efficiently (Wikipedia, n. d.). There are some approaches to green computing, include the way of algorithmic efficiency, virtualization of computer resources, terminal servers, power management, power supply, storage, Video Card, Display, Operating System issues, Materials recycling, Telecommuting, and etc. (Raftery, 2008). We are moving into a new era of technology with all of these concepts. I think that companies should start taking advantage of it and start developing applications both green and cloud hosted. I have always believed that IT is the engine of an efficient economy; it also can drive a greener one.

Energy efficiency is increasingly important for future information and communication technologies (ICT), because the increased usage of ICT, together with increasing energy costs and the need to reduce green house gas emissions call for energy-efficient technologies that decrease the overall energy consumption of computation, storage and communications. Cloud computing has recently received considerable attention, as a promising approach for delivering ICT services by improving the utilization of data centre resources. In principle, cloud computing can be an inherently energy-efficient technology for ICT provided that its potential for significant energy savings that have so far focused on hardware aspects, can be fully explored with respect to system operation and networking aspects.

Cloud providers are going to make Green claims. I can easily see Google and Amazon competing on that basis. The potential of ICT technologies and cloud computing to drive low-carbon economic growth underscore the importance of building cloud infrastructure in places powered by clean renewable energy. Companies like Facebook, Google, and other large players in the cloud computing market must advocate for policy change at the local, national and international levels to ensure that, as their appetite for energy increases, so does the supply of renewable energy. More cloud-computing companies are pursuing design

and siting strategies that can reduce the energy consumption of their data centres, primarily as a cost containment measure. For most companies, the environmental benefits of green data design are generally of secondary concern. There are two aspects in any discussion of Cloud Computing and Green IT. The first perspective is that of energy savings which there is some debate about the perspective that we will examine is how Cloud Computing can enable and enhance the operations of 'Green' Energy networked devices such as smart meters and wind turbines (Raftery, 2008).

CONCLUSION

Cloud computing appears poised to further increase its profile among enterprise users over the next few years. CIO's and other tech decision makers are interested in expanding their use of the cloud, and IT vendors and solution providers are increasing their ability to meet this demand. Clearly there is growing momentum behind cloud computing, evidenced by climbing adoption rates and greater awareness. But cloud computing adoption is still nascent. The year ahead will be one of evaluation, trial and error and, most importantly, opportunity. The market will sort through the role IT channel companies will play, as well as best business models, sales and marketing strategies and most relevant technologies. The need to identify better ways to deliver IT services requires a serious look at increasingly strategic uses of cloud services. We believe working with educational institutions will support new applications for cloud computing and offer the next generation more options for deploying their scientific or business applications. The new concept of green computing with cloud adds a new dimension in our progress towards saving earth as well as our self.

REFERENCES

Aggarwal, B. B., & Barnes, M. (2010). *The case for cloud CRM in India*. Beijing, China: Springboard Research.

Barry, D. K. (n. d.). *Database concepts and standards*. Retrieved from http://www.service-architecture.com/database/articles/index.html

Browne, C. (n. d.). *NonRelational database systems*. Retrieved from http://linuxfinances.info/info/nonrdbms.html

Browne, J. (2009). *Brewer's CAP theorem*. Retrieved from http://www.julianbrowne.com/article/viewer/brewers-cap-theorem

Chinnici, R. (2008). *Profiles in the Java EE 6 platform*. Retrieved from http://weblogs.java.net/blog/robc/archive/2008/02/profiles_in_the_1.html

Churcher, C. (2007). *Beginning database design: From novice to professional*. New York, NY: Apress.

D'Monte, L. (2010). *Good opportunity for Indian IT firms*. Retrieved from http://business.rediff.com/column/2010/jan/05/guest-cloud-computing-good-opportunity-for-indian-firms.htm

D'Monte, L. (2010). *Work in the cloud*. Retrieved from http://www.business-standard.com/india/news/work-incloud/381619/

Emerge Forum. (2009). *India market and the SaaS/cloud computing landscape*. Retrieved from http://emerge.nasscom.in/2009/08/india-market-and-the-saascloud-computing-landscape/

Geelan, J. (2009). *The top 150 players in cloud computing*. Retrieved from http://cloudcomputing.sys-con.com/node/770174

Goncalves, A. (2010). *Beginning Java™ EE 6 Platform with GlassFish™ 3: From novice to professional*. New York, NY: Apress. doi:10.1007/978-1-4302-2890-5

Hamilton, M. (2009). *How green is cloud computing?* Retrieved from http://blogs.sun.com/marchamilton/entry/how_green_is_cloud_computing

Hogan, M. (2008). *Cloud computing & databases: How databases can meet the demands of cloud computing*. Menlo Park, CA: ScaleDB Inc.

Lewis, C. (2008). *Databases in the cloud*. Retrieved from http://clouddb.blogspot.com/

Marks, E. A., & Lozano, B. (2010). *Executive's guide to cloud computing*. New York, NY: John Wiley & Sons.

Murty, J. (2010). *Programming Amazon Web Services: S3, EC2, SQS, FPS, and SimpleDB*. Sebastopol, CA: O'Reilly Media.

Oppel, A. (2004). *Databases DeMYSTiFieD* (2nd ed.). New York, NY: McGraw-Hill.

Popli, G. S., & Rao, D. N. (2009). *An empirical study of SMEs in electronics industry in India: Retrospect & prospects in post WTO era*. Retrieved from http://papers.ssrn.com/sol3/papers.cfm?abstract_id=1459809

Raftery, T. (2008). *How green is cloud computing?* Retrieved from http://greenmonk.net/how-green-is-cloud-computing/

Rao, M. L. N. (2010). *SaaS opportunity for Indian channels*. Santa Clara, CA: Jamcracker Inc.

Rittinghouse, J. W., & Ransome, J. F. (2009). *Cloud computing implementation, management and security*. Boca Raton, FL: Taylor & Francis Group/CRC Press.

Sasikala, P. (2011). Cloud computing in higher education: Opportunities and issues. *International Journal of Cloud Applications and Computing, 1*(2), 1–13.

Sasikala, P. (in press). Cloud computing: Present status and future implications. *International Journal of Cloud Computing*.

Sasikala, P. (in press). Research challenges and potential green technological applications in cloud computing. *International Journal on Cloud Computing.*

Shachor, G. (2001). *Working with mod_jk.* Retrieved from http://tomcat.apache.org/tomcat-3.3-doc/mod_jk-howto.html

Sharma, P. (2009). *Is India ready for cloud computing?* Retrieved from http://dqchannels.ciol.com/content/space/109102902.asp

Springboard Research. (2010a). *Cloud computing in Asia Pacific 2010.* Beijing, China: Springboard Research.

Springboard Research. (2010b). *IT market in India 2009- 2013.* Beijing, China: Springboard Research.

Varley, I. T. (2009). *No relation: The mixed blessings of non-relational databases.* Austin, TX: University of Texas.

Wagner, J. (2010). *Cloud computing: Preparing for the next 20 years of IT.* Retrieved from http://www.worldsystems-it.com/?p=209

Wikipedia. (n. d.). *Green computing.* Retrieved from http://en.wikipedia.org/wiki/Green_computing

Wikipedia. (n. d.). *Service management facility.* Retrieved from http://en.wikipedia.org/wiki/Service_Management_Facility

This work was previously published in the International Journal of Cloud Applications and Computing (IJCAC), Volume 1, Issue 4, edited by Shadi Aljawarneh & Hong Cai, pp. 1-24, copyright 2011 by IGI Publishing (an imprint of IGI Global).

Chapter 17

A Framework for Analysing the Impact of Cloud Computing on Local Government in the UK

Jeffrey Chang
London South Bank University, UK

ABSTRACT

Cloud computing is hailed as the next-revolution of computing services. Although there is no precise definition, cloud computing refers to a scalable network infrastructure where consumers receive IT services such as software and data storage through the Internet like a utility on a subscription basis. With an increasing number of data centres hosted by large companies such as Amazon, Google, and Microsoft cloud computing offers potential benefits including cost savings, simpler IT, and reduced energy consumption. Central government and local authorities, like commercial organisations, are considering cloud-based services. However, concerns are raised over issues such as security, access, data protection, and ownership. This paper develops a framework for analysing the likely impact of cloud computing on local government and suggests an agenda for research in this emerging area.

INTRODUCTION

Cloud computing is held as the next revolution in the delivery of computing services. It offers enormous potential advantages to organisations such as cost savings, scalable computing services, simpler IT infrastructure, reduced energy consumption, and so on. In response to declining IT budgets and the lack of adequate skills, and as part of the e-Government agenda, cloud-based delivery models are rapidly gaining the attention of government IT leaders.

The primary driving factor compelling government organisations to consider cloud computing is the potential for capital and operational expenditure savings. Faced with the ongoing economic downturn public sector organisations are seriously considering cost saving strategies to

DOI: 10.4018/978-1-4666-1879-4.ch017

optimise computing resource. However, software applications; hardware, infrastructure, platforms, services and storage, or whether the government should develop its own cloud, are key issues that require careful consideration.

Concerns about the use of cloud computing in private organisations have already been raised. But this has not deterred the government from continuing developing the government cloud (G-Cloud). In the UK government has already begun the process of developing the G-Cloud which will be rolled out over the next few years. Key concerns relate to the security and ownership of data, the potential impact on employment and the structural and cultural implications of moving to cloud provision.

As yet, little research has been carried out on the implications for local government. This paper provides a framework for analysing the likely impact of cloud computing use within local authorities and suggests an agenda for research in this emerging area.

THE CONCEPT OF CLOUD COMPUTING

Although there is no universally agreed definition of cloud computing the term refers to a computing service provided via internet connections, a service that can be scaled up and down. It can mean a storage service; or it can be seen as a platform or as a software service. According to a group of researchers at Gartner cloud computing has five key attributes: service-based, scalable and elastic, shared, metered by use and using Internet Technology (Plummer *et al.,* 2009). Customers and providers of cloud services will consider any of above attributes or a combination of attributes to determine the expected services. Essentially it is a style of computing where IT capabilities are provided as a service delivered over the Internet to a customer's workplace, similar to utilities like water and electricity which are 'piped' to the customer's premises.

It can be argued that cloud computing evolves from and integrates a number of IT practices both private and public organisations have experienced over past years: outsourcing, software as service (SaaS), web-based storage etc. The development of cloud-based services has accelerated its pace in the last few years due to improved technologies and faster internet speed.

For service users cloud computing is an attractive alternative to building their own computing infrastructure, which can be cost efficient (Korri, 2009). The key advantages of cloud computing are held to be greatly reduced costs, increased efficiency and a significant reduction in energy consumption leading to cost savings and greener IT (Foster *et al.*, 2008; Luis *et al.,* 2008; Aymerich *et al.*, 2009; Grossman, 2009; Korri, 2009; Maggiani, 2009; Nelson, 2009). In the *Digital Britain* (2009) report, the UK government sees the adoption of cloud computing as critical to the success of its plans to increase efficiency in the public sector and is working with various suppliers to develop a dedicated G-Cloud for the delivery of all government services.

In the private sector, concerns have been expressed both about the security of data management and loss of organisational control of a key resource (Buyya *et al.*, 2009; Grossman, 2009). The confidential and sensitive nature of data stored in the public sector has made this issue particularly sensitive (Nelson, 2009). There is also concern about possible effects on employment caused by the introduction of centrally-run computer services. So far, in the public sector, there has been limited adoption of cloud computing.

THE CONCEPTUAL FRAMEWORK AND ITS APPLICATION

The conceptual framework (Figure 1) and application of cloud computing in local government draw

Figure 1. A conceptual framework of implementing cloud computing in local government

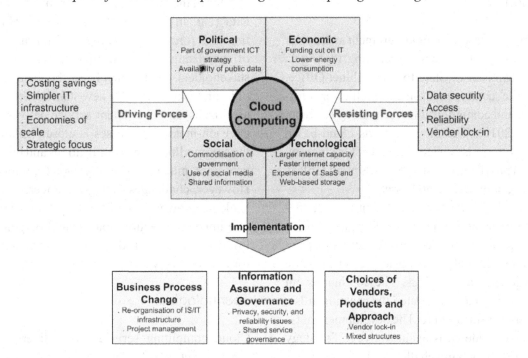

on a number of models for analysing the change process including Lewin's model (Lewin, 1947) and PEST. Key issues of cloud computing are outlined using the PEST framework (political, economical, social and technological) to give the general background and the relevance of cloud computing for government organisations. Through applying Kurt Lewin's change process model driving forces and resisting forces are identified.

The implementation of cloud computing involves business process change, information assurance and governance, and choices of vendors, products, platform and approach. These issues will be discussed to raise awareness for local authorities when considering cloud-based services.

Background

Political

The move towards cloud-based services is largely driven by the UK government's intention to cen-

tralise ICT resources. This policy was established when the Cabinet Office announced that all central government and agency websites would be routed through open.gov.uk in 1994. According to official records (Cabinet Office, 2010) 100% of citizen-based services in the UK are now fully online. Such transformation is a response to demand by citizens and businesses to gain access to public service data in the same way they would when dealing with private organisations.

As part of the ICT strategy for the 21st century focusing on common infrastructure, standards and capabilities the UK government is developing a government cloud infrastructure that will enable pubic organisations to select and host ICT services from a "secure, resilient and cost-effective shared environment". The adoption and use of shared services are in line with and a key part of Transformational Government Agenda. The implementation of G-Cloud, by the government's own prediction, would result in some £3.2 billion savings (Cabinet Office, 2010).

Economic

An obvious reason for the government to consider cloud computing is cost. According to Angela Smith, Minister of State for the Cabinet Office, the public sector spends some £16 billion on information and communication technology (Cabinet Office, 2010). The move towards cloud-based services promises significant cost reduction, through provision of a single access point for ICT services, applications and assets.

The global economic downturn of 2008/9 has significant implications for government funded public IT services, market structures and investment models, leading to greater pressure for efficiency and savings. The continuous cut in IT budgets has accelerated the pace towards a simpler IT infrastructure. The development of G-Cloud is considered as an innovative ICT strategy in meeting economic challenges in the next few decades. This is in line with e-Government projects which require government agencies to prioritise IT services and focus on portfolio management and the reuse of existing solutions.

In particular, it is important for government agencies to explore public cloud services in order to achieve savings and economies of scale on less 'mission-critical' workloads (Di Maio, 2009). Apart from the financial cost of technology one also needs to consider the environmental cost. Cloud computing is claimed to reduce energy consumption significantly providing an environmentally friendly approach to ICT.

Social

A key driver that leads to the implementation of cloud computing in local authorities is commoditisation of government; that is, using public cloud services to host public data and to address the commoditisation of government channels through the use of social media (Di Maio, 2009). This is an important social trend which is impacting citizen service delivery and the government workplace.

One of the initiatives about 'open government data' (e.g., in the UK with the Power of Information task force) suggests the need for and value of socialising public information with external stakeholders. On the other hand, the socialisation of information to involve new stakeholders (such as citizens) in government service delivery and decision-making processes is closely related to the commoditisation of applications and business processes made possible by cloud computing. However, challenges facing the government IT leaders are ownership of assets, perceived negative impact on security and reliability, potential loss of control of data location, and accountability for data recovery and discovery.

Technological

Cloud computing services are delivered over the Internet which, for government agencies, makes them a viable alternative to building their own computing infrastructure. Larger internet capacity and faster internet speed ensure reliant and appropriate cloud service delivery. Private organisations, as well as government agencies, have experienced a range of IT innovations over recent years including outsourcing, grid computing, SaaS, virtualisation and Web-based storage. These form a foundation for the development of cloud computing.

One of the benefits of cloud computing is its flexibility and speed of change. Owing to constantly expanding storage and processing capabilities cloud service providers can offer a level of flexibility and regular software updates. For example, users can opt for more hardware-focused cloud solutions, where the user would own back-end software but use cloud computing for processing or storage (such as Amazon's Elastic Compute Cloud); more software-focused cloud solutions where SaaS is accessed through a software cloud (such as Google Apps); or data-focused clouds, providing data and information which are stored on a hardware cloud (such as YouTube or Google Health).

Driving Forces

The key driving forces that prompt local government to consider cloud computing are cost savings, simpler IT infrastructure, economies of scale and strategic focus. When companies or individuals launch new products and services, they do not need to have a big initial investment in computer architecture. Cloud computing offers the hardware and makes sure that the platforms hardware functions as supposed.

Cost savings, according to Vaquero *et al.* (2008) are on two sides, investments in hardware and upkeep costs. Both of these costs are reduced when building a service on a cloud computing platform. Hardware is rented from the computing service provider and the price is based on usage. In this way the users can enjoy the economies of scale that data centres provide (Grossman, 2009).

Second and probably the most desired advantage is scalability. Cloud computing platforms have the ability to scale computing power up automatically according to usage (McEvoy & Schulze, 2008). Usage can be increased through the instructions of the user and can scale up and down according to requirements. This means that there does not always need to be a massive infrastructure behind the service running it, only when the usage volume requires it. This also means that the user as well as the computing platform can utilise the hardware more efficiently (Aymerich *et al.,* 2008) by freeing up internal resources.

Another attractive feature of cloud computing is that vendors have typically been reputable enough to offer customers reliable service and large enough to deliver huge data centres with seemingly endless amounts of storage and computing capacity. Cloud computing minimises energy consumption in IT systems (AbdelSalam *et al.,* 2009). The freed resources will enable the government to commoditise infrastructure and related

services and focus its attention on strategic areas and make better use of the limited IT and human resources (Vining & Di Maio, 2009). In addition, with cloud-based software services clients do not need to worry about updating or maintaining software, which eliminates the need to worry about licensing issues and related installations (Vining & Di Maio, 2009).

Resisting Forces

Security and reliability of data and information and concerns over privacy and access are perhaps some of the most worrying issues for government to consider in cloud computing. The problem is that the user does not know where the machines are; they are somewhere in the cloud. However, there are privacy laws which governments and companies must comply with. This means that they have to know exactly where the data is being stored and who can access it (Buyya *et al.,* 2008).

The successful implementation of cloud computing relies largely on the trust between buyers and vendors. With respect to the security of data the vendor has to provide some assurance in service level agreements (SLA) to convince the customer on security issues (Kandukuri *et al.,* 2009). However, guaranteeing the security of corporate data in the cloud is difficult, if not impossible, as they provide different services like Software as a service (SaaS), Platform as a service (PaaS) and Infrastructure as a service (IaaS). Each service has its own security issues.

Cloud computing uses the Internet as the communication medium. There will be issues related to multiple customers sharing the same piece of hardware and bandwidth-related issues (Grossman, 2009), whereas the major concerns lie in the reliability and the security of the service provider. At the moment most cloud computing vendors use Secure Sockets Layer (SSL) to protect data. However, a majority of cloud offerings store data in shared environments (Vining & Di Maio, 2009).

Data information and classification is critical and the government will need to ensure that all public data being stored and backed up in the cloud be encrypted and the vendor needs to demonstrate and certify their security capabilities such as disaster recovery, intrusion protection and access controls. To ensure the security, integrity and compliance of data the researcher suggests that the service providers need to include a test encryption schema to ensure the shared storage environment safeguards all data, stringent access controls to prevent unauthorised access to the data, and scheduled data backup and safe storage of the backup media.

Implementation Issues

Government IT leaders who are evaluating the use of cloud computing should look at cloud computing as a 'sourcing option' (Di Maio, 2009). They should carefully consider the benefits, costs and risks of different cloud services models. In addition, they should implement cloud computing as a business process change project applying project management skills.

The implementation of cloud computing will have effects beyond technical issues and will have considerable impact on IT infrastructure, service users and how the government provides its services well into the future.

Business Process Change

The implementation of cloud computing will inevitably involve changes in business processes and the way local authorities provide public services. Technological change has always catalysed organisational change. There is a body of literature that focuses on identifying the factors that influence the ICT-enabled business process change. For example, Scholl's work (Scholl, 2003) suggests that e-Government projects require a holistic view of the public organisation, its culture, systems, processes and stakeholders.

Apart from a detailed workflow analysis successful implementation of major e-Government projects requires flexible planning for disruptive change process, realistic goals and objectives, an understanding of business processes, skills and senior management support (Scholl, 2003). This implies that the local authorities need to have adequate project management skills. This would be a concern as local government is not used to running large IT projects and the high failure rates are evident in Gilbert's work (Gilbert, 2009).

Information Assurance and Governance

To ensure low cost and scalability cloud-based services do not generally provide any assurance about where client data is being hosted on a server network. The government needs to consider a suitable cloud to host public information. Government needs to evaluate and explore cloud-based services for pubic data, similar to how private organisations host web sites, images, videos, documents and so on (Vining & Di Maio, 2009).

Cloud-based services are provided based on trust between the user and service provider. Not only does cloud computing raise issues regarding privacy, security, liability, reliability and government surveillance, relevant existing laws do not appear to be applicable to this innovative practice. This is particularly the case when government needs to comply with regulations on data protection, individual privacy, intellectual property and security. Data needs to be classified according to its nature and level of sensitivity in order for government to decide whether it should be put on a public cloud or private cloud.

When a cloud is made available in a pay-as-you-go manner to the general public, it is called Public Cloud (Armbrust *et al.,* 2009). In contrast a Private Cloud implementation has a bounded membership that is exclusive. It is likely that government would adopt a hybrid type of government cloud services. This model is not unusual. For example, the US government has launched a new cloud storefront that features applications called Apps.gov to be used on non-sensitive data (Di Maio, 2009). According to Di Maio (2009) public or private cloud computing are just sourcing options for governments that want to use a scalable and elastic virtualised infrastructure or application as a service. Cloud computing services used by government can be managed by government, by external providers or by a combination of both. It has been suggested that the success or failure depends on the appropriateness of the shared-service governance more than on the technical adequacy of the solution. Given the common risks in areas such as data privacy, access, loss and security and fears of vendor lock-in government-owned/controlled cloud computing may have greater chance of short-term viability (Vining & Di Maio, 2009).

Choices of Vendors, Products and Approach

There are currently plenty of cloud services providers out in the marketplace and pricing modes differ from one provider to another; some of them are quite complex. For example, the price of Amazon's Web Services comprises of used hours, used storage and the amount of data transferred (Amazon, 2010). The total costs of use may be difficult to predict.

The biggest disadvantage of cloud computing is vender lock-in. Every service provider has their own proprietary application programming interfaces (API). It would be difficult to change the interface when considering moving from one provider to another (Buyya *et al.*, 2008).

Another concern is the reliability of the service provider. There are a number of big cloud service providers and many other small vendors. One may question whether they are going to be around in five or ten years. Buyers would have to consider the cost to change the technology if the company discontinues the service (Foster *et al.*, 2008). What if the costs become very expensive? These are questions the customer should think about carefully before they choose a cloud computing provider.

In terms of the implementation of cloud computing in government Di Maio (2009a, 2009b) proposes a staged approach and a mixed structure of alternative clouds. In the short term government should work on greater availability of public data, while in the medium term the government will be faced with choices concerning whether and how to use cloud-based services to host systems and applications preparing for the establishment of government-driven private cloud platforms for selected services and data storage. Given the sensitivity of citizens' data, different requirements and concerns over privacy, security and integrity issues it is predicted the government is likely to adopt a hybrid approach to government cloud services. This could be the use of public clouds and less-public clouds, a mixture of different cloud services or a combination of cloud services with G-Cloud.

CONCLUSION

Cloud computing represents an innovative but major change in the way information services are delivered in the public sector (Thomson, 2009). If all the benefits are realised, users can certainly look forward to enhanced service provision. There may, however, be major risks with the Cloud. A key issue is security. Highly sensitive public data will be stored in data centres around the world. Such an arrangement raises concerns about ownership and control of public data, possible infringement of privacy and information warfare.

A recent report by the Society of IT Managers (Socitm) has drawn attention to the security dangers but argues that public service IT managers will simply have to accept the risk and urges them to start drawing up plans to embrace the new technologies (Thomson, 2009). Risk assessment, information assurance and governance, service level agreements and policies are areas that will require further investigation.

Apart from security concerns, cloud computing will entail fundamental changes in organisation and service provision. As yet, these are poorly understood. Presumably, if services are managed centrally, there will be a need for fewer IT staff. Those who remain in-house are likely to occupy different roles related to managing the outsourced service provision and technology strategy. A key constraint on the implementation of cloud computing, therefore, may be fear around potential job losses and changes in job roles. A related issue is competence in introducing and managing the new form of service delivery.

Widespread ignorance about, and confusion over, cloud computing is likely to stymie efforts to draw up plans to move to cloud computing. Some local authorities have experienced difficulties managing shared IT services, which raises concerns about the implementation of cloud provision (Gilbert, 2009). Further research of cloud computing in local government will need to take a holistic view, addressing the culture, systems, processes and people elements of the IT-enabled business process change. Views and comments need to be gathered in order to gauge reactions to cloud computing amongst those likely to be affected by it, and to consider how local authorities perceive the mandated arrival of cloud computing and the implications of its implementation. The research will provide an opportunity for those most affected by this momentous change to voice constructive dissent while at the same time providing opportunities to reflect on the possibilities presented by global service provision.

REFERENCES

AbdelSalam, H., Maly, K., Mukkamala, R., Zubair, M., & Kaminsky, D. (2009). Towards energy efficient change management in a cloud computing environment. In R. Sadre & A. Pras (Eds.), *Proceedings of the Third International Conference on Autonomous Infrastructure, Management and Security* (LNCS 5637, pp. 161-166).

Amazon. (2010). *Amazon elastic compute cloud.* Retrieved from http://aws.amazon.com/ec2/

Aymerich, F. M., Fenu, G., & Surcis, S. (2008, August). An approach to a cloud computing network. In *Proceedings of the 1st International Conference on the Applications of Digital Information and Web Technologies* (pp. 113-118).

Buyya, R., Yeo, C. S., & Venugopal, V. (2008, September). Vision, hype and reality for delivering IT services as computing utilities. In *Proceedings of the 10th IEEE International Conference on High Performance Computing and Communications* (pp. 5-13).

Cabinet Office. (2010). *Government ICT Strategy: Smarter, cheaper, greener.* Norwich, UK: Cabinet Office.

Di Maio, A. (2009). *Cloud computing in government: Private, public, both or none?* Stamford, CT: Gartner.

Di Maio, A. (2009). *Government in the cloud: Much more than computing.* Stamford, CT: Gartner.

Di Maio, A. (2009). *GSA launches Apps.gov: What it means to government IT leaders?* Stamford, CT: Gartner.

Digital Britain. (2009, June). *Final report presented to Parliament.* London, UK: Stationery Office.

Foster, I., Zhao, Y., Raicu, I., & Lu, S. (2008, November). Cloud computing and grid computing 360-degree compared. In *Proceedings of the Grid Computing Environments Workshop* (pp. 1-10).

Grossman, R. L. (2009). The case for cloud computing, computer.org/ITPro. *IT Professional, 11*(2), 23-27.

Kandukuri, B. R., Ramakrishna, P. V., & Rakshit, A. (2009). Cloud security issues. In *Proceedings of the IEEE International Conference on Services Computing* (pp. 517-520).

Korri, T. (2009, April). *Cloud computing: Utility computing over the Internet.* Paper presented at the Seminar on Internetworking, Helsinki, Finland.

Lewin, K. (1947). Frontiers in group dynamics: Concept, method, and reality in social science. *Human Relations, 1*(1), 5–42. doi:10.1177/001872674700100103

Luis, M. V., Luis, R, Caceres, J., & Lindner, M. (2008). A break in the clouds: Towards a cloud definition. *SIGCOMM Computer Communication Review, 39*(1).

Maggiani, R. (2009, November). Cloud computing is changing how we communicate. In *Proceedings of the International Conference on Professional Communication* (pp. 1-4).

McEvoy, G. V., & Schulze, B. (2008). Using clouds to address grid limitations. In *Proceedings of the 6th International Workshop on Middleware for Grid Computing* (pp. 1-6).

Nelson, M. R. (2009). The cloud, the crowd and public policy. *Issues in Science and Technology,* 71–76.

Plummer, D. C., Smith, D. M., Bittman, T. J., Cearley, D. W., Cappuccio, D. J., & Scott, D. (2009). *Five refining attributes of public and private cloud computing.* Stamford, CT: Gartner.

Scholl, H. J. (2003). E-government: A special case of ICT-enabled business process change. In *Proceedings of the 36th Hawaii International Conference on System Sciences.*

Thomson, R. (2009, February 24). Socitm: Cloud computing revolutionary to the public sector. *Computer Weekly.*

Vaquero, L. M., Rodero-Merino, R., Caceres, J., & Lindner, M. (2008). A break in the clouds: Towards a cloud definition. *SIGCOMM Computer Communication Review, 39*(1).

Vining, J., & Di Maio, A. (2009). *Cloud computing for government is cloudy.* Stamford, CT: Gartner.

Chapter 18
Survey of the State-of-the-Art of Cloud Computing

Sanjay P. Ahuja
University of North Florida, USA

Alan C. Rolli
University of North Florida, USA

ABSTRACT

Cloud computing as a computational model has gathered tremendous traction. It is not completely clear what this term represents though it generally is thought to include a pay-as-you model for computation and storage. This paper explains what Cloud Computing is and contrasts it with Grid computing. It describes the major cloud services offered, discusses architectural details, and gives details about the infrastructure of Cloud Computing. This paper surveys the state of cloud computing and associated research and discusses the probable directions to the future of this evolving field of computing.

1. INTRODUCTION

In the early 1990s, Ian Foster and Carl Kesselman came up with the idea of "The Grid"; applying the same concept of plugging into a grid for metered utility service to computing (Wallis, 2008; Douglis, 2009). This concept has transformed into what has been coined Cloud Computing. With Cloud Computing, data is stored in the "cloud" of the internet where web-based applications are utilized to access the data and perform various

tasks. Amazon was one of the first to provide service, the Elastic Compute Cloud (EC2). EC2 charged for their resources by the hour, like the electric company charges their clients per kilowatts an hour of usage. Presently, there are three major categories of Cloud Computing systems offered: software-as-a-service (SaaS), platform-as-a-service (PaaS), and infrastructure-as-a-service (IaaS) (Viega, 2009). Each of the aforementioned services as well as the emerging Database-as-a-Service provides customers a variety of choices to meet their specific needs.

DOI: 10.4018/978-1-4666-1879-4.ch018

1.1. Comparing and Contrasting Grid and Cloud Computing

On the outside there is little difference between a Grid and a Cloud environment. They both are a collection of machines such as servers, network devices, and computers that, to the user, appear as a single resource. This resource provides services such as Web infrastructure, databases, application support and more. So what truly sets these two environments apart?

Grids were originally designed to help provide a way for researchers to allocate large amounts of resources in order to perform some complex computation that would take a significant amount of time longer to accomplish with a single machine. In the Grid, scientists could allocate as many nodes in the environment as necessary to complete their task, and leave the remaining nodes available for other such computations. The nodes are allocated only for as long as it takes to complete the task. Once the task is completed, the nodes are leased back into the Grid to be allocated to another process.

In contrast to Grids, Cloud environments are geared towards smaller requests for resources. Instead of requesting 2000 nodes in order to perform some difficult task, one might request enough resources in order to meet their individual, business, or personal, needs. The resources in a Cloud are more semi-permanent as opposed to a Grid. In other words, the allocated resources in a Cloud will remain persistent until the client decides to cancel the service.

Another difference is that the environment in a Cloud will generally be a virtual environment. What this means is that if somebody were to request a resource such as a Web server, they would not be given an entire server in which to host their environment. Instead they would be allocated a virtual environment, there could be several such environments running on each physical server, where the user's Web server would run.

While research on Grid systems has been expanding on working towards being able to tie resources from different administrative domains, Cloud Computing systems are currently restricted to a single domain.

An additional key difference in cloud computing is how data is stored and distributed. Relational databases, as typically used in an Enterprise environment, are not very scalable because the data cannot easily be distributed across the domain. A relational database takes a huge performance hit if the files are distributed across a large number of nodes. If the workload doubles or triples overnight, you would have to upgrade the server quickly, which is not an easy task.

By using a distributed hash table you can distribute your data across thousands of nodes without having to have one or even several centralized database servers. Not only is this good for redundancy, it also allows for faster insertion and retrieval for things such as Web services.

1.2. Benefits of Cloud Computing

As you can see from the comparing and contrasting the Grid to Cloud Computing, there are clear advantages to Cloud Computing (Figure 1). One of the immediate benefits is that of time-to-application deployment. Cloud platforms allow you to develop and deploy new applications on existing infrastructure as quickly as you can write the code. Companies are also benefiting from the ability to scale their IT budget as demand grows, freeing up equity.

1.2.1. Operational Benefits

Server uptime and availability are very important to any company, regardless of the size of the business. The cloud-based infrastructure provides better uptime and availability by using technologies that have been developed for distributed systems for redundancy of resources. As discussed earlier, a Cloud environment is very similar to a grid

Figure 1. Comparing on site computing with cloud computing

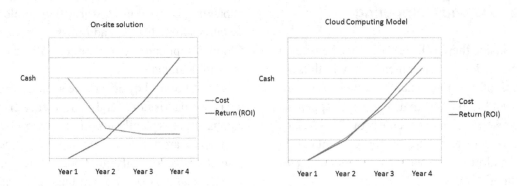

in that the processes are generally distributed throughout the network. A properly configured Cloud environment will be fault tolerant and have the ability to recover from lost resources by reallocating new ones. In order to accomplish this Amazon allows users to place their work in what they call "Availability Zones" (Thorsten, 2009). An Availability Zone is essentially a separate data center where a user can have their system replicated for fail-over. If one data center goes (or simply one of the servers crash), then there is a replicated server in another location in which traffic can be routed to.

Besides having reliable and stable servers, the ability for deployment teams to launch new projects and bring new virtual servers up almost instantaneously is a largely attractive benefit of Cloud Computing. Having an internal infrastructure team, combined with an in house system support team, combined with an in house architectural review team can create a lot of red tape and delays in getting new material deployed. Having the ability to stand up a new virtual server, preconfigured possibly replicating another existing one, within a few seconds is a great time saver.

1.2.2. Financial Benefits

One of the largest benefits of Cloud Computing is the cost savings. Having the ability to pay-as-you go instead of needing to invest a large sum upfront

gives the company a lot more flexibility in their cash availability. Not having a large upfront cost allows companies to easily scale according to demand and gives them the ability to have more cash on hand in order to fund future projects. This model also leverages more annual capital by lowering operational costs associated with overhead.

The resources of an average company's data center are vastly underutilized. Servers will often spend 85% of their time in an idle state (Perry, 2008). According to the white paper published in American Power Conversion (2003), "The Total Cost of Ownership of a rack in a data center is approximately $120K over the data center lifetime."

Another cost associated with a data center that is often not thought of is the cost to actually build a proper data center in the first place. A financial institution has very strict requirements on how it must secure and store its data. There are also governing regulations covering the ability for a data center to be functional within a certain time after a natural disaster. This means that the data center must be able to be self contained, and powered by some type of backup power source such as a generator. Or that there must be a second site available should the main one go down. The average cost to build such a data center in 2008 was around $250 per square foot. If you had a data center of 22,500 feet the cost would be over $5,600,000 just to build the data center (http://www.reedconstructiondata.com/rsmeans/

models/data-center/). This cost does not even take into account all of the costs of the actual servers themselves as well as the salaried people necessary to maintain them. Now, the choices are to either build a second data center in a remote location to be used in case of an emergency, or to create a data center with such structural integrity, and power backup that it could withstand anything thrown at it. Either way, the cost goes way beyond the standard cost for a physical server.

According to research done at Hewlet Packard Laboratories, the annual cost to maintain a data center with 100 racks of servers would be approximately \$4,200,000 (Patel & Shah, 2005). Comparing this with the price of Amazon's EC2, while estimating 10 servers per rack (therefore 1000 servers), we would have an annual cost of \$1,820,000 (at \$1,800 per extra large CPU instance) in server cost. In order to calculate our cost for data transfer, the average data usage for a typical Web site is multiplied that by the number of servers in the system. The typical usage for a website is around 2-3 gigabytes per month (Good Web Hosting, 2006). At a data transfer cost of \$0.10 per GB of data transfer, and 2.5 GB per month usage for each of our 1000 servers, this gives us the following calculation:

Bandwidth Cost = 1000 * 2.5 * 0.10

This enables only paying \$250 per month, or \$3,000 per year.

1.3. Services Offered

There is not only a variety players in the Cloud Computing arena, there is also, as noted above, a variety of service choices provided to customers to meet their specific needs; Software-as-a-Service (SaaS), Platform-as-a-Service (PaaS), Infrastructure-as-a-Service (IaaS), and Database-as-a-Service (DBaaS).

1.3.1. SaaS

With SaaS, clients can purchase a piece of software that is accessed online, or possibly data that will be stored online. While SaaS can almost generally be applied to all types of software, in the context of Cloud Computing we take it to represent software that is accessed from a machine residing on the Internet as opposed to software that is downloaded and installed locally.

1.3.2. IaaS

When we talk about PaaS, IaaS is certainly involved. In order for a vendor at a Cloud Computing company to allocate a client the resources to host a Web site, they invariably need to allocate some resources to the client. However, they do have the ability to allocate resources for different purposes. Suppose that a client wanted to perform some Data Warehousing or Data Mining but didn't have the servers necessary to perform such compute intensive tasks. The client could acquire resources from a Cloud vendor that could be used to perform such tasks.

1.3.3. PaaS

The PaaS type systems are especially useful for Web servers. Instead of having to purchase their own infrastructure and host that hardware somewhere, clients can purchase the rights to host their Web sites in the Cloud, on someone else's infrastructure. Google App Engine, GoDaddy and Amazon EC2 are some popular examples of the many online hosting services available.

1.3.4. DBaaS

There is a relatively new concept to the cloud computing community being termed as Database-as-a-Service (DBaaS). Microsoft recently released a beta version of SQL Server called SQL Server Data Services (SSDS). They intend this to be used in cloud environments so that customers of a specific cloud provider can still use the mainstream database they have come to know and love without having to purchase full licenses that would have to be installed by the cloud provider.

While RDBMSs are being developed for cloud environments, they certainly are not the norm when it comes to storing data in the cloud. One of the main driving forces for cloud computing is for the on-demand capability for the environment to be highly scalable. If the data within the system were to triple over night, the typical RDBMS might not be able to handle the sudden increase in data and activity. Relational databases scale well when running on a single server. However, when the capacity of that single node is reached, we need to either distribute that load among various other servers or quickly replace the current storage capacity. Replacing the storage is costly and time consuming, and trying to distribute a relational system across hundreds or thousands of nodes within a network is near impossible and will degrade the performance of the relational system.

The need for a scalable database system in cloud environments that would allow for a highly scalable environment led to a new kind of database management system that is being referred to as a key/value store (Bain, 2009). This kind of database is not an entirely new concept, but has gone mostly unused in general, being more of a small niche market. The benefit of this type of system is that it scales well across a large domain. Trying to retrieve large chunks of data from multiple file stores in a relational system is a costly procedure and indexing such a schema can be very difficult to implement. The main downside to this new kind of database approach is that relational systems provide data type integrity whereas key / value stores depend on the application developer to handle these types of constraints.

1.4. Cloud Providers

According to *Cloud Computing Journal* there are over 150 "players" in the Cloud Computing arena (Geelan, 2009). A detailed overview of a few of the top providers from their list is provided:

- **Adaptivity:** Tony Bishop founded Adaptivity in 2007. Adaptivity focuses its services around "IT Blueprinting" (http://adaptivity.com/blueprint4it/downloads/news). The company's business platform pushes the ideal that "IT must invest in architecture and engineering." One such architecture is via the Cloud, and Tony Bishop sees the advantages of Cloud Computing as broad: "IaaS type resources managed externally from the enterprise do provide value; however, the larger opportunity is enabling enterprises to change how they deliver and consume IT resources" (Geelan, 2010).

- **Agathon Group:** The Agathon Group is owned by Joel Boonstra. The company offers Cloud Hosting via a dedicated grid environment. The company's target clientele is charitable and non-profit campaigns and includes local high schools, regional telecommunications firms and worldwide mission organizations (http://www.agathongroup.com).

- **Akami:** Akamai claims it has been optimizing the cloud for over ten years. Its cofounder and Chief Scientist is Tom Leighton. Their services include virtualization, IaaS, PaaS, SaaS, and cloud optimization services. Their cloud optimization services provide performance, scale and reliability for their virtualization, IaaS, PaaS and SaaS services (Akamai, 2009).

- **Amazon:** Amazon offers IaaS through its EC2 services offerings as well as DBaaS. Amazon was one of the first to provide services in the cloud computing arena. EC2 charges for their resources by the hour, like the electric company charges their clients per kilowatts an hour of usage; offering a pay-as-you go for only the space you consume web services model (Amazon, n. d.).

- **AT&T:** AT&T offers SaaS and Synaptic Compute as a Service, which is essentially IaaS services. Its pay-as-you go for only the space you consume web services model is nearly identical to that of Amazon (http://www.business.att.com/enterprise/Family/application-hosting-enterprise/synaptic-compute-as-a-service-enterprise/)
- **Microsoft's Azure:** Microsoft offers PaaS by providing a cloud computing operating system, as well as IaaS and DBaaS. They are adding more services and have announced their goal to focus on cloud computing services in the future (http://www.microsoft.com/windowsazure/)
- **Google:** Google offers PaaS and claims to be a pioneer in SaaS by offering many services such as Google Docs, Gmail, Picasa, etc., but in fact it is a pioneer in the industry as a whole. They have multiple patents pending for various cloud computing related services (http://thecloudtutorial.com/googlecloud.html).

1.5. Cloud Providers

All of the players in the cloud computing arena offer various services and pricing structures. As noted above, Amazon charges by the hour. They charge between $0.085 and $8,650/hour depending on whether a reserved instance is required, what OS is desired (Linux or Windows) and how much space is needed (Amazon, n. d.) Google, for its PaaS services charges per application, $8 per user, per month up to a maximum of $1000 a month. For their IaaS offerings, they charge $50 per year per user account (http://thecloudtutorial.com/googlecloud.html, http://www.google.com/apps/intl/en/business/features.html). It appears

that the pricing structures are as unique as the varying services that each player offers, which could make choosing a vendor hard, but the good news is that the goal of cloud computing as a whole, seems to be to provide a financial benefit.

2. CLOUD COMPUTING ARCHITECTURE

When talking about the architecture of a cloud environment people generally refer to the architecture as having two major divisions: the front end and the back end (Sun Microsystems, 2009; Strickland, n. d.; Wikipedia, n. d.). The front end is the various ways that an end user connects to the cloud, whether it is via a Web browser or some developed application. Interaction with the cloud is usually done via web service calls.

The back end is where things get a bit more technical. Cloud environments basically consist of computers and a variety of servers that can be configured in many different ways. The way in which these servers are configured can vary depending on the infrastructure model chosen, whether it is going to be a public cloud, a private cloud or maybe some type of hybrid. The configuration also depends on the type of services being offered (SaaS, PaaS, IaaS).

Starting from the ground up there are physical servers. There will generally be one (or possibly more, depending on traffic and whether or not there is an external facing storage server) server configured to be the main point of access into the cloud. This server acts as the proxy for all communication in and out of the cloud. In the case of (http://www.eucalyptus.com/), it also will act as the Cloud Controller which is responsible for making decisions about tasks and scheduling of cloud components.

Inside of the cloud environment there will be computers and servers working together to provide the guts of the system. The key to them working together is the middleware agent installed on each machine. The architecture of any cloud center is dependent upon virtualization, which is explained further in Section 2.1 of this article. On top of the middleware we have the virtual Operating systems (OS) that are installed per virtual instance. Depending on the type of services offered, there may be application software installed as well on these virtual instances.

2.1. Hypervisors

Hypervisor technology is the technology that paved the way for cloud computing and virtualization. A major problem in application development is being able to write software that will run on multiple operating systems with multiple versions. Operating system programmers also suffer from this problem when trying to interface with the underlying hardware that is being developed by the many different companies out there. Up until 2006, you could not even install the Mac operating system on an Intel x86 architecture (Kerris, Dowling, & Beermann, 2005). So how does one go about placing multiple OSes on a single server in a virtual environment? Hypervisors were developed to answer this question by placing a layer of software on the physical hardware that sits between the hardware and the operating systems wishing to interact with it. These hypervisors create an abstraction layer that allows each piece of software wishing to interact with the physical hardware to act as if it was running directly on the hardware. With most of the major players in the OS market embracing the Intel x86 architecture, the difficult task faced by those developing hypervisors was eased.

Within the hypervisor community there are two different flavors: native (known as bare metal) and hosted. The native hypervisors run on the physical hardware and are able to interface with directly with the components. Here are a few of the more popular native hypervisors: VMware, Xen, Oracle VM, Microsoft Hyper-V, TRANGO,

IBM's POWER Hypervisor, KVM and Sun's Logical Domains Hypervisor.

Hosted hypervisors run inside of a standard operating system and rely on the host system to provide the interactions with the underlying OS. VMware provides several different hosted environments for their hypervisor, which makes it a bit confusing as to whether it is a native or hosted hypervisor. While they do provide a native version, there are a few such as VMware Server and VMware Workstation that allow you to run the virtual environment on either a Windows or Linux OS. Some of the other hosted VM's include Citrix XenServer, Microsoft's Virtual PC, Sun's VirtualBox and Parallels Workstation.

3. SUMMARY OF CLOUD COMPUTING RESEARCH

The easiest thing for someone to do is to point out, research and write about the flaws in technology, especially a relatively new technology. Accordingly, while cloud computing is definitely moving forward in a positive direction, and more and more people are moving their applications and data into the cloud every year, there are still many issues that have yet to be clearly defined or worked out. While there are several concerns with using cloud services, there are a few major issues that seem to repeatedly appear in research.

3.1. Data Governance and Security

Most people would agree that data is probably the most important thing to a company, whether it is a bank's transactional data or a manufacture's inventory and accounting data or even patient records. Ensuring that the data is secure and that only authorized people have access to it is a large concern. Who owns the intellectual property once it is in a cloud environment is another data governance topic much greater than this paper. The basic concern though stems from the concept that traditionally, data governance models have called for individuals companies to maintain the information and records. With Cloud Computing, companies must rely on their vendors to ensure the safety of their data and that they are following all the applicable IT governance and rule sets as well as their governing laws (Wikipedia, n. d.; Zhen, 2008; Doelitzcher, Reich, Sulistio, & Furtwangen, 2010).

Cloud computing differs from the traditional way that data and resources are stored. Even when outsourcing a server to somebody else's data center you know exactly where the data is stored and what resources may be shared. However, Cloud computing obscures such low-level details by decoupling the actual data from the physical infrastructure. It is possible that your data could

be spread across multiple physical servers that also happen to be storing data for other clients on the same machine.

There are three main security issues to consider: security and privacy, regulatory compliance, and legal or contractual issues discussed by Wang (2009). The authors are credited with providing a security issue checklist (Figure 2) of items to consider before choosing a vendor (or whether to go with cloud computing at all).

Security and privacy involves ensuring that only those authorized to view data are able to. In order to provide this type of security you must ensure that your applications do not allow access to data without proper authorization, users are not able to access data directly, data is not accessible from other sources that may share resources on

Figure 2. Security issue checklist (adapted from Wang, 2009)

Area	Topics
Security and privacy	Data segregation and protection
	Vulnerability management
	Identity management
	Physical and personnel security
	Data leak prevention
	Availability
	Application security
	Incident response
	Privacy
Compliance	Business continuity and disaster recovery
	Logs and audit trail
	Specific requirements (e.g., PCI, HIPAA, EU privacy, Basel II, FFIEC)
Other legal and contractual issues	Liability
	Intellectual property
	End of service support
	Auditing agreement

the same physical network (i.e., other customers whose data is stored at the same physical location or within the same physical network), and users must be able to report on possible security violations.

Depending on the type of business, or the type of data you being stored in the cloud, users may fall under certain regulatory compliance laws. If, for example, a company is a bank, it will have to ensure that there is an adequate disaster recovery plan, data history that meets the audit trail compliances and other federal requirements that govern banking.

The last major security issue to consider is that of legal or contractual obligations. If a company X is providing a service for a third party (say it is developing a web site for company Y) and it wants to host their Web site in a cloud environment, company X needs to be able to meet the contractual agreement for uptime, even though it doesn't have control over the physical infrastructure. Company X may also be considered liable for intellectual property that the client company Y may store. Even though company X is not the one ultimately responsible for what happens to the physical network and servers, it can still be held responsible for the outcome of its decision to go with a specific vendor. It is also possible that there is not a clear definition of who actually controls the intellectual property stored in the cloud. There is no clear definition when it comes to this, and it is possible that the vendor could use the old adage "ownership is 9 / 10s of the law." It is advised that a user ensures that there is an intellectual property clause in the contract when signing with a specific vendor.

Questions arise such as "who is responsible if security is compromised?" In order to identify proper security methods for cloud computing vendors the US National Institute of Standards and Technology (NIST) has decided to create a section that is responsible for outlining security standards (Kaufman, 2009). Their goal is to ensure that vendors are properly implementing their

security guidelines in order to protect both the vendor and their clients.

3.2. Manageability

Hypervisor technology is now being open sourced, therefore management capabilities becomes the main difference in cloud services. A lot of Iaas and PaaS service providers are simply providing raw infrastructures and platforms, which ultimately can limit or require full management capabilities. For example, only some companies are offering auto-scaling or load balancing. Others require users to manage this on their own, or at least recognize when they need to request for things to be rebalanced or space to be expanded and request assistance in doing such (Zhen, 2008). For small businesses or even larger companies trying to keep their IT infrastructure to a minimum, manageability can quickly become a key determining factor in which service provider they choose.

3.3. Reliability

T-Mobile's Sidekick embarrassment, where T-Mobile's Sidekick's complete loss of contacts and other synced information as well as Google's Gmail and News outages demonstrate the causes of concern over the reliability of cloud services (Bradley, 2009a, 2009b; Coursey, 2009). There are many reasons and causes for the reliability, security and data governance and manageability concerns.

In our opinion, the concerns require reflection and assessment prior to enterprise implementation. In any IT Infrastructure, these issues are a concern, but relying on a third-party, rather than a company's own employees or data centers for security and data retention measures raises some red flags when it comes to sensitive information.

It is likely that this emerging technology will see more laws come into play as it continues to become a more widespread medium. Data Security and Management models will be enacted soon, and are already being developed. It is expected that these issues will be worked out in the long run.

4. CONCLUSION AND THE FUTURE OF CLOUD COMPUTING

PC World said it best, "Today's weather report for the internet: increasing cloudiness with a chance of strong winds (of change)" (Bradley, 2009a). Various researchers from Merrill Lynch, Morgan Stanley and various Cloud Computing Journals suggest all suggest that Cloud Computing is expected to grow into a $150 billion plus market opportunity and that the current trend supports this (Buyya, Yeo, & Venugopal, 2008; Stanley, 2008; http://www.financialpost.com/money/).

In this volatile market, companies are looking for ways to save money, and technology seems to be a target for savings. So, it is natural that with the cost-benefits of Cloud Computing, and continuing efforts to get all the kinks are worked out, even more enterprises will move towards cloud solutions.

As noted, Cloud Computing, as any relatively new type of technology, has its drawbacks, but this isn't anything that technology, especially the Iinternet hasn't faced before. We believe we can expect to see changes in the way of laws being drafted, management and security models being created and Cloud Computing consuming the majority of the market in the very near future.

REFERENCES

Akamai. (2009). *Solutions for cloud computing: Accelerate, scale and fortify applications and platforms running in the cloud.* Retrieved from http://www.akamai.com/dl/brochures/Cloud_Computing_Brochure.pdf

Amazon. (n. d.). *EC2 pricing.* Retrieved from http://aws.amazon.com/ec2/pricing/

Amazon. (n. d.). *Elastic compute cloud.* Retrieved from http://aws.amazon.com/ec2/

American Power Conversion. (2003). *Determining total cost of ownership for data center and network room infrastructure.* Retrieved from http://www.lamdahellix.com/%5CUserFiles%5CFile%5Cdownloads%5C6_whitepaper.pdf

Bain, T. (2009). *Is the relational database doomed?* Retrieved from http://www.readwriteweb.com/enterprise/2009/02/is-the-relational-database-doomed.php?p=2

Bradley, T. (2009a). *IBM and AT&T unveil cloud computing services.* Retrieved from http://www.pcworld.com/businesscenter/article/182238/ibm_and_atandt_unveil_cloud_computing_services.html

Bradley, T. (2009b). *Sidekick foul-up is not a failure of the cloud.* Retrieved from http://www.pcworld.com/businesscenter/article/173485/sidekick_foulup_is_not_a_failure_of_the_cloud.html

Buyya, R., Yeo, C. S., & Venugopal, S. (2008). Market-oriented cloud computing: Vision, hype, and reality for delivering IT services as computing utilities. In *Proceedings of the 10th IEEE International Conference on High Performance Computing and Communications.*

Coursey, D. (2009). *Google outages defy customer confidence.* Retrieved from http://www.pcworld.com/businesscenter/article/172575/Google_Outages_Defy_Customer_Confidence.html

Doelitzcher, F., Reich, C., Sulistio, A., & Furtwangen, H. (2010). Designing cloud services adhering to government privacy laws. In *Proceedings of the 10th International Conference on Computer and Information Technology* (pp. 930-935).

Douglis, F. (2009). Staring at clouds. *IEEE Internet Computing*, *13*(3), 4–6. doi:10.1109/MIC.2009.70

Geelan, J. (2009). *The top 150 players in cloud computing.* Retrieved from http://cloudcomputing. sys-con.com/node/770174

Geelan, J. (2010). *IT must invest in architecture and engineering: Adaptivity CEO.* Retrieved from http://cloudcomputing.sys-con.com/ node/1061561

Good Web Hosting. (2006). *How much data transfer do you really need?* Retrieved from http:// www.goodwebhosting.info/article.py/58

Kaufman, L. M. (2009). Data security in the world of cloud computing. *IEEE Security & Privacy, 7*(4), 61–64. doi:10.1109/MSP.2009.87

Kerris, N., Dowling, S., & Beermann, T. (Eds.). (2005). *Apple to use Intel microprocessors beginning in 2006.* Retrieved from http://www.apple. com/pr/library/2005/jun/06intel.html

McMillan, R. (2009). *Researchers hack into Intel's vPro: Invisible things labs discovers a way to compromise the vPro architecture.* Retrieved from http://www.infoworld.com/d/security-central/ researchers-hack-intels-vpro-604

Patel, C. D., & Shah, A. J. (2005). *Cost model for planning, development and operation of a data center.* Retrieved from http://www.hpl.hp.com/ techreports/2005/HPL-2005-107R1.pdf

Perry, G. (2008). *How cloud & utility computing are different.* Retrieved from http://gigaom. com/2008/02/28/how-cloud-utility-computing- are-different/

Stanley, M. (2008). *Technology trends.* Retrieved from http://www.morganstanley.com/institu- tional/techresearch/pdfs/TechTrends062008.pdf

Strickland, J. (n. d.). *How cloud computing works.* Retrieved from http://communication.howstuff- works.com/cloud-computing1.htm

Sun Microsystems. (2009). *Introduction to cloud computing architecture.* Retrieved from http:// webobjects.cdw.com/webobjects/media/pdf/ Sun_CloudComputing.pdf

Thorsten. (2008). *Setting up a fault-tolerant site using Amazon's availability zones.* Retrieved from http://blog.rightscale.com/2008/03/26/ setting-up-a-fault-tolerant-site-using-amazons- availability-zones/

Viega, J. (2009). Cloud computing and the common man. *IEEE Computer, 42*(8), 106–108.

Wallis, P. (2008). *A brief history of cloud comput- ing: Is the cloud there yet?* Retrieved from http:// cloudcomputing.sys-con.com/node/581838

Wang, C. (2009). *How secure is your cloud: A close look at cloud computing security issues.* Retrieved from http://www.forrester.com/rb/ Research/how_secure_is_cloud/q/id/45778/t/2

Wikipedia. (n. d.). *Cloud computing.* Retrieved from http://en.wikipedia.org/wiki/Cloud_com- puting

Wikipedia. (n. d.). *Information lifecycle manage- ment.* Retrieved from http://en.wikipedia.org/ wiki/Information_Lifecycle_Management

Wojtczuk, R. (2008). *Track: Virtualization.* Re- trieved from http://blackhat.com/html/bh-usa-08/ bh-usa-08-speakers.html#Wojtczuk

Zhen, J. (2008). *Five key challenges of enterprise cloud computing.* Retrieved from http://cloudcom- puting.sys-con.com/node/659288

This work was previously published in the International Journal of Cloud Applications and Computing (IJCAC), Volume 1, Issue 4, edited by Shadi Aljawarneh & Hong Cai, pp. 34-43, copyright 2011 by IGI Publishing (an imprint of IGI Global).

Chapter 19
Cloud Computing Towards Technological Convergence

P. Sasikala
Makhanlal Chaturvedi National University of Journalism and Communication, India

ABSTRACT

With the popularization and improvement of social and industrial IT development, information appears to explosively increase, and people put much higher expectations on the services of computing, communication and network. Today's public communication network is developing in the direction that networks are widely interconnected using communication network infrastructure as backbone and Internet protocols; at the same time, cloud computing, a computing paradigm in the ascendant, provides new service modes. Communication technology has the trend of developing towards computing technology and applications, and computing technology and applications have the trend of stepping towards service orientation architecture. Communication technology and information technology truly comes to a convergence. Telecom operators are planning to be providers of comprehensive information services in succession. To adopt cloud computing technology not only facilitates the upgrade of their communication network technology, service platform and supporting systems, but also facilitates the construction of the infrastructure and operating capacity of providing comprehensive information services. In this paper, the development processes of public communication network and computing are reviewed along with some new concepts for cloud computing.

DOI: 10.4018/978-1-4666-1879-4.ch019

1. INTRODUCTION

1.1. The Development of Communication Network

1.1.1. The Development of Voice Network

The development of communication network has mainly experienced five stages, electromechanical switch, stored program control switch (SPC switch), value-added service (VAS), softswitch and IP multimedia subsystem (IMS). The development itself is the increasing process of computing elements. The switch evolvement from electromechanical to SPC is a significant change of telephone communication in the 20th century, which has immensely heightened the communication quality and efficiency. It provided more convenient service to clients. The third-generation telephone network evolving from SPC to softswich is another big change. It is essentially distinguished from current telephone network in the aspect of operation mode and profit model. IMS references the irrelevance of accessing in the softswitch, and also its functional improvement made the network architecture more open and more flexible. IMS made the seamless integration between all-IP network and public service telephone network possible, and made the resources more unifiedly manageable.

• Electromechanical Switch

The pre-stage of SPC is mainly telephone communication, i.e., the manual operator services. Operators should finish each call connection manually.

• SPC switch

Control the consecutive automatic electronic switch using the pre-compiled program. The system adopts SPC is called the program controlled system, which complete the control to switch through the interaction of software (program) and hardware. The evolvement of SPC includes three stages of space-division, simulation technique and digital technique.

• Value-added service platform

Beginning in the 1990s, many value-added services was introduced into SPC exchange, bringing a further functional improvement in it (Ali, 1986). By means of resources and other communication facilities in the public telecommunication network, VAS could provide consumers more highly and diversified information basing on the guarantee of basic communication needs. VAS is an integrated service combined voices, images and words. To provide VAS, specific service logic and service environment running on computer servers were gradually introduced, including SMS, CRBT, MMS, WAP and other varied VAS groups (Houssos et al., 2002).

• Softswitch and IMS

The basic principal of softswitch: soft the traditional exchange equipment component, and realize call control and multimedia processing using computing technique. Basing on the IP network, IMS is a network architecture which provides voice and multimedia service. The main feature of IMS includes uniform user data, uniform and open application/service layer, uniform call session control. IMS further separates service/application and control. Moreover, it concentrates user data, and is irrelevant to access technology (Loreto, Mecklin, Opsenica, & Rissanen, 2010). The application of IMS makes the control to system environment much easier, and the user experience is much better because of QoS, single sign-on and customer service. As a new service form, IMS could now satisfy terminal users with novel and diversified multimedia service. Softswitch and IMS network have totally relied on computing technique. They employ standard protocol and software, and introduce large numbers of calculation servers into network entity. The VAS capability and application have been totally a computer system platform.

1.1.2. The Development of Data Communication Network

The purpose of data communication is to complete information transfer between two computers, and between computer and data terminals. Data transfer should be done to realize data communication. Data information, which generate from one data source, are transferred to another data receiving set via data communication network. The development of this switching technology has gone through the stages of circuit mode, group mode, frame mode and cell mode.

Circuit mode transfers information from one point to another, and fixes the occupied circuit bandwidth resources, such as DDN special line. Group mode divides information into packets with certain length, then store-and-forwards those packets. Early typical packet switching network is "X.25 Network". Frame mode simplifies the process of error recovery and congestion avoidance between packet switches, and its typical technology is frame relay. ATM, which was the development peak of circuit switch and packet switch, is sometimes called cell relay. Ethernet has the advantage of cheap cost and flexibility. With the acceleration of global networking, widespread PC use and emergence of electronic commerce, Internet technique basing on the Ethernet was commercialized rapidly. Entered for the 21st century, industrial circles have been trying to build a uniform network platform one by one. It could provide multi-service on a MPLS-based platform simultaneously, including the service of voice, traditional data (frame relay, ATM, circuit simulation), Internet, video conference, video phone, and mobile data. The platform maintains the flow control technology, and solves the problem of

QoS which can't be guaranteed by IP technology. Meanwhile, by utilizing the technique of virtual router, it provides a practicable technological approach to the Next Generation Network (NGN).

It can be seen that the development of data communication plays a predominant role in promoting computer technology, while the computer communication network, computing-based application and service also promotes the up-to-date data communication network.

1.2. The Development of Public Service

1.2.1. The Development of Telephone Switch

The development of telephone switch can be divided into three steps:

Step 1: User sets up Private Branch Exchange (PBX). It realizes internal call through manual switching or automatic connection. Then, it uses public telephone network user account to call external via manual relay.

Step 2: Client side does not use PBX anymore. It is necessary to prepare customer exchange board PBX. It provides trunk circuit by public telephone network and internal call is provided through automatic connection. Then, it call external via manual switching or automatic second tempt connection. It is shown in Figure 1(a).

Step 3: Withdraw customer small branch exchange. It takes Centrex (Central Exchange) technology. Customer intranet merges into public telephone network. Then, telecom departments provide services with internal China Netcom and public telephone network functions. It is shown in Figure 1(b).

Up to today, era which needs client to prepare internal telephone network already ends. Telephone communication is provided by public

Figure 1. Revolution of telephone exchange

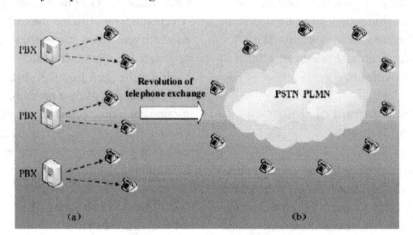

telephone network operators, which can make service abundant. Public telephone network covers communication network of entire region or country. It consists of public exchange telephone network PSTN and public terrestrial mobile communication network PLMN.

1.2.2. The Development of Data Service

Early data network provides basic transmission service, including data, telegraph text and image. Today, it is developing towards the way of broad band, integrating totalization, differentiation.

Service broad band: common applications such as video application should achieve the bandwidth of 2M, while higher-end application should achieve 8M.

Service integrating totalization: provide services of voice, video, data and interconnected enterprises on MAN.

Service differentiation: provide different QoS service according to different client needs. Improve service quality by reliability improvement.

Meanwhile, an important development trend appears. Enterprises satisfy their own needs via public communication network. Video conferencing system, for example, using operator's broad band video service, is not investing MCU switch

equipment necessarily, not maintenance necessarily and not investing network update necessarily.

For another, market change also drives the development of Information and Communication Technology (ICT). The operators used to focus on the communication technology and information technology. Now, they pay close attention to an overall solution about client information and communication. Using public communication, operators provide services such as circuits renting, information software renting, ICT integration and IDC.

1.3. PCN Digitization and Business Computerization

It can be seen from the above that the development of PCN and business itself is a process of evolving from analog to digitization and computerization. From switching manually to today's rise of value-added business and generation-upgrade of softswitch and IMS core network technology; a lot of computing technologies are inducted. The differences in hardware, software and protocol are gradually reduced. Now the only thing end-users need to do is figuring out what resources they need and how to get the related services from network. At the same time, communication network itself becomes easier to extend dynamically.

It also shows from the development of telephone switching, data switching, video conferencing and service platform that communication network is getting more help from computing technologies to meet users' needs and to provide convenient, high-quality and abundant services for end-users. In short, from the beginning of digital communication technology, the progress of computing technologies keeps promoting the improvement of communication technology. Nowadays, the characteristic of cloud computing is more and more remarkable from communication networking itself to its bearing services. Therefore, PCCN which is based on cloud computing technology and combines the characteristics of telecommunication network and computer network emerges. The definition of PCCN will be proposed in the next chapter, and further explanations will be given.

2. PUBLIC COMPUTING COMMUNICATION NETWORKS

2.1. Public Communication and Cloud Computing

2.1.1. The Revolutions of Computing

IT industry has experienced two major revolutions, and there is a qualitative leap between them.

The first revolution is the popularity of personal computers. The equipment for personal work and home entertainment win more favor than that for special scientific computing. Personal computers enable individuals to have their own capacity for computing and storage. Since then, with the popularity of the network, Client/Server pattern begins to be widely used. High-performance servers provide users with the required services such as WWW, Email and so on. It is shown in Figure 2(a).

However, with the number of applications and users increasing, the server-side load is increasingly heavier, which has become a bottleneck to its development. That is the IT industry's second revolution. Distributed computing (Dai, Xie, & Poh, 2003; Yang, Hu, & Guo, 2009; Dai, Pan, & Raje, 2007) grid computing (Dai, Pan, & Raje, 2007; Dai, Pan, & Zou, 2007; Dai, Levitin, & Trivedi, 2007) P2P technology, Web2.0 and so-on have been widely studied and applied, and each person is not only the user of resources but also the provider. So a large number of users share the huge calculation, transmission and storage requirements.

Since then, with the explosive growth of information services, system operation efficiency gets lower, and energy consumption, emission rapidly increase. Cloud computing emerges as the times require; it is shown in Figure 2(b). In the cloud

Figure 2. Revolutions of computing

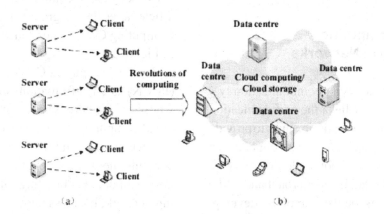

computing model, the server-side hardware and software are integrated into shared resources, in accordance with the "sharing, on-demands, flexibility, services" principle, users only need to connect to the servers via the network to obtain a variety of services and capabilities. There is no need for users to have exclusive computing, storage capacities. Cloud computing has been applied in the Internet, and new business models, new industries, new economy, new lifestyle are appearing.

2.1.2. The Concept of Cloud Computing

Cloud computing originates from telephone network, the industry called the transparent and dark telephone network cloud. In the past, cloud was on behalf of the telephone network, but later it was used in computer network diagrams as the Internet's underlying infrastructure. Actually, the term of cloud is referenced to telecom's telephone network. At the beginning telecommunications company were providing point to point line services, until the 20th century 90 years it started to provide the same quality of virtual private network services whose cost is much lower by reasonably balancing the use flow for more efficient use of overall network bandwidth. In addition, clouds can also be used to distinguish the responsibilities of suppliers and users. Today, the scope of cloud computing have expanded, and it includes not only the network infrastructure but also the servers (Wikipedia, n. d.).

2.2. Public Computing Communication Networks

Communication network is a communication system which organically links the communication endpoints, nodes and transmission links to provide connected or disconnected transmissions between two or more communication endpoints (Gao & Guo, 2009). As the "all IP" and broadband of the telecommunications network rapidly develop, there is a trend between telecommunications networks and computer networks to penetrate and fuse in terminals, access, transmission and business applications and so on. Recently cloud computing has been concerns of the telecommunications industry. The CPU operation speed has been improved, but the I/O speed is still a bottleneck; with the development of communication technology, the speed of network transmission becomes faster and faster and gradually tends to realize the communication between computers through the network and breakthrough the bottleneck of I/O Speed. On the other hand, traditional communication technology is service-oriented, and it provides the public with voice and data transmission services. Now computing technology is also going to be service-oriented, and many new computing models are emerging such as Utility Computing, Service Computing and so on. In short, communication technology and its business are developing towards computer technology and its applications; at the same time, computing technology and its applications are developing towards network and services provided. CT, IT are really moving towards integration and Cloud computing is just the product of the integration. Cloud computing is to integrate the IT resources by the public communication network, and provide users with computing capabilities and applications. Taking example by cloud computing development ideas of Internet, the new trends of telecommunications is to organize the resources of telecommunication network by cloud computing technology, and the article called the integration model PCCN (Public Computing Communication Network), is shown in Figure 3.

Public Computing Communication Network (PCCN) is an information processing network whose architecture is based on virtualization and cloud computing, it uses cloud computing as its core technology, and fuses telecom network and computer network. Using the visualization technology of cloud computing can create a supporting network, business network and unify the in-

Figure 3. PCCN and telecommunication network, clouding computing, computer network

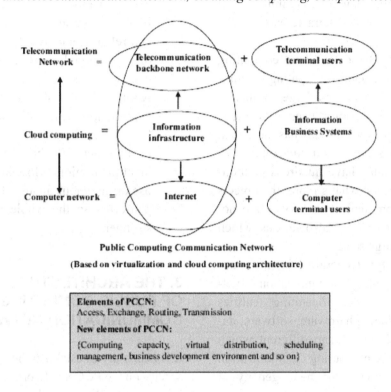

frastructure resource pools; what's more, we can organize and utilize the infrastructure resource pools by the idea of cloud computing. Based on the access, switching, routing, transmission factors of original public communication network, PCCN can also realize the technologies about computing capacity, virtual distribution, scheduling management, and business development and so on. Thus, Public Computing Communication Network will be based on the public communications network which will provide platforms, software and applications in the form of Internet services, it can enhance the ability of the telecommunications business, reduce the investment cost (CAOEX) and operating costs (OPEX), and lower end requirements, but also help operators expand new markets. As network services are developing from information services to the information management services, now public communications network are changing to public computing communication network (PCCN), which will build

distributed data centers corresponding to the core network level, and provide the public information management services, namely, cloud computing. Cloud computing will make people use computers as like using telephone, all the computing and storage capacity in need are just like the communication access, transmission, switching and routing capabilities, and the basic services will be provided by the public basic service providers.

2.3. The Characteristics of PCCN

Inheriting the operation quality assurance of public communication network and learning from the development ideas of cloud computing about Internet, the public computing communication network will organizes the resources of telecommunications networks (including access, scheduling, management, storage, security and services, etc.) using the technology of cloud computing, it will become the new development trend of telecom business. The public computing communication network has the following characteristics:

- **Large scale:** Based on Cloud computing technology, the public computing communication network needs a large number of servers as information processing centers, for example Google already has more than 100 million servers, and Amazon, IBM, Microsoft, yahoo have hundreds of thousands of servers; what's more, the objects of public communication service are not only large in scale but also various, which decides its large scale.

- **Heterogeneity:** Generally, in the public computing communication network, the various operating entities systems(including hardware, software, databases, platforms, etc) are different, which requires public computing communication network to support the heterogeneity of systems, data, networks and different manufacturers, etc.

- **Cooperative:** The public computing communication network provides strong computing capability and rich applications through communication network, and it needs strong interconnection, interoperability and synchronization / asynchronization capabilities between networks, systems, data, interfaces, terminals and so on.

- **Virtualization:** Virtualization is the basis of cloud computing, but also one of the characteristics of the public computing communication network.It completely virtualizes the underlying hardware (including server storage and network equipment), and build a shared basic resource pool which is selected with the needs, so it will form a service-oriented, scalable IT infrastructure.

- **Reliability:** Credible information service is important for the public computing communication network, so the operators must ensure that the operation and storage of reliable user's applications and data is secure and reliable, and avoid the data and codes of commercial secrets, personal privacy lost.

- **Robust:** In the abnormal and dangerous condition, robustness is the key for system to survive. The public computing communication network should be able to implement the minimum loss in the case of misuse or malicious attacks through the cloud platform. So the whole network will be stronger.

3. THE ARCHITECTURE OF PUBLIC COMPUTING COMMUNICATION NETWORK

It is necessary to establish the public computing service on the PCCN network which is described in the paper. Based on the idea, it analyses the architecture of public computing communication and related entities.

3.1. Requirements of Public Computing Services

In recent years, as there are lots of telecom value-added services, the requirements of public computing services for users are provided and appear to be developed diversely. This trend promotes the development of cloud computing in the field of computing and network technology. And it satisfies the diversification of requirements for public computing services. It will be described from three aspects as following.

3.1.1. From the Point of Whether has Third-Party Application Interface

- The cloud computing platform without third-party extension, such as software as a service. Those services are provided for telecom operators and don't provide third-party extension interface. So those own the performance of strong specific application and weak scalability. And Users only use cloud computing services and business model provided by telecom operators completely.

- The cloud computing platform with third-party extension, such as Microsoft cloud OS-Azure. It can provide trusteeship services for third-party. Those services own the performance of strong scalability and users' customization. However, as more and more advanced virtualization technology is required, there have not many telecom operators for this type platform now. This kind of platform can quickly expand the own ability of cloud computing services based on itself platform according to the software store model.

3.1.2. From the Point of Operating Environment

- Private cloud: It belongs to some companies, organization, departments and so on. It only can be used themselves and operated in the private LAN. It aims at lowering the enterprise operation and maintenance costs, reducing hardware resources waste. So this type cloud will compete with public business cloud.
- Public cloud: It mainly belongs to a kind of business model for public and general cloud computing. This cloud will support wide kinds of business and services. It will include many aspects of cloud computing services and provide the public services for lots of ordinary users, especially for some departments and enterprise.

3.1.3. From the Point of Business Model

- Cloud for community services: This cloud mainly provides cloud services for community users, such as blog group and city network business area and so on. As the saying goes, birds of a feather flock together. In the future, cloud computing will provide more and more wide community cloud services for users.
- Cloud for different services: Different application area will produce different type of cloud, e.g. online CRM, ERP service, etc. In the future, there have many SaaS industry services.
- Cloud for basic services: For example, services for demand and elasticity storage management, supercomputing, search engine. These services with the characters of cloud computing will mine users' information fully and provide more high quality cloud computing business model. Based on the performance, it can process to deliver the exact advertising.
- E-marketplace: such as Apple company software store. This kind of platform can provide basic transaction model and some management and marketing method for users' funds and merchandise. It will be the one of most important cloud computing business model.
- Cloud development environment: It can provide quick, convenient, simple, intelligent development interface, such as script and natural language. By way of changing cloud environment on the back stage and matching specific system environment, it can lower the development limitation and attract more SP, AP and person to develop the application.

3.2. The Framework of Public Computing Communication Network (PCCN)

In order to meet the above business requirements, we need to construct Service Open Operational Oriented Intelligent Architecture (SO³IA). The core of NGN is soft exchange and IMS. Similarly, the core of the future computing communication network is SO³IA. Hence, we will describe its three-level framework from coarse to fine, in terms of overall framework, system framework and functional modules, respectively.

3.2.1. L1 Framework

With a reference to ITU-T, L1 framework brings out NGN framework, including factors of network payload and access, computing resources, protocol stack, business analysis, development environment, business applications, operation support, intelligent terminal, guarantee of trustworthiness and etc. It provides computing power, development ability, application ability, business delivery and support ability through network. We illustrate the framework of Pubic Computing Communication Network (PCCN) from the two angles of functionality and region, as in Figure 4.

From the angle of functionality, Operation Support System can be divided into three layers from top to down, namely, Application Layer, Payload Layer and Delivery Layer.

From the angle of region, Public Computing Communication Network (PCCN) can be divided into Backbone Network, Metropolitan Area Network, Access Network and Customer Premises Network. Backbone Network is entity set fulfilling the main transmission function in the communication network. As for telecommunications network, Backbone Network (Takeda, Oki, Inoue, Shiomoto, Fujihara, & Kato, 2008) includes the huge network architecture consisting of communication link between province centre and county centre.

3.2.2. L2 Framework

As in Figure 5, with a reference to NGN, L2 framework brings out the framework of IMS sub-system. Network Attachment Sub-System (NASS) has the ability of resources virtualization, in addition to the ability of access layer registration and the ability of user terminal initialization. Resource and Admission Control Sub-System (RACS) has the ability of resource scheduling, resources virtualization control and business adaptive match, in addition to the ability of resources

Figure 4. Overall concepts chart of Public Computing Communication Network (PCCN) framework

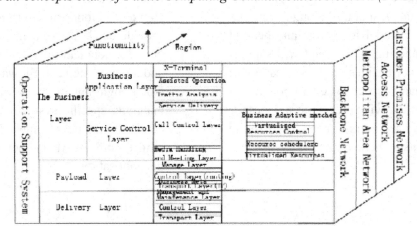

Figure 5. Network element chart of Public Computing Communication Network (PCCN)

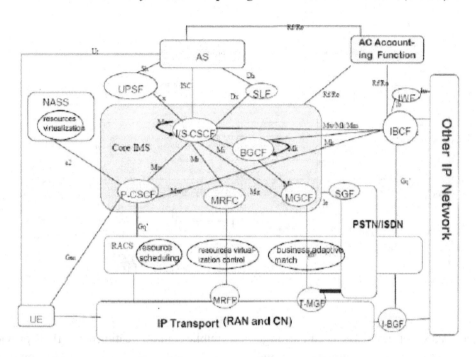

transmission and control, related to the ability of Qos transmission and business control.

3.2.3. L3 Framework Computing Related Entity

As in Figure 6, the related entity of Public Computing Communication Network (PCCN) consist of five layers, including hardware and infrastructure layer, virtualization and cloud management layer, application cloud platform service layer, cloud business service layer and terminal, business delivery and support system, security and guarantee of trustworthiness system.

Hardware and infrastructure layer is at the bottom of the entity, including physical devices of hosts, network and storage. On one hand, virtualization and cloud management layer virtualizes the hardware totally from the physical layer, constructs structured mass data resources management and distributed file system, for the sake of storage and access of mass data. On the other hand, virtualization and management (such as

Xen, VirtualBox, KVM, VMware) realizes functions of on demand dynamic supply, resource configuration, task scheduling, load balance and etc., by virtue of virtualization technology, carries out security administration in aspects of safety and supervisory control, in order to detect the real-time working status of different resources and guarantee the security of data.

Virtualization and cloud management layer provides Infrastructure as a Service (IaaS). It supplies processing ability, storage ability, network and other basic computing resources for the users. Users can use these resources to arrange and run their own software, such as operation systems and application programs. They can control operating systems, storage, arranged application programs, or have the limited control right of network element. However, they cannot manage or control the bottom cloud infrastructure.

Application cloud platform service layer and cloud business service layer are on top of virtualization and cloud management layer. Application cloud platform service layer provides Platform

Figure 6. Computing function entity chart of Public Computing Communication Network (PCCN)

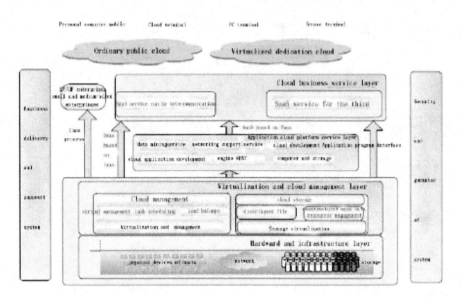

as a Service (PaaS) for users. Users can arrange own or purchased application program onto cloud infrastructure, directly carry out the intelligent and simplified development, arrangement, and web application management on the platform. They can control the arranged application program and application configuration environment, but cannot manage or control the bottom cloud infrastructure. Cloud business service layer is the service layer on top of cloud platform service layer, providing SaaS service for the third part, including self-owned SaaS service and enterprise support system.

On the left, it is the interface of business delivery and support system, providing the end to end penetration of system, business and service process, such as service handling, charging, closing, customer service, etc.

On the right, it is the security and guarantee of trustiness system, providing functions of high quality, safe and trustworthy computing communication business security environment, as well as automatic detection of malfunction and etc.

Public Computing Communication Network (PCCN) uses ordinary public cloud and virtual dedication cloud (such as MPLS, IP virtual dedication cloud, SDH, ATM leased line and etc. in

public communication network) to provide service and control for terminals (personal computer, mobile, cloud terminal, Internet for Things sensors and etc.). Among them, ordinary public cloud provides the services from current network of communication, internet, IT support system for public users. Virtual dedication cloud constructs a virtual private cloud for enterprises, using public computing network resources from Internet service provider, to meet the own development desires of enterprises.

4. KEY TECHNIQUES OF PUBLIC COMPUTING COMMUNICATION NETWORK (PCCN)

4.1. Computing Related Techniques

Distributed Networking Hierarchical Structure Techniques

Computer framework is crucial for the networking of Public Computing Communication Network (PCCN), because servers and components of network are physical carriers which accomplish bottom communication and implement services. Network architecture is a framework formulating different physical components of network, function organization and configuration, network operation rules, data format of program and transmission. It is the premise to practice technologies and services.

Distributed File Systems and Storage Techniques

Since the quantity scale of resources is huge, it is necessary to research and design large-scale file systems, for realizing the classification and efficient management of resources. Hence, it enables the guarantee in essence for providing users in-time and accurate technique support of corresponding services. It is of equal importance to study techniques of server high-speed memory and Solid-State Drive (SSD), which is the key to break through the bottleneck of the times of implementing I/O operations per -second (IOPS). In addition, techniques of energy conservation and cloud container provide a mobile data centre for users. It is convenient to extend computing resources and storage resources. It brings out higher processing speed and less cost than creating the whole data centre.

Virtualization and Cloud Management Layer Techniques

Virtualization and cloud management layer is the core of the whole pubic computing communication network (PCCN). Many kinds of techniques are adopted on this layer. They are the key of implementing the functions of the whole pubic computing communication network (PCCN). The applied primary techniques are heterogeneous resources virtualization and management, data concentration framework, resources matching algorithms, resources scheduling algorithms, heterogeneous virtual-machine live-migration techniques, virtual fault tolerance, disaster preparation techniques

and etc. Heterogeneous resources virtualization (Garbacki & Naik, 2007) is necessary at present when various kinds of systems coexist. It becomes a powerful technique to successfully provide customized resources under the circumstance of grid and data centre. It provides a logical view for data, computing power, storage resources and other resources, rather than a physical view. Therefore, it shields many physical and structural details. In addition, this layer undertakes the important tasks of managing network resources. In other words, pubic computing communication network (PCCN) needs to implement resources matching and resources scheduling, parallel execution/concurrent execution, according to the complex practical applications, for the maximized use of resources. In the meanwhile, it improves the implementing competence and runtime performance of the whole network system, for supplying users higher quality services. As a result, it is necessary to intensively study resources scheduling algorithms. In order to guarantee that pubic computing communication network (PCCN) is able to provide efficient and reliable services, fault tolerance mechanisms are essential and appear to be more important under the circumstance of heterogeneous environment. At present, there are mainly four strategies commonly used for fault tolerance: (1) simply retry; (2) dual-system backup; (3) passive backup; (4) setting checkpoint (Chen & Ren, 2009).

Cloud Terminal Techniques

Beyond doubt, the final end of constructing pubic computing communication network (PCCN) is to provide safe and convenient services for terminal users. However, at present many users face a big security problem that is the terminal security. "Cloud terminal" can help users to resolve the terminal security problem, by way of centralized and unified management to better manage enterprise terminals. Cloud terminal is an ingenuity and extraordinaire network computer with very high price quality. In fact, physical hosts are virtualized into several computers, and each cloud terminal is used as a virtual terminal. Compared with traditional network computers (NC), it has great advantages of prices. Compared with the so called thin clients, it the advantage of saving the expensive software license.

Cloud Test Techniques

Cloud test is a new test technique based on cloud computing. It provides convenience employ for users. However, at present this technique hasn't become a prototype yet. It needs to be further studied. In the whole system implementation and maintenance of pubic computing communication network (PCCN), cloud security needs to be prudent considered. Like the promises of cloud computing, systems can be used as the most reliable and safe data storage centre for users.

4.2. Network Related Techniques

The implementation of pubic computing communication network (PCCN) involves several network related techniques as follows:

Novel Network Devices and Networking Techniques

Similar to the appearance of the new kind based on the previous integration of payload and transmission, the integration of routing and switching in communication network, pubic computing communication network will probably bring out a new kind of network device based on the integration of computing and communication network resources, like integrated communication gateway. Since the changes of traffic modeling, business granule and business logic, we need to revise, upgrade or even re-design the bottom exchange, the contents, algorithms, protocols and theories of routing. In the meanwhile, we also need to explore and research the hierarchy and framework of networking.

High Speed Network Adaptor Techniques

With the exponential increasing speed of Internet link circuit, a large number of network servers will adopt 10Gb/s Ethernet to connect with Internet. However, a large number of high speed network adaptors supporting 10Gb/s will be used on these network servers. Based on this, it becomes a hot research point to design high speed network adaptor (Druschel, Peterson, & Davie, 1994) and optimize its corresponding performances. The technique of Peripheral Component Interconnect Express (PCI-E) deserves a mention. Its main advantage is the high rate of data transmission. Gigabit Interface Converter (GBIC) is a signal converter to be commonly used on Gigabit Ethernet and Fiber Channel. By the standard specification of the converter, the ports of Gigabit Ethernet (Seifert, 1998) devices can directly map to various kinds of entity transmission interfaces, including copper wire, multi-mode fiber and single mode fiber. Small Form-factor Pluggable (SFP) network adaptor with the speed as 10GB/S is a miniaturized version of GBIC. Its size is only half of GBIC. In this way, it saves costs and the ports density of network devices becomes twice.

Network Adaptor Virtualization Techniques

It is necessary to configure network adaptor on the terminal. At present, the already21 released network adaptors supporting virtualization techniques, include SR-IOV and MR-IOV from PCI Special Interest Group (PCI-SIG), VMDq from Intel. Network adaptor techniques can propel the virtualization process.

Hardware Virtual Switches Techniques

Now people have already accepted and begun implementing virtualization techniques step by step. In order to provide ideal performance and reliable security, hardware virtualization techniques are necessary. Hardware virtualization begins with processor, then proceeds to chip set, and proceeds to IO devices. Its development is step by step, because each stage is based on the previous stage. Virtual Machine Device Queues (VMDq) is a dedicated technique to improve the virtual I/O (Seelam & Teller, 2007) performance of network adaptor. In fact, it realizes a half-software and half-hardware virtual machine. Compared to the original pure software scheme, it provides better performance and lower resources occupancy rate.

Protocol Offload and Improving Performance Techniques

High performance computing needs communication delay as short as possible. Communication delay mainly falls into two kinds: the delay of processing messages and the delay of network. Modern high speed Internet achieves very high transmission speed, so the bottleneck of communication transfers from the former Internet to software handling the receiving and dispatching of messages. TCP/IP offload engine (ToE) (Wu & Chen, 2006) is a technique to speed the connection of TCP/IP. It transfers the TCP processing from hosts CPU to dedicated TCP accelerators. In this way, the bottleneck of communication is solved by the speed advantage of hardware.

Network Live Migration and Automatic Configuration Techniques

Live migration techniques are hot topics in the research of virtualization techniques. It enables the transfer of a virtual machine from one physical server to another physical server. It is used for dual-computer fault tolerance or load balance.

4.3. Operation Management Related Techniques

Operation Support Techniques

The operation support of computing communication network mainly focuses on the end to end penetration of business process, such as service handling, charging and closing, customer service, etc. Service handling includes the activation, variation, remove of service, etc. Charging and closing technique mainly includes techniques of charging subscribers, subscribers account management, subscribers credit control, middle-wares collecting telephone fares, computing rules engine, charging based on data granularity and etc. Customer service techniques mainly include handling customer explains, customer explains feedback and tracking, customer retention, customer loyalty management and etc. In addition, we also need to study charging business model, such as paying monthly or making a price according to traffic and talking time, according to the period of time based on data peak and valley, etc.

Security and Guarantee of Trustiness Techniques

The security key of computing communication network focuses on: data protection, identity management, security holes management, physical and personal security, application program security, time corresponding and privacy measures. The main security techniques are virtual machine separation techniques, sensitive data encryption techniques, the security authentication techniques of data access, audit techniques, automatic data backup and recovery techniques, intrusion detection techniques and etc. In addition, we also need to study the institution and procedure of security management. We need to guarantee security and trustworthy in aspects of techniques and management.

5. CONCLUSION

With the popularization and improvement of social and industry informatization, information appears to increase explosively. People need urgently not only rapid promotion of the ability of providing high bandwidth and mobile basic communication network services, but also resources which support comprehensive information services. At the same time, they are trending towards getting "information management services". The information resources of the whole society should be integrated to the max and also should be constructed or operated as public infrastructure so that the efficiency can be improved, the resources can be saved and the requirement of the development of society and economy can be satisfied. For that reason, PCN should evolve to PCCN and provide cloud computing services (Li, 2009) to the public by creating data center correspond to core network switch level. At present, the architecture and key technology of PCCN is understudy and development. So the following works should be done pressingly: (1) Turn the computing resources into network elements, form a optimal architecture of network, platform, terminal and service; (2) Formulate the compatible and interoperable technical standard of computing resources; (3) Integrate the research, manufacture, network constructing, application development and service supply, build a more open business model.

REFERENCES

Ali, S. R. (1986). Analysis of total outage data for stored program control switching systems. *IEEE Journal on Selected Areas in Communications*, 4(7), 1044–1046. doi:10.1109/JSAC.1986.1146441

Chen, N., & Ren, S. (2009). Adaptive optimal checkpoint interval and its impact on system's overall quality in soft real-time applications. In *Proceedings of the 24th ACM Symposium on Applied Computing* (pp. 1015-1020).

Dai, Y. S., Levitin, G., & Trivedi, K. S. (2007). Performance and reliability of tree-structured grid services considering data dependence and failure correlation. *IEEE Transactions on Computers*, 56(7), 925–936. doi:10.1109/TC.2007.1018

Dai, Y. S., Pan, Y., & Raje, R. (2007). *Advanced parallel and distributed computing: Evaluation, improvement and practices*. New York, NY: Nova Science.

Dai, Y. S., Pan, Y., & Zou, X. K. (2007). A hierarchical modeling and analysis for grid service reliability. *IEEE Transactions on Computers*, 56(5), 681–691. doi:10.1109/TC.2007.1034

Dai, Y. S., Xie, M., & Poh, K. L. (2003). A study of service reliability and availability for distributed systems. *Reliability Engineering & System Safety*, 79(1), 103–112. doi:10.1016/S0951-8320(02)00200-4

Druschel, P., Peterson, L. L., & Davie, B. S. (1994). Experiences with a high-speed network adaptor: A software perspective. *Computer Communication Review*, 24(4), 2. doi:10.1145/190809.190315

Gao, H., & Guo, J. (2009). Application of vulnerability analysis in electric power communication network. In. *Proceedings of the International Conference on Machine Learning and Cybernetics*, 4, 2072–2077. doi:10.1109/ICMLC.2009.5212210

Garbacki, P., & Naik, V. K. (2007). Efficient resource virtualization and sharing strategies for heterogeneous grid environments. In *Proceedings of the 10th IFIP/IEEE International Symposium on the Integrated Network Management* (pp. 40-49).

Houssos, N., Gazis, E., Panagiotakis, S., Gessler, S., Schuelke, A., Quesnel, S., et al. (2002). Value added service management in 3G networks. In *Proceedings of the IEEE/IFIP Network Operations and Management Symposium* (pp. 529-544).

Li, L. (2009). An optimistic differentiated service job scheduling system for cloud computing service users and providers. In *Proceedings of the 3rd International Conference on Multimedia and Ubiquitous Engineering* (pp. 295-299).

Loreto, S., Mecklin, T., Opsenica, M., & Rissanen, H.-M. (2010). IMS service development API and testbed. *IEEE Communications Magazine, 48*(4), 26–32. doi:10.1109/MCOM.2010.5439073

Seelam, S. R., & Teller, P. J. (2007). Virtual I/O scheduler: A scheduler of schedulers for performance virtualization. In *Proceedings of the 3rd International Conference on Virtual Execution Environments* (pp. 105-115).

Seifert, R. (1998). *Gigabit ethernet*. Reading, MA: Addison-Wesley.

Takeda, T., Oki, E., Inoue, I., Shiomoto, K., Fujihara, K., & Kato, S. (2008). Implementation and experiments of path computation element based backbone network architecture. *IEICE Transactions on Communications, 91*(8), 2704–2706. doi:10.1093/ietcom/e91-b.8.2704

Wikipedia. (n. d.). *Cloud computing*. Retrieved from http://en.wikipedia.org/wiki/Cloud_computing

Wu, Z. Z., & Chen, H. C. (2006). Design and implementation of TCP/IP offload engine system over gigabit Ethernet. In *Proceedings of the 15th International Conference on Computer Communications and Networks* (pp. 245-250).

Yang, H., Hu, S., & Guo, J. (2009). Cost-oriented task allocation and hardware redundancy policies in heterogeneous distributed computing systems considering software reliability. *Computers & Industrial Engineering, 56*(4), 1687–1696. doi:10.1016/j.cie.2008.11.001

This work was previously published in the International Journal of Cloud Applications and Computing (IJCAC), Volume 1, Issue 4, edited by Shadi Aljawarneh & Hong Cai, pp. 44-59, copyright 2011 by IGI Publishing (an imprint of IGI Global).

Chapter 20

Fault Tolerant Architecture to Cloud Computing Using Adaptive Checkpoint

Ghalem Belalem
University of Oran (Es Senia), Africa

Said Limam
University of Oran (Es Senia), Africa

ABSTRACT

Cloud computing refers to both the applications delivered as services over the Internet and the hardware and systems software in the datacenters that provide those services. Failures of any type are common in current datacenters, partly due to the number of nodes. Fault tolerance has become a major task for computer engineers and software developers because the occurrence of faults increases the cost of using resources and to meet the user expectations, the most fundamental user expectation is, of course, that his or her application correctly finishes independent of faults in the node. This paper proposes a fault tolerant architecture to Cloud Computing that uses an adaptive Checkpoint mechanism to assure that a task running can correctly finish in spite of faults in the nodes in which it is running. The proposed fault tolerant architecture is simultaneously transparent and scalable.

1. INTRODUCTION

Cloud computing can be defined as a new style of computing in which dynamically scalable and often virtualized resources are provided as a services over the Internet. Cloud computing has become a significant technology trend, and many experts expect that cloud computing will reshape information technology (IT) processes and the IT marketplace. With the cloud computing technology, users use a variety of devices, including PCs, laptops, smartphones, and PDAs to access programs, storage, and application-development platforms over the Internet, via services offered by cloud computing providers. Advantages of the

DOI: 10.4018/978-1-4666-1879-4.ch020

cloud computing technology include cost savings, high availability, and easy scalability (Borko & Armando, 2010).

With the development of virtualization technology, more and more companies are using virtual machines to improve the utilization of resources. In a cloud computing environment, companies may purchase computing and storage capability from cloud providers, such as Amazon EC2, and then use the purchased resources to provide services, such as web application services, to users. Technically, Amazon EC2 (Kaushal & Anju, 2011) assigns virtual machines to those companies which deploy the application instances on those VMs. It is expected that many instances of the same application can run on different virtual machines in the same cloud environment (http://aws.amazon.com/ec2/).

Cloud computing refers to both the applications delivered as services over the Internet and the hardware and systems software in the datacenters that provide those services. The services themselves have long been referred to as Software as a Service (SaaS), so we use that term. The datacenter hardware and software is what we will call a Cloud (Armbrust et al., 2009).

For cloud service, two issues are of great importance, the reliability and the performance of cloud service. Cloud service reliability is concerned with how probable that the cloud can successfully provide the service requested by users. Cloud service performance, on the other hand, is concerned with how fast the cloud can provide the requested service.

As a rule of thumb, all computer engineers know that when a computer system becomes more complex, the system's susceptibility to faults increases. Such rule is fully applicable to Cloud Computing.

Failures of any type are common in current datacenters, partly due to the number of nodes. In spite of the quality improvement in each single component of the nodes, one may argue that, as certain as the sun will rise tomorrow, any com-

ponent will suffer some kind of failure: a disk can break down or a link can fail anytime, and this can lead to computation time wasting when executing medium and long-running tasks and the crash occurs before finishing the task execution. For instance, if a task that takes 24 hours is executing in a node, and this node crashes 5 minutes before the task execution ends, almost one day of execution will be wasted (Goiri et al., 2010).

Dealing with fault tolerance has become a major task for computer engineers and software developers because the occurrence of faults increases the cost of using resources. However, the inclusion of fault tolerance in a system increases the complexity of such system from the user point of view.

In this paper, we proposed a fault tolerant architecture to Cloud Computing that uses a Checkpoint mechanism to assure that a task running can correctly finish in spite of faults in the nodes in which it is running. This mechanism records the system state periodically to establish recovery points. Upon a node crash, the last checkpoint can be restored, and task execution can be resumed from that point. In the development of such architecture, we assumed that, in order to attend the user expectations, the first user expectation is, of course, that his/her application correctly finishes independent of faults in the node. This is obviously the main obligation of any fault-tolerant architecture for Cloud Computing and demands no further discussion. The next requirement is that the fault-tolerant architecture must be easy to use. A fault-tolerant architecture for Cloud Computing must simultaneously be transparent, and scalable.

The rest of this paper is organized as follows. Section 2, the Checkpoint mechanism is presented. In Section 3 we summarize important related works in Cloud Computing. Section 4 presents our fault tolerant architecture, Section 5 shows experiments and Section 6 concludes the paper.

2. CHECKPOINTING

The checkpoint and rollback technique has been widely used in distributed systems. In distributed systems components may be geographically or logically remote. They cooperate by exchanging messages. Checkpointing of such systems poses a number of challenging problems.

Checkpointing protocols require the processes to take periodic checkpoints with varying degrees of coordination. At one end of the spectrum, co-ordinated checkpointing requires the processes to coordinate their checkpoints to form global consistent system states. Coordinated checkpointing generally simplifies recovery and garbage collection, and yields good performance in practice. At the other end of the spectrum, uncoordinated checkpointing does not require the processes to coordinate their checkpoints, but it suffers from potential domino effect, complicates recovery, and still requires coordination to perform output commit or garbage collection. Between these two ends are communication-induced checkpointing schemes that depend on the communication patterns of the applications to trigger checkpoints. These schemes do not suffer from the domino effect and do not require coordination. Recent studies, however, have shown that the nondeterministic nature of these protocols complicates garbage collection and degrades performance (Elnozahy et al., 2002).

Without checkpointing upon failure of the system typically all work performed so far is lost and the computation has to start anew. Checkpointing aims at reducing the amount of work that is lost upon failure of the system by intermediately saving the whole state of the system. Saving the system state usually comes with some cost, or overhead. The checkpointing cost can be the time needed for checkpointing. It can also be some cost incurred by system operation. The checkpoint latency is the time needed to establish a checkpoint. A checkpoint is also called recovery point. If no failure happens checkpointing is seen purely as retarding system operation. If a failure happens during computation, on the other hand, then the system does not have to roll back to the initial state, instead it can roll back to the most recent checkpoint (Wolter, 2010).

There exist three types of implementations of checkpointing. Checkpointing can be integrated into applications by the application programmer. Checkpointing can also be implemented as a function of the operating system. This is called system-initiated checkpointing. Cooperative checkpointing combines both application-initiated as well as system initiated checkpointing.

In general, a checkpoint is a snapshot of the entire state of the process at the moment it was taken. It represents all the information that we would need to restart the process from that point. We record the checkpoint on stable storage, i.e., storage in whose reliability we have sufficient confidence.

Different solutions using checkpoints have been proposed in the literature. For example, (Vallee et al., 2006) proposes using a union file system in order to save time storing VM checkpoints. It also introduces a remote storage in order to store the checkpoints and introduces the storage of VM disk space during the checkpoint phase. In addition, it also proposes to store only the differences in the disk.

In Ta-Shma et al. (2008) authors present a CDP (Continuous Data Protection) with live-migration-based checkpoint mechanism. They use a central repository approach and intercept migration data flow to create the checkpoint images. Although the authors say that it has good performance, no experimentation is presented.

Parallax developed by Warfield et al. (2005) is a storage subsystem for Xen to be used in cluster Xen Virtual Machines. The solution proposed by the authors makes coupled checkpoints of both memory and disk using a Copy-on-Write mechanism (CoW) to maintain the remote images.

Goiri et al. (2010) propose a smart checkpoint for virtualized service providers. It uses Another

Union File System to differentiate read-only from read-write parts in the virtual machine image. In this way, read-only parts can be checkpointed only once, while the rest of checkpoints must only save the modifications in read-write parts, thus reducing the time needed to make a checkpoint. The checkpoints are stored in a Hadoop Distributed File System.

3. FAULT-TOLERANT IN CLOUD

Cloud computing is expected to be the platform for next generation computing, in which users carry thin clients such as smart phones while storing most of their data in the cloud and submitting computing tasks to the cloud. A web browser serves as the interface between clients and the cloud. Operating system in web browsers allows the users to manage their data and computation tasks. One of the main drivers for the interest in cloud computing is cost and reliability (Deng et al., 2010). Providing highly reliable cloud service is a challenging and critical research problem. It is too expensive to provide redundant alternative components for all the cloud components to reduce the cost and to develop highly reliable cloud applications within the limited budget.

Cloud computing and virtualization have opened a new window in the failure management. Pausing, resuming, and migrating VMs are powerful mechanisms to manage failures in such environments. A VM can be easily migrated to another node when a failure is predicted or detected. Although migration is a good solution in order to deal with failures, it has two main problems. First, it has considerable overhead, especially if entire VM images have to be migrated. Second, the prediction and advance detection of failures is key issue. We cannot start the migration of a VM if its running node has already failed (Goiri et al., 2010).

Deng et al. (2010) propose techniques to improve the fault-tolerance and reliability of a rather general scientific computation: matrix multiplication. Matrix multiplication serves as the foundation for many complex problem solving and optimization. They investigate a cloud selection strategy to decompose the matrix multiplication problem into several tasks which will be submitted to different clouds and they demonstrate that fault-tolerance and reliability against faulty and even malicious clouds in cloud computing can be achieved.

In Cloud Storage Bonvin et al. (2010) propose a self-managed key-value store that dynamically allocates the resources of a data cloud to several applications in a cost efficient and fair way. The proposed approach offers and dynamically maintains multiple differentiated availability guarantees to each different application despite failures. We employ a virtual economy, where each data partition (i.e., a key range in a consistent-hashing space) acts as an individual optimizer and chooses whether to migrate, replicate or remove itself based on net benefit maximization regarding the utility offered by the partition and its storage and maintenance cost.

Zheng et al. (2010) propose FTCloud which is a component ranking based framework for building fault-tolerant cloud applications. FTCloud employs the component invocation structures and the invocation frequencies to identify the significant components in a cloud application. An algorithm is proposed to automatically determine optimal fault tolerance strategy for these significant components.

4. FAULTS TOLERANT ARCHITECTURE

This section describes our architecture and implementation for supporting fault-tolerance in Cloud Computing. Our proposed architecture can be divided into four phases. The phases are shown in Figure 1.

Figure 1. Operational phases

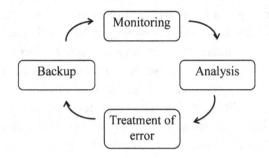

A. Backup Phase (State save)

In this phase we record the system state periodically to establish recovery points. Upon a node crash, the last checkpoint can be restored, and task execution can be resumed from that point. This phase is the major responsible for resources consumed by the fault tolerance mechanism as well as for the enlargement in the execution time in the absence of failures.

B. Monitoring Phase

This phase corresponds to the supervision of the application, it consists in collecting information about existing resources (nodes failing, Data Center Overload) to provide to the following steps.

C. Analysis Phase

From the previous phase with the aim of an efficient decision making and correct. It consists in analyzing the information produced by the monitoring phase in order to identify system status.

D. Treatment of Error (Error Handling)

This is the last step; it corresponds to the implementation of measures to restore the system. This phase is triggered when the analysis phase detects a failure, it is to study the reports gener-

ated during this last phase at the end to take the necessary decisions.

Thus, this phase is to change some settings to adjust the system behavior based on the analysis conducted during the previous phase (Figure 2). The error handling is guided by rules for selecting the most appropriate mechanism:

1. Monitor agent
    ◦ Control the load balancing of Datacenter;
    ◦ Failure detection using a heartbeat watchdog mechanism.
2. Treatment of error agent:
    ◦ Identifies the state of the Datacenter;
    ◦ Change the checkpoint interval and watchdog cycle;
    ◦ Selects the least loaded Datacenter and select an available node;
    ◦ Restart failed processes from their most recent checkpoints.

To execute tasks, provider uses Virtual Machines (VM) that are created on demand. The creation of a VM image involves dealing with a large amount of data, including the installation of the guest operating system and the deployment of the required software.

Making a checkpoint of task running within a VM must include all the information needed to resume the task execution in another node: the task context, the memory content, and the disks.

In order to make a checkpoint, we need to save the task current status (basically its memory and disk contents). Taking a checkpoint increases the application execution time: this increase is defined as the checkpoint overhead. The checkpoint interval also defined the number of checkpoints taken during the executions. If the checkpoint interval has expired, the checkpoint program suspends virtual machine to be checkpointed.

The interaction between the application and the checkpoints determines the enlargement of the

Figure 2. Fault tolerant architecture

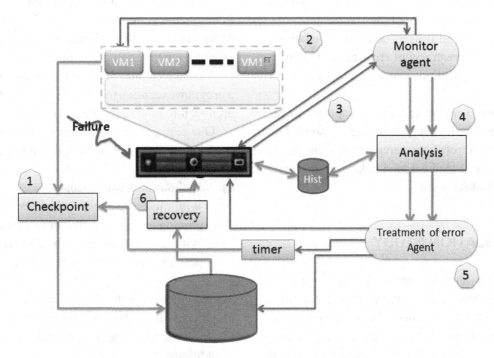

Figure 3. Checkpoint algorithm

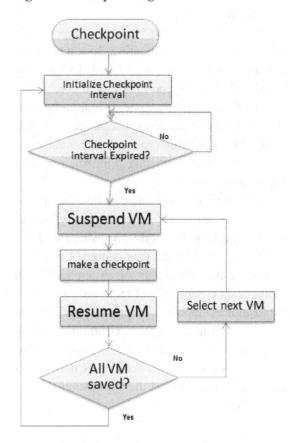

application execution time. An optimal checkpoint interval minimizes the checkpoint overhead.

Failure detection is through a heartbeat mechanism: the nodes periodically emit heartbeats or "I am alive" messages. If a sufficiently long sequence of heartbeat messages is missed from a node, it is declared to have failed.

A good choice of the heartbeat/watchdog cycle value in the monitoring process is essential when optimizing failure detection. The heartbeat/watchdog cycle determines how fast the monitor agent will detect a failure. Short cycles reduce the response time, but also increase the interference over the communication channel, while if the timeout is too long reduces the sensibility of the failure detection mechanism; because the watchdog will take more time to detect a heartbeat missing.

We have used an adaptive checkpoint interval and heartbeat cycle (Figure 3). The checkpoint interval and watchdog cycle vary according to the state of the Datacenter. The number of free resources, the occurrence of a new fault and failure history defined the state of the Datacenter.

When the monitor agent detects a node failure, it informs the treatment of error agent who must restart the applications that were running on that node.

The recovery procedure corresponds to restarting a task since its previous checkpoint. The treatment of error agent searches a host with lesser computational load for load balance and selects the recent checkpoint.

5. SIMULATION STUDY

To test our fault-tolerant architecture we need tools that allow us to evaluate the hypothesis prior to real deployment in an environment where one can reproduce tests. Access to the infrastructure in Cloud Computing incurs payments in real currency; simulation-based approaches offer significant benefits.

Currently there is no Simulator support fault tolerance for Cloud Computing. To test our fault tolerance architectures we extended CloudSim (version 2.1.1) to support fault tolerance.

CloudSim is a new, generalized, and extensible simulation framework that enables seamless modeling, simulation, and experimentation of emerging Cloud computing infrastructures and application services.

In CloudSim, Cloudlet models the Cloud-based application services (content delivery, social networking, business workflow), which are commonly deployed in the data centers. CloudSim represents the complexity of an application in terms of its computational requirements. Every application component has a pre-assigned instruction length (inherited from GridSim's Gridlet component) and amount of data transfer (both pre and post fetches) that needs to be undertaken for successfully hosting the application (Buyya et al., 2009).

Task completion time is considered in general in unreliable systems that are subject to failures. Mechanisms can be evaluated by how much they improve the task completion time at what cost.

We have created many fault scenarios during a program execution to evaluate our fault-tolerant architecture in the presence of failures.

Our test procedure consisted in measuring the Cloudlet runtimes in the following scenarios:

a) Without fault tolerance;
b) With fault tolerance, with different checkpoint intervals, in different fault scenarios.
c) With fault tolerance, with adaptive checkpoint, with different fault scenarios.

We calculated the total execution time for cloudlets without fault tolerance and compared the results against the total execution time with fault tolerance.

The first group of tests served to evaluate the behavior of system in the presence of failures and without fault tolerance.

The Figure 4 contains the results for four tests, and we calculated the number of failure for each test. The simulation environment consisted of a data center with 50 hosts, where each host was modeled to have four CPU core (1000MIPS), 2GB of RAM memory. We modeled the user (through the DatacenterBroker) to request creation of 100 VMs having following constraints: 512MB of physical memory, 1 CPU core. We have varied the number of cloudlets (100,150,200 and 300) with each Cloudlet unit requiring 120000 million instructions to be executed on a host

The second experiment consisted to study the influence of the checkpoint interval. We experiment the system with different checkpoint intervals (40, 80, 120 and 240 seconds) and compared the results with the executions without fault tolerance.

The checkpoint interval had more influence over the behavior of the system. This influence

Figure 4. Influence of faults over the number of cloudlets

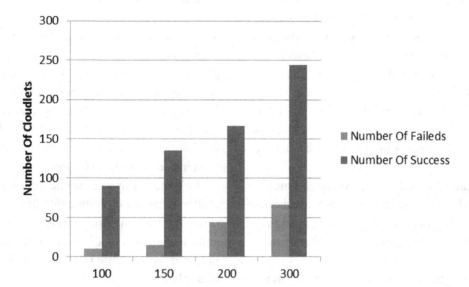

of the checkpoint over the system's behavior is consequence of the time needed for checkpointing.

The checkpoint interval also defined the number of checkpoints taken during the executions.

In this simulation we created one data center. A data center in the system is modeled to have 50 computing hosts, each host have 2GB of memory, 4 processors with 1000 MIPS of capacity. Data center broker on behalf of the user requests instantiation of a VM that requires 512 MB of memory, 1GB of storage, 1 CPU. The broker requests instantiation of 100 VMs and associates tow Cloudlet to each VM to be executed. Each Cloudlet is modeled to be having 120000 MIs.

In Figure 5, we present a summary of these tests. The figure represents the total execution time (in seconds) for cloudlets.

Figure 5. Impact of checkpoint interval over the execution time

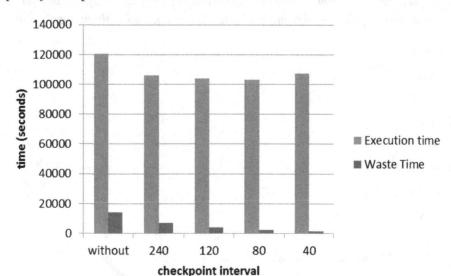

The enlargement in the execution times increases when the checkpoint interval is shorter because more checkpoints occur during the program execution. The waste times increases when the checkpoint interval is larger in the presence of multiple failures.

The Figure 6 contains the results for three scenarios: without fault tolerance and with fault tolerance using checkpoint interval and adaptive checkpoint.

The figure shows the total execution time (in seconds) of cloudlets and the time overheads caused by the operation of the fault tolerance mechanism; we have varied the number of fault per 5 for each three scenarios.

Analyzing the curves, we see that the architecture using adaptive checkpoint is better than the architecture using periodic checkpoint. This occurred because the checkpoint interval had more influence over the behavior of the system.

The results obtained with different scenarios shown that the interference of the fault tolerance operation have not imposed a strong overhead over the execution time.

6. CONCLUSION

With the cloud computing technology, users use a variety of devices, including PCs, laptops, smartphones, and PDAs to access programs, storage, and application development platforms over the Internet, via services offered by cloud computing providers. The size and the complexity of machines in cloud computing environments will continue to increase in the near future.

Fail inevitably; an application will fail, no matter what its environment. In this scenario, users and system administrators will need tools and mechanisms that help them to manage failures and with as little influence on the execution of application.

In this paper, we have proposed a fault tolerant architecture to Cloud Computing. Our architecture implements an adaptive checkpoint to assure that an application running can correctly finish in spite of faults in the nodes in which it is running.

We have performed a series of experiments to evaluate the cost of proposed fault tolerant architecture in many scenarios. Results show that the use of our fault tolerant architecture reduces the waste time caused by faults and have not imposed a strong overhead over the execution time.

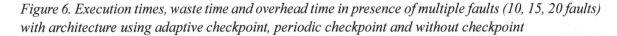

Figure 6. Execution times, waste time and overhead time in presence of multiple faults (10, 15, 20 faults) with architecture using adaptive checkpoint, periodic checkpoint and without checkpoint

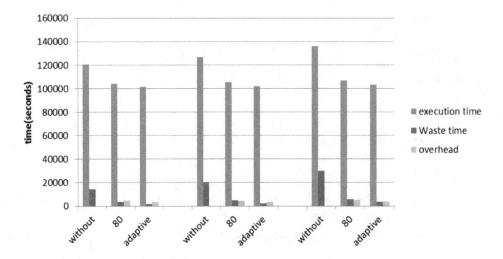

REFERENCES

Armbrust, M., Fox, A., Griffith, R., Joseph, A. D., Katz, R. H., Konwinski, A., et al. (2009). *Above the clouds: A Berkeley view of cloud computing* (Tech. Rep. No. UCB/EECS-2009-28). Berkeley, CA: University of California at Berkeley.

Bonvin, N., Papaioannou, T. G., & Aberer, K. (2010). A self-organized, fault-tolerant and scalable replication scheme for cloud storage. In *Proceedings of the ACM Symposium on Cloud Computing*, Indianapolis, IN.

Borko, F., & Armando, E. (2010). Cloud computing fundamentals. In Furht, B., & Escalante, A. (Eds.), *Handbook of cloud computing*. New York, NY: Springer.

Buyya, R., Ranjan, R., & Calheiros, R. N. (2009, June 21-24). Modeling and simulation of scalable cloud computing environments and the CloudSim toolkit. In *Proceedings of the International Conference on High Performance Computing & Simulation*, Leipzig, Germany (pp. 1-11).

Deng, J., Huang, S. C.-H., Han, Y. S., & Deng, J. H. (2010). Fault-tolerant and reliable computation in cloud computing. In *Proceedings of the IEEE Globecom Workshop on Web and Pervasive Security*, Miami, FL.

Elnozahy, M., Alvisi, L., Wang, Y.-M., & Johnson, D. B. (2002). A survey of rollback-recovery protocols in message-passing systems. *ACM Computing Surveys*, *34*(3), 375–408. doi:10.1145/568522.568525

Goiri, I., Julià, F., Guitart, J., & Torres, J. (2010, April 19-23). Checkpoint-based fault-tolerant infrastructure for virtualized service providers. In *Proceedings of the 12th IEEE/IFIP Network Operations and Management Symposium*, Osaka, Japan (pp. 455-462).

Kaushal, V., & Anju, B. (2011). Autonomic fault tolerance using HAProxy in cloud environment. *International Journal of Advanced Engineering Sciences and Technologies*, *7*(2), 222–227.

Ta-Shma, P., Laden, G., Ben-Yehuda, M., & Factor, M. (2008). Virtual machine time travel using continuous data protection and checkpointing. *ACM SIGOPS Operating Systems Review*, *42*, 127–134. doi:10.1145/1341312.1341341

Vallee, G., Naughton, T., Ong, H., & Scott, S. (2006, October 17). Checkpoint/restart of virtual machines based on Xen. In *Proceedings of the High Availability and Performance Computing Workshop*, Santa Fe, NM.

Warfield, A., Ross, R., Fraser, K., Limpach, C., & Hand, S. (2005, June 12-15). Parallax: Managing storage for a million machines. In *Proceedings of the 10th Workshop on Hot Topics in Operating Systems*, Santa Fe, NM (pp. 1-11).

Wolter, K. (2010). *Stochastic models for fault tolerance - Restart, rejuvenation and checkpointing*. New York, NY: Springer.

Zibin, Z., Zhou, T. C., Lyu, M. R., & King, I. (2010). FTCloud: A component ranking framework for fault-tolerant cloud applications. In *Proceedings of the 21th IEEE International Symposium on Software Reliability Engineering* (pp. 398-407).

This work was previously published in the International Journal of Cloud Applications and Computing (IJCAC), Volume 1, Issue 4, edited by Shadi Aljawarneh & Hong Cai, pp. 60-69, copyright 2011 by IGI Publishing (an imprint of IGI Global).

Compilation of References

Abdelsalam, H. S., Maly, K., Mukkamala, R., Zubair, M., & Kaminsky, D. (2009). Analysis of energy efficiency in clouds. In *Proceedings of the Computation World: Future Computing, Service Computation, Cognitive, Adaptive, Content, Patterns* (pp. 416-421).

AbdelSalam, H., Maly, K., Mukkamala, R., Zubair, M., & Kaminsky, D. (2009). Towards energy efficient change management in a cloud computing environment. In R. Sadre & A. Pras (Eds.), *Proceedings of the 3rd International Conference on Autonomous Infrastructure, Management and Security: Scalability of Networks and Services* (LNCS 5637, pp. 161-166).

Abdi, H. (2009). *The methods of least squares*. Dallas, TX: The University of Texas.

Aerts, A. T. M., Szirbik, N. B., & Goossenaerts, J. B. M. (2002). A flexible agent-based ICT for virtual enterprises. *Journal of Computers in Industry, 49*(3). doi:10.1016/S0166-3615(02)00096-9

Aggarwal, B. B., & Barnes, M. (2010). *The case for cloud CRM in India*. Beijing, China: Springboard Research.

Ahonena, H., Alvarengaa, A. G., Provedela, A., & Paradab, V. (2010). A client-broker-server architecture of a virtual enterprise for cutting stock. *International Journal of Computer Integrated Manufacturing, 14*(2), 194–205. doi:10.1080/09511920150216314

Aiftimiei, C., Andreozzi, S., Cuscela, G., Donvito, G., Dudhalkar, V., Fantinel, S., et al. (2007). Recent evolutions of gridice: A monitoring tool for grid systems. In *Proceedings of the Workshop on Grid Monitoring* (pp. 1-8). New York, NY: ACM Press.

Akamai. (2009). *Solutions for cloud computing: Accelerate, scale and fortify applications and platforms running in the cloud*. Retrieved from http://www.akamai.com/dl/brochures/Cloud_Computing_Brochure.pdf

Alexandre, R., Prata, P., & Gomes, A. (2009). A grid infrastructure for online games. In *Proceedings of the 2nd International Conference on Interaction Sciences* (pp. 670–673). New York, NY: ACM Press.

Ali, S. R. (1986). Analysis of total outage data for stored program control switching systems. *IEEE Journal on Selected Areas in Communications, 4*(7), 1044–1046. doi:10.1109/JSAC.1986.1146441

Aljawarneh, S. (2011). A web engineering security methodology for e-learning systems. *Network Security Journal, 2011*(3), 12-16.

Amazon Web Services. (n. d.). *Amazon's CloudWatch monitoring system*. Retrieved from http://aws.amazon.com/cloudwatch/

Amazon. (2010). *Amazon elastic compute cloud*. Retrieved from http://aws.amazon.com/ec2/

Amazon. (2010). *Amazon web services: Overview of security processes*. Retrieved from awsmedia.s3.amazonaws.com/pdf/AWS_Security_Whitepaper.pdf

Amazon. (n. d.). *EC2 pricing*. Retrieved from http://aws.amazon.com/ec2/pricing/

American Power Conversion. (2003). *Determining total cost of ownership for data center and network room infrastructure*. Retrieved from http://www.lamdahellix.com/%5CUserFiles%5CFile%5Cdownloads%5C6_whitepaper.pdf

Amir, M. S. (2010). It's written in the cloud: the hype and promise of cloud computing. *Journal of Enterprise Information Management, 23*(2), 131–134. doi:10.1108/17410391011019732

Amit, G., Heinz, S., & David, G. (2010). Formal models of virtual enterprise architecture: Motivations and approaches. In *Proceeding of the Pacific Climate Information System Workshop* (pp. 1207-1217).

Amrhein, D., Anderson, P., & de Andrade, A. (2010). *Cloud computing use cases white paper.* Retrieved from http://opencloudmanifesto.org/Cloud_Computing_Use_Cases_Whitepaper-4_0.pdf

Armbrust, M., Fox, A., Griffith, R., Joseph, A. D., Katz, R. H., Konwinski, A., et al. (2009). *Above the clouds: A Berkeley view of cloud computing* (Tech. Rep. No. UCB/EECS-2009-28). Berkeley, CA: University of California at Berkeley.

Armbrust, M., Fox, A., Griffith, R., Joseph, A. D., Katz, R., & Konwinski, A. (2010). A view of cloud computing. *Communications of the ACM, 53*, 4. doi:10.1145/1721654.1721672

Arshad, J. (2009). *Integrated intrusion detection and diagnosis for clouds.* Paper presented at the 39th Annual IEEE International Conference on Dependable Systems and Networks, Lisbon, Portugal.

Arshad, J., & Townend, P. (2009). Quantification of security for compute intensive workloads in clouds. In *Proceedings of the 15th International Conference on Parallel and Distributed Systems* (pp. 479-486). Washington, DC: IEEE Computer Society.

Arshad, J., Townend, P., & Xu, J. (2010a). An automatic approach to intrusion detection and diagnosis for clouds. *International Journal of Automation and Computing.*

Arshad, J., Townend, P., & Xu, J. (2010b). An intrusion diagnosis perspective to cloud computing. *International Journal of Automation and Computing.*

Arshad, J., Townend, P., Xu, J., & Jie, W. (2010). Cloud computing security: Opportunities and pitfalls. *International Journal of Cluster Computing.*

Arthur, C. (2010). *Google's ChromeOS means losing control of data, warns GNU founder Richard Stallman.* Retrieved from http://www.guardian.co.uk/technology/blog/2010/dec/14/chrome-os-richard-stallman-warning

Assuncao, M. D., Costanzo, A., & Buyya, R. (2010). A cost-benefit analysis of using cloud computing to extend the capacity of clusters. *Journal of Cluster Computing, 13*, 335–347. doi:10.1007/s10586-010-0131-x

Averitt, S., Bugaev, M., Peeler, A., Schaffer, H., Sills, E., Stein, S., et al. (2007, May 7-8). The virtual computing laboratory. In *Proceedings of the International Conference on Virtual Computing Initiative*, Triangle Park, NC.

Avetisyan, A., Campbell, R., Gupta, I., Heath, M., Ko, S., & Ganger, G. (2010). Open cirrus: A global cloud computing testbed. *IEEE Computer, 43*(4), 42–50.

Aymerich, F. M., Fenu, G., & Surcis, S. (2008, August). An approach to a cloud computing network. In *Proceedings of the 1st International Conference on the Applications of Digital Information and Web Technologies* (pp. 113-118).

Bain, T. (2009). *Is the relational database doomed?* Retrieved from http://www.readwriteweb.com/enterprise/2009/02/is-the-relational-database-doomed.php?p=2

Barnard, K., Duygulu, P., Forsyth, D., De Freitas, N., Blei, D. M., & Jordan, M. I. (2003). Matching words and pictures. *Journal of Machine Learning Research*, 1107–1135. doi:10.1162/153244303322533214

Barnett, W., Presley, A., Johnson, J., & Liles, D. H. (1994). An architecture for the virtual enterprise. In *Proceedings of the IEEE International Conference on Systems, Man, and Cybernetics*, San Antonio, TX (Vol. 1, pp. 506-511).

Barry, D. K. (n. d.). *Database concepts and standards.* Retrieved from http://www.service-architecture.com/database/articles/index.html

Baun, C., & Kunze, M. (2009). Building a private cloud with Eucalyptus. In *Proceedings of the IEEE International Conference on E-Science Workshops* (pp. 33-38).

BBC. (2009). *The sidekick cloud disaster.* Retrieved from http://bbc.co.uk./blogs/technology/2009/10/the_sidekick_cloud_disaster.html

Beaty, K., Kochut, A., & Shaikh, H. (2009, May 23-29). Desktop to cloud transformation planning. In *Proceedings of the IEEE International Symposium on Parallel and Distributed Processing*, Rome, Italy (pp. 1-8).

Behnia, K. (2009). *Time to rethink IT service delivery & bring the clouds down to earth*. Virtual Strategy Magazine.

Bell, M. (2008). *Introduction to service-oriented modeling, service-oriented modeling: Service analysis, design, and architecture*. New York, NY: John Wiley & Sons.

Beloglazov, A., Buyya, R., Lee, Y. C., & Zomaya, A. (2011). *A taxonomy and survey of energy-efficient data centers and cloud computing systems*. Advances in Computers.

Bengtsson, P., Lassing, N., Bosch, J., & Vliet, H. V. (2004). Architecture-level modifiability analysis. *Journal of Systems and Software*, *69*(1-2), 129–147. doi:10.1016/S0164-1212(03)00080-3

Berral, J. L., Goiri, Í., Nou, R., Julià, F., Guitart, J., Gavaldà, R., et al. (2010, April 13). Towards energy-aware scheduling in data centers using machine learning. In *Proceedings of the 1st International Conference on Energy-Efficient Computing and Networking*, Passau, Germany (pp. 215-224).

Biggs, S., & Vidalis, S. (2010). Cloud computing storms. *International Journal of Intelligent Computing Research*, *1*(1).

Birge, L., & Massart, P. (2001). Gaussian model selection. *Journal of the European Mathematical Society*, *3*(3), 203–268. doi:10.1007/s100970100031

Blogspot. (2008). *The blue pill*. Retrieved from http://theinvesiblethings.blogspot.com/2008/07/0wning-xen-invegas.html

BonFire. (2010). *Building service testbeds on fire*. Retrieved from http://www.bonfire-project.eu/

Bono, S. C., Green, M., Stubblefield, A., Juels, A., Rubin, A. D., & Szydlo, M. (2005). Security analysis of a cryptographically-enabled RFID device. In *Proceedings of the 14th Conference on USENIX Security*, Berkeley, CA.

Bonvin, N., Papaioannou, T. G., & Aberer, K. (2010). A self-organized, fault-tolerant and scalable replication scheme for cloud storage. In *Proceedings of the ACM Symposium on Cloud Computing*, Indianapolis, IN.

Borko, F., & Armando, E. (2010). Cloud computing fundamentals. In Furht, B., & Escalante, A. (Eds.), *Handbook of cloud computing*. New York, NY: Springer.

Bradley, T. (2009a). *IBM and AT&T unveil cloud computing services*. Retrieved from http://www.pcworld.com/businesscenter/article/182238/ibm_and_atandt_unveil_cloud_computing_services.html

Bradley, T. (2009b). *Sidekick foul-up is not a failure of the cloud*. Retrieved from http://www.pcworld.com/businesscenter/article/173485/sidekick_foulup_is_not_a_failure_of_the_cloud.html

Brandic, I., Music, D., Leitner, P., & Dustdar, S. (2009, August 25-28). VieSLAF framework: Enabling adaptive and versatile SLA-management. In *Proceedings of the 6th International Workshop on Grid Economics and Business Models*, Delft, The Netherlands.

Brand, M. (2007). *Forensics analysis avoidance techniques of malware*. Perth, Australia: Edith Cowan University.

Brewer, E. A. (2000). Towards robust distributed systems. In *Proceedings of the 19th ACM Symposium on Principles of Distributed Computing*, Portland, OR (p. 7).

Briscoe, G., & Marinos, A. (2009, June 1-3). Digital ecosystems in the clouds: Towards community cloud computing. In *Proceedings of the 3rd IEEE International Conference on Digital Ecosystems and Technologies*, New York, NY (pp. 103-108).

Bronk, C. (2008). Hacking the nation-state: Security, information technology and policies of assurance. *Information Security Journal: A Global Perspective*, *17*(3), 132-142.

Browne, C. (n. d.). *NonRelational database systems*. Retrieved from http://linuxfinances.info/info/nonrdbms.html

Browne, J. (2009). *Brewer's CAP theorem*. Retrieved from http://www.julianbrowne.com/article/viewer/brewers-cap-theorem

Bulkeley, W. M. (2007). IBM, Google, Universities combine 'Cloud' forces. Retrieved from http://online.wsj.com/article/SB119180611310551864.html

Burchard, L., Hovestadt, M., Kao, O., Keller, A., & Linnert, B. (2004). The virtual resource manager: An architecture for SLA-aware resource management. In *Proceedings of the IEEE International Symposium on Cluster Computing and the Grid* (pp. 126-133). Washington, DC: IEEE Computer Society.

Burton Group. (2010). *Comprehensive research and advisory solution.* Retrieved from http://www.burtongroup.com/research/

Buscmann, F., Meunier, R., Rohnert, H., Sommerlad, P., & Stal, M. (1996). *Pattern-oriented software architecture- A system software of patterns.* Chichester, UK: John Wiley & Sons.

Buyya, R., Beloglazov, A., & Abawajy, J. (2010, July 12-15). *Energy-efficient management of data center resources for cloud computing: A vision, architectural elements, and open challenges.* Paper presented at the International Conference on Parallel and Distributed Processing Techniques and Applications, Las Vegas, NV.

Buyya, R., Ranjan, R., & Calheiros, R. N. (2009, June 21-24). Modeling and simulation of scalable cloud computing environments and the CloudSim toolkit. In *Proceedings of the International Conference on High Performance Computing & Simulation*, Leipzig, Germany (pp. 1-11).

Buyya, R., Yeo, C. S., & Venugopal, S. (2008). Market-oriented cloud computing: Vision, hype, and reality for delivering IT services as computing utilities. In *Proceedings of the 10th IEEE International Conference on High Performance Computing and Communications.*

Buyya, R., Yeo, C. S., & Venugopal, V. (2008, September). Vision, hype and reality for delivering IT services as computing utilities. In *Proceedings of the 10th IEEE International Conference on High Performance Computing and Communications* (pp. 5-13).

Buyya, R., Yeo, C. S., Venugopal, S., Broberg, J., & Brandic, I. (2009). Cloud computing and emerging IT platforms: Vision, hype, and reality for delivering computing as the 5th utility. *Journal of Future Generation Computer Systems*, *25*(6), 559–616. doi:10.1016/j.future.2008.12.001

Cabinet Office. (2010). *Government ICT Strategy: Smarter, cheaper, greener.* Norwich, UK: Cabinet Office.

Cáceres, J., Vaquero, L. M., Rodero-Merino, L., Polo, A., & Hierro, J. (2010). Service scalability over the cloud. In Furht, B., & Escalante, A. (Eds.), *Handbook of cloud computing.* New York, NY: Springer. doi:10.1007/978-1-4419-6524-0_15

Cai, H., Zhang, K., Zhou, M. J., Cai, J. J., & Mao, X. S. (2009). An end-to-end methodology and toolkit for fine granularity SaaS-ization. In *Proceedings of the IEEE International Conference on Cloud Computing* (p. 101).

Calheiros, R. N., Ranjan, R., Beloglazov, A., De Rose, C. A. F., & Buyya, R. (2011). ClouSim: A toolkit for modeling and simulation of cloud computing environments and evaluation of resource provisioning algorithms. *Software, Practice & Experience*, *41*, 23–50. doi:10.1002/spe.995

Camarinha-Matos, L. M., & Afsarmanesh, H. (1998). Towards and architecture for virtual enterprises. *Journal of Intelligent Manufacturing*, *9*(2), 189–199. doi:10.1023/A:1008880215595

Cameron, K. W. (2010). The challenges of energy-proportional computing. *IEEE Computer*, *43*, 82–83.

Cappelli, D. M., Trzeciak, R. F., & Moore, A. B. (2006). *Insider threats in the SLDC: Lessons learned from actual incidents of fraud: Theft of sensitive information, and IT sabotage.* Pittsburgh, PA: Carnegie Mellon University.

Cárdenas, R. G., & Sanchez, E. (2005). Security challenges of distributed e-learning systems. In F. F. Ramos, V. A. Rosillo, & H. Unger (Eds.), *Proceedings of the 5th International School and Symposium on Advanced Distributed Systems* (LNCS 3563, pp. 538-544).

Cardosa, M., Korupolu, M. R., & Singh, A. (2009). Shares and utilities based power consolidation in virtualized server environments. In *Proceedings of the IFIP/IEEE International Symposium on Integrated Network Management* (pp. 327-334).

Carr, N. (2008). *The big switch: Our new destiney.* New York, NY: W.W. Norton.

Carr, N. (2008). *The big switch: Rewiring the world, from Edison to Google.* New York, NY: Norton & Company.

Carter, J., & Rajamani, K. (2010). Designing energy-efficient servers and data centers. *IEEE Computer, 43,* 76–78.

Cassandra (n. d.). *The apache cassandra project.* Retrieved from http://cassandra.apache.org/

Cent, O. S. (2009). *The community ENTerprise operating systems.* Retrieved from http://www.centos.org/

Cent, O. S. Wiki. (2009). *Creating and installing a CentOS 5 domU instance.* Retrieved from http://wiki.centos.org/HowTos

Cervone, H. F. (2010). An overview of virtual and cloud computing. *OCLC Systems & Services, 26*(3), 162–165. doi:10.1108/10650751011073607

Cesare, S., & Xiang, Y. (2010). Classification of malware using structured control flow. In *Proceedings of the Eighth Australasian Symposium on Parallel and Distributed Computing,* Brisbane, Australia. (Vol. 107).

Chang, F., Dean, J., Ghemawat, S., Hsieh, W. C., Wallach, D. A., Burrows, M., et al. (2006). Bigtable: A distributed storage system for structured data. In *Proceedings of the 7th USENIX Symposium on Operating Systems Design and Implementation.* (Vol. 7, pp. 205-218).

Chang, V., Mills, H., & Newhouse, S. (2007, September). *From open source to long-term sustainability: Review of business models and case studies.* Paper presented at the UK e-Science All Hands Meeting, Nottingham, UK.

Chang, V., Wills, G., & De Roure, D. (2010a, July 5-10). A review of cloud business models and sustainability. In *Proceedings of the Third IEEE International Conference on Cloud Computing,* Miami, FL.

Chang, V., Wills, G., & De Roure, D. (2010b). Case studies and sustainability modelling presented by cloud computing business framework. *International Journal of Web Services Research.*

Chang, V., Wills, G., & De Roure, D. (2010c, September 13-16). *Cloud business models and sustainability: Impacts for businesses and e-research.* Paper presented at the UK e-Science All Hands Meeting Software Sustainability Workshop, Cardiff, UK.

Chang, V., Wills, G., De Roure, D., & Chee, C. (2010, September 13-16). *Investigating the cloud computing business framework - modelling and benchmarking of financial assets and job submissions in clouds.* Paper presented at the UK e-Science All Hands Meeting on Research Clouds: Hype or Reality Workshop, Cardiff, UK.

Chen, N., & Ren, S. (2009). Adaptive optimal checkpoint interval and its impact on system's overall quality in soft real-time applications. In *Proceedings of the 24th ACM Symposium on Applied Computing* (pp. 1015-1020).

Cherkasova, L. (1999). *FLEX: Design and management strategy for scalable web hosting service (Tech. Rep. No. HPL 1999□64R1).* Palo Alto, CA: Hewlett-Packard Laboratories.

Chickowski, E. (2007). Safely eliminating e-waste. *IEEE Processor, 29,* 12.

Chinnici, R. (2008). *Profiles in the Java EE 6 platform.* Retrieved from http://weblogs.java.net/blog/robc/archive/2008/02/profiles_in_the_1.html

Chong, F., & Carraro, G. (2006). *Architecture strategies for catching the long tail.* Retrieved from http://msdn.microsoft.com/en-us/library/aa479069.aspx

Choudhury, A. R., King, A., Kumar, S., & Sabharwal, Y. (2008). Optimisations in financial engineering: The least-squares Monte Carlo method of Longstaff and Schwarts. In *Proceedings of the IEEE International Symposium on Parallel and Distributed Computing* (pp. 1-11).

Chou, T. (2009). *Seven clear business models.* Active Book Press.

Chow, G. J. (2009). Controlling data in the cloud: Outsourcing computation without outsourcing control. In *Proceedings of the ACM Workshop on Cloud Computing and Security,* Chicago, IL (pp. 85-90).

Chow, R., Golle, P., Jakobsson, M., Shi, E., Staddon, J., Masuoka, R., et al. (2009). Controlling data in the cloud: Outsourcing computation without outsourcing control. In *Proceedings of the ACM Workshop on Cloud Computing Security* (pp. 85-90). New York, NY: ACM Press.

Churcher, C. (2007). *Beginning database design: From novice to professional.* New York, NY: Apress.

Cirstoiu, C., Grigoras, C., Betev, L., Costan, A., & Legrand, I. C. (2007). Monitoring, accounting and automated decision support for the alice experiment based on the monalisa framework. In *Proceedings of the Workshop on Grid Monitoring* (pp. 39-44). New York, NY: ACM Press.

City, A. M. (2010). *Business with personality.* Retrieved from http://www.cityam.com

Clarke, R. (2010). User requirements for cloud computing architecture. In *Proceedings of the 10th IEEE/ACM International Conference on Cluster, Cloud and Grid Computing* (pp. 623-630).

Clayman, S., Galis, A., Chapman, C., Toffetti, G., Rodero-Merino, L., & Vaquero, L. M. (2010). Monitoring service clouds in the future internet. In Tselentis, G., Galis, A., Gavras, A., Krco, S., Lotz, V., & Simperl, E., (Eds.), *Towards the future internet - emerging trends from European research.* Amsterdam, The Netherlands: IOSPress.

Cloud Computing World. (2010). *Why your small business needs cloud computing.* Retrieved from http://www.cloud-computingworld.org/cloud-computing-for-businesses/why-your-small-business-needs-cloud-computing.html

Cloud Portal. (2010). *Cloud computing portal.* Retrieved from http://cloudcomputing.qrimp.com/portal.aspx

Cloud Security Alliance. (2009). *Security guidance for critical areas of focus in cloud computing V2.1.* Retrieved from http://www.privatecloud.com/2010/01/26/security-guidance-for-critical-areas-of-focus-in-cloud-computing-v2-1/?fbid=uCRTvs9w3Cs

Computer Technology Documentation Project. (2010). *Network and computer security tutorial 2010.* Retrieved from http://www.comptechdoc.org/

Cooper, B. F., Ramakrishnan, R., Srivastava, U., Silberstein, A., Bohannon, P., & Jacobsen, H.-A. (2006). PNUTS: Yahoo!'s hosted data serving platform. *Very Large Data Base Endowment, 1*(2), 1277–1288.

Couch, A. L., & Kumar, K. (2008, December). *Workshop on power aware computing and systems.* Retrieved from http://www.usenix.org/publications/login/2009-04/open-pdfs/hotpower08.pdf

Coursey, D. (2009). *Google outages defy customer confidence.* Retrieved from http://www.pcworld.com/businesscenter/article/172575/Google_Outages_Defy_Customer_Confidence.html

CPNI. (2010). *Information security briefing 01/2010 cloud computing.* Retrieved from www.cpni.gov.uk/Documents

CSA. (2010). *Top threats to cloud computing, v1.0.* Retrieved from https://cloudsecurityalliance.org/topthreats/csathreats.v1.0.pdf

D'Monte, L. (2010). *Good opportunity for Indian IT firms.* Retrieved from http://business.rediff.com/column/2010/jan/05/guest-cloud-computing-good-opportunity-for-indian-firms.htm

D'Monte, L. (2010). *Work in the cloud.* Retrieved from http://www.business-standard.com/india/news/work-incloud/381619/

Dai, W. (2009). The impact of emerging technologies on small and medium enterprises (SMEs). *Journal of Business Systems. Governance and Ethics, 4*(4), 53–60.

Dai, Y. S., Levitin, G., & Trivedi, K. S. (2007). Performance and reliability of tree-structured grid services considering data dependence and failure correlation. *IEEE Transactions on Computers, 56*(7), 925–936. doi:10.1109/TC.2007.1018

Dai, Y. S., Pan, Y., & Raje, R. (2007). *Advanced parallel and distributed computing: Evaluation, improvement and practices.* New York, NY: Nova Science.

Dai, Y. S., Pan, Y., & Zou, X. K. (2007). A hierarchical modeling and analysis for grid service reliability. *IEEE Transactions on Computers, 56*(5), 681–691. doi:10.1109/TC.2007.1034

Dai, Y. S., Xie, M., & Poh, K. L. (2003). A study of service reliability and availability for distributed systems. *Reliability Engineering & System Safety, 79*(1), 103–112. doi:10.1016/S0951-8320(02)00200-4

Danchev, D. (2009). *Zeus crimeware using Amazon's EC2 as command and control server.* Retrieved from http://www.zdnet.com/blog/security/zeus-crimeware-using-amazons-ec2-as-command-and-control-server/5010

Danielson, K. (2008). *Distinguishing cloud computing from utility computing.* Retrieved from http://www.ebizq.net/blogs/saasweek/2008/03/distinguishing_cloud_computing/

D'Arcy, J., & Hovav, A. (2007). Deterring internal information systems misuse. *Communications of the ACM, 50*(10), 113–117. doi:10.1145/1290958.1290971

Dean, J., & Ghemawat, S. (2008). MapReduce: Simplified data processing on large clusters. *Communications of the ACM, 51*(1), 107–113. doi:10.1145/1327452.1327492

DeCandia, G., Hastorun, D., Jampani, M., Kakulapati, G., Lakshman, A., & Pilchin, A. (2007). Dynamo: Amazon's highly available key-value store. In *Proceedings of the 21st ACM SIGOPS Symposium on Operating Systems Principles* (Vol. 21, pp. 205-220).

Deng, J., Huang, S. C.-H., Han, Y. S., & Deng, J. H. (2010). Fault-tolerant and reliable computation in cloud computing. In *Proceedings of the IEEE Globecom Workshop on Web and Pervasive Security*, Miami, FL.

Deng, Y., & Wang, F. A. (2007). Heterogeneous storage grid enabled by grid service. *Operating Systems Review, 41*(1), 7–13. doi:10.1145/1228291.1228296

Di Maio, A. (2009). *Cloud computing in government: Private, public, both or none?* Stamford, CT: Gartner.

Di Maio, A. (2009). *Government in the cloud: Much more than computing.* Stamford, CT: Gartner.

Di Maio, A. (2009). *GSA launches Apps.gov: What it means to government IT leaders?* Stamford, CT: Gartner.

Digital Britain. (2009, June). *Final report presented to Parliament.* London, UK: Stationery Office.

Diver, S. (2007). *Information security policy – a development guide for large and small companies* (p. 43). Reston, VA: SANS Institute.

Doelitzcher, F., Reich, C., Sulistio, A., & Furtwangen, H. (2010). Designing cloud services adhering to government privacy laws. In *Proceedings of the 10ᵗʰ International Conference on Computer and Information Technology* (pp. 930-935).

Douglis, F. (2009). Staring at clouds. *IEEE Internet Computing, 13*(3), 4–6. doi:10.1109/MIC.2009.70

Druschel, P., Peterson, L. L., & Davie, B. S. (1994). Experiences with a high-speed network adaptor: A software perspective. *Computer Communication Review, 24*(4), 2. doi:10.1145/190809.190315

Dumitrescu, C., Raicu, I., & Foster, I. (2005). Di-gruber: A distributed approach to grid resource brokering. In *Proceedings of the ACM/IEEE Conference on Supercomputing* (p. 38). Washington, DC: IEEE Computer Society.

Duy, T. V. T., Sato, Y., & Inoguchi, Y. (2010). *Performance evaluation of a Green Scheduling Algorithm for energy savings in cloud computing.* Paper presented at the IEEE International Symposium on Parallel and Distributed Processing.

Dyachuk, D., & Mazzucco, M. (2010). On allocation policies for power and performance. In *Proceedings of the 11th IEEE/ACM International Conference on Grid Computing*, Belgium.

Elnozahy, E., Kistler, M., & Rajamony, R. (2003). Energy-efficient server clusters. In B. Falsafi & T. N. Vijaykumar (Eds.), *Proceedings of the Second International Workshop on Power-Aware Computer Systems* (LNCS 2325, pp. 179-197).

Elnozahy, M., Alvisi, L., Wang, Y.-M., & Johnson, D. B. (2002). A survey of rollback-recovery protocols in message-passing systems. *ACM Computing Surveys, 34*(3), 375–408. doi:10.1145/568522.568525

Emerge Forum. (2009). *India market and the SaaS/cloud computing landscape.* Retrieved from http://emerge.nasscom.in/2009/08/india-market-and-the-saascloud-computing-landscape/

ENISA. (2010). *Cloud computing: Benefits, risks and recommendations for information security.* Retrieved from http://www.coe.int/t/dghl/cooperation/economic-crime/cybercrime/cy-activity-interface-2010/presentations/Outlook/Udo%20Helmbrecht_ENISA_Cloud%20Computing_Outlook.pdf

Erdogmus, H. (2009). Cloud computing: Does nirvana hide behind the nebula? *IEEE Software, 26*(2), 4–6. doi:10.1109/MS.2009.31

European Network and Information Security Agency. (2009). *Cloud computing - benefits, risks and recommendations for information security*. Retrieved from http://itlaw.wikia.com/wiki/Cloud_Computing:_Benefits,_Risks,_and_Recommendations_for_Information_Security

Evoy, G. V. M., & Schulze, B. (2008). Using clouds to address grid limitations. In *Proceedings of the 6th International Workshop on Middleware for Grid Computing* (p. 11).

F5 Networks. (2009). *Cloud computing: Survey results*. Seattle, WA: F5 Networks.

Fan, B., Tantisiriroj, W., Xiao, L., & Gibson, G. (2009). DiskReduce: RAID for data-intensive scalable computing. In *Proceedings of the 4th Annual Workshop on Petascale Data Storage*, Portland, OR (pp. 6-10).

Feiman, J., & Cearley, D. W. (2009). *Economics of the cloud: Business value assessments*. Stamford, CT: Gartner RAS Core Research.

Ferguson, T. (2009). *Outage hits thousands of businesses*. Retrieved from http://news.cnet.com/8301-1001_3-10136540-92.html

Ferreiro, D. S. (2010). *Guidance on managing records in cloud computing environments*. Retrieved from http://www.egov.vic.gov.au/focus-on-countries/north-and-south-america-and-the-caribbean/united-states/government-initiatives-united-states/culture-sport-and-recreation-united-states/archives-and-public-records-united-states/guidance-on-managing-records-in-cloud-computing-environments.html

Financial Times. (2009). *Interview with Lord Turner, Chair of Financial Services Authority*. Retrieved from http://www.ft.com/cms/s/0/d76d0250-9c1f-11dd-a42e-000077b07658.html#axzz1Iqssz7az

Findlaw.com. (n. d.). *U.S. vs. Robert Johnson - Child pornography indictment*. Retrieved from http://news.findlaw.com/hdocs/docs/chldprn/usjhnsn62805ind.pdf

Foster, I. (2002). *What is the grid? A three point checklist*. Retrieved from http://dlib.cs.odu.edu/WhatIsTheGrid.pdf

Foster, I., Zhao, Y., Raicu, I., & Lu, S. (2008, November). Cloud computing and grid computing 360-degree compared. In *Proceedings of the Grid Computing Environments Workshop* (pp. 1-10).

Foster, I. T. (2005). Globus toolkit version 4: Software for service-oriented systems. *Journal of Computer Science and Technology*, *21*(4), 513–520. doi:10.1007/s11390-006-0513-y

Foster, I., & Kesselman, C. (2004). *The grid 2: Blueprint for a new computing infrastructure* (2nd ed.). San Francisco, CA: Morgan Kauffman.

Free Software Foundation. (2010). *The free software definition*. Retrieved from http://www.gnu.org/philosophy/free-sw.html

FreeBSD Wiki. (2010). *FreeBSD/Xen: FreeBSD/Xen port*. Retrieved from http://wiki.freebsd.org/FreeBSD/Xen

Freeman, L. (2009). *Reducing data center power consumption through efficient storage*. Retrieved from http://www.gtsi.com/eblast/corporate/cn/06_2010/PDFs/NetApp%20Reducing%20Datacenter%20Power%20Consumption.pdf

Freeman, T., LaBissoniere, D., Marshall, P., Bresnahan, J., & Keahey, K. (2009). *Nimbus elastic scaling in the clouds*. Retrieved from http://www.nimbusproject.org/files/epu_poster4.pdf

Gain, B. (2010). Cloud computing & SaaS in 2010. *IEEE Processor*, *32*, 12.

Galán, F., Sampaio, A., Rodero-Merino, L., Loy, I., Gil, V., Vaquero, L. M., et al. (2009). Service specification in cloud environments based on extensions to open standards. In *Proceedings of the Fourth International Conference on Communication System Software and Middleware*, Dublin, Ireland (p. 19) New York, NY: ACM Press.

Gandhi, A., Harchol-Balter, M., Das, R., & Lefurgy, C. (2009). Optimal power allocation in server farms. In *Proceedings of the 11th International Joint Conference on Measurement and Modeling of Computer Systems* (pp. 157-168).

Gao, H., & Guo, J. (2009). Application of vulnerability analysis in electric power communication network. In. *Proceedings of the International Conference on Machine Learning and Cybernetics, 4*, 2072–2077. doi:10.1109/ICMLC.2009.5212210

Garbacki, P., & Naik, V. K. (2007). Efficient resource virtualization and sharing strategies for heterogeneous grid environments. In *Proceedings of the 10th IFIP/IEEE International Symposium on the Integrated Network Management* (pp. 40-49).

Garfinkel, S. (2007). Anti-forensics: Techniques, detection and countermeasures. In *Proceedings of the 2nd International Conference in i-Warefare and Security* (p. 77).

Garfinkel, T., & Rosenblum, M. (2005). When virtual is harder than real: Security challenges in virtual machine based computing environments. In *Proceedings of the 10th Workshop on Hot Topics in Operating Systems* (p. 20).

Gartner. (2010). *Gartner research.* Retrieved from http://blogs.gartner.com/

Gartner. (2010). *Gartner's expertise in a variety of ways.* Retrieved from http://www.gartner.com/

Geelan, J. (2009). *The top 150 players in cloud computing.* Retrieved from http://cloudcomputing.sys-con.com/node/770174

Geelan, J. (2009, May 18-19). *Deploying virtualization in the enterprise.* Paper presented at the Virtualization Conference, Prague, Czech Republic.

Geelan, J. (2010). *IT must invest in architecture and engineering: Adaptivity CEO.* Retrieved from http://cloudcomputing.sys-con.com/node/1061561

Gehling, B., & Stankard, D. (2005). eCommerce security. In *Proceedings of the Information Security Curriculum Development Conference*, Kennesaw, GA (pp. 32-37). New York, NY: ACM Press.

Geiger, M. (2005). *Evaluating commercial counter-forensic tools.* Paper presented at the Digital Forensic Research Workshop, Pittsburgh, PA.

Gelbman, M. (2010). *Highlights of web technology surveys, July 2010: CentOS is now the most popular Linux distribution on web servers.* Retrieved from http://w3techs.com/blog/entry/highlights_of_web_technology_surveys_july_2010

Glisson, W., & Welland, R. (2005). Web development evolution: The assimilation of web engineering security. In *Proceedings of the Third Latin American Web Congress* (p. 49). Washington, DC: IEEE Computer Society.

Goel, A., Schmidt, H., & Gilbert, D. (2009). Towards formalizing virtual enterprise architecture. In *Proceedings of the Distributed Object Computing Conference Workshops* (pp. 238-242).

GoGrid. (2010). *Scalable load-balanced windows and linux cloud-server hosting.* Retrieved from http://www.gogrid.com/

Goiri, I., Julià, F., Guitart, J., & Torres, J. (2010, April 19-23). Checkpoint-based fault-tolerant infrastructure for virtualized service providers. In *Proceedings of the 12th IEEE/IFIP Network Operations and Management Symposium*, Osaka, Japan (pp. 455-462).

Golden, B. (2009). *Capex vs. Opex: Most people miss the point about cloud economics.* Retrieved from http://www.cio.com/article/484429/Capex_vs._Opex_Most_People_Miss_the_Point_About_Cloud_Economics

Goncalves, A. (2010). *Beginning Java™ EE 6 Platform with GlassFish™ 3: From novice to professional.* New York, NY: Apress. doi:10.1007/978-1-4302-2890-5

Good Web Hosting. (2006). *How much data transfer do you really need?* Retrieved from http://www.goodweb-hosting.info/article.py/58

Google. (2010). *Indian version of this popular search engine.* Retrieved from http://www.google.co.in/

Google. (2011a). *Google trends: Cloud computing, Amazon ec2.* Retrieved from http://www.google.de/trends?q=cloud+computing%2C+amazon+ec2&ctab=0&geo=all&date=all

Google. (2011b). *Google trends: Private cloud, public cloud.* Retrieved from http://www.google.de/trends?q=private+cloud%2C+public+cloud

Greengard, S. (2010). Cloud computing and developing nations. *Communications of the ACM, 53*(5), 18–20. doi:10.1145/1735223.1735232

Grigoriu, A. (2009). *The cloud enterprise.* Retrieved from http://www.bptrends.com/publicationfiles/TWO_04-09-ART-The_Cloud_Enterprise-Grigoriu_v1-final.pdf

Grobauer, B., Walloscheck, T., & Stocker, E. (2010). Understanding cloud-computing vulnerabilities. *IEEE Security and Privacy, 9*(2), 50–57. doi:10.1109/MSP.2010.115

Gross, D. (2010). *WikiLeaks cut off from Amazon servers.* Retrieved from http://articles.cnn.com/2010-12-01/us/wikileaks.amazon_1_julian-assange-wikileaks-amazon-officials?_s=PM:US

Grossman, R. L. (2009). The case for cloud computing, computer.org/ITPro. *IT Professional, 11*(2), 23-27.

Guenin, B. (2002). The many flavors of ball grid array packages. *Electronics Cooling, 8*(1), 32–40.

Guidance Software. (n. d.). *Computer forensics solutions and digital investigations.* Retrieved from http://www.guidancesoftware.com/

Guo, C. J., Sun, W., Huang, Y., Wang, Z. H., & Gao, B. (2007). A framework for native multi-tenancy application development and management. In *Proceedings of the 9th IEEE International Conference on E-Commerce Technology and the 4th IEEE International Conference on Enterprise Computing, E-Commerce, and E-Services* (pp. 551-558).

Gurav, U., & Shaikh, R. (2010). Virtualization – A key feature of cloud computing. In *Proceedings of the International Conference and Workshop on Emerging Trends in Technology* (pp. 227-229).

Hadoop (n. d.). *The apache hadoop project.* Retrieved from http://hadoop.apache.org/

Hadoop ZooKeeper. (n. d.). *The apache hadoop zookeeper project.* Retrieved from http://hadoop.apache.org/zookeeper/

Hamilton, M. (2009). *How green is cloud computing?* Retrieved from http://blogs.sun.com/marchamilton/entry/how_green_is_cloud_computing

Hamnett, C. (2009). *The madness of mortgage lenders: Housing finance and the financial crisis.* London, UK: King's College.

Harris, R. (2006). Arriving at an anti-forensics consensus: Examining how to define and control the anti-forensics problem. *Digital Investigation, 3,* 44–49. doi:10.1016/j.diin.2006.06.005

Hartley, M. W. (2007). *Current and future threats to digital forensics.* ISSA Journal.

Hata, H., Kamizuru, Y., Honda, A., Shimizu, K., & Yao, H. (2010). Dynamic IP-VPN architecture for cloud computing. In *Proceedings of the Information and Communication Technologies Conference* (pp. 1-5).

Hayes, B. (2008). Cloud computing. *Communications of the ACM, 51*(7), 9–11. doi:10.1145/1364782.1364786

HBase. (n. d.). *The apache hbase project.* Retrieved from http://hbase.apache.org/

Heath, N. (2010). *How cloud computing will help government save taxpayer £3.2bn.* Retrieved from http://www.silicon.com/management/public-sector/2010/01/27/how-cloud-computing-will-help-government-save-taxpayer-32bn-39745389/

Hickey, A. R. (2008). *Gartner: Green IT needs to be on midsize CIOs' radar screens.* Retrieved from http://www.crn.com/hardware/208700292

Hinchcliffe, D. (2009). *Eight ways that cloud computing will change business.* Retrieved from http://www.major-cities.org/pics/download/1_1247671028/ZDNET_Eight_ways_that_cloud_computing_will_change_business.pdf

Hobson, D. (2009). *Global secure systems: Into the Cloud we go....have we thought about security issues?* Retrieved from http://www.globalsecuritymag.com/David-Hobson-Global-Secure-Systems,20090122,7110

Hogan, M. (2008). *Cloud computing & databases: How databases can meet the demands of cloud computing.* Menlo Park, CA: ScaleDB Inc.

Höne, K., & Eloff, J. H. O. (2002). Information security policy — what do international information security standards say? *Computers & Security, 21*(5), 382–475. doi:10.1016/S0167-4048(02)00504-7

Houssos, N., Gazis, E., Panagiotakis, S., Gessler, S., Schuelke, A., Quesnel, S., et al. (2002). Value added service management in 3G networks. In *Proceedings of the IEEE/IFIP Network Operations and Management Symposium* (pp. 529-544).

Huai, J., Li, Q., & Hu, C. (2007). CIVIC: A hypervisor based computing environment. In *Proceedings of the International Conference on Parallel Processing Workshops* (p. 51).

Huerta-Canepa, G., & Lee, D. (2010). A virtual cloud computing provider for mobile devices. In *Proceedings of the 1st ACM Workshop on Mobile Cloud Computing & Services: Social Networks and Beyond.*

Hull, J. C. (2009). *Options, futures, and other derivatives* (7th ed.). Upper Saddle River, NJ: Pearson/Prentice Hall.

Huong Ngo, H. (1999). Corporate system security: towards an integrated management approach. *Information Management & Computer Security*, 7(5), 217–222. doi:10.1108/09685229910292817

Iakobashvili, R. M. M. (2007). *Welcome to curl-loader.* Retrieved from http://curl-loader.sourceforge.net/

IBM. (2009). *Power systems: Introduction to virtualization* (Tech. Rep. No. 5733-SSI). Armonk, NY: IBM Corporation.

Imamagic, E., & Dobrenic, D. (2007). Grid infrastructure monitoring system based on nagios. In *Proceedings of the Workshop on Grid Monitoring* (pp. 23-28). New York, NY: ACM Press.

India Online. (2010). *Population of India.* Retrieved from http://www.indiaonlinepages.com/population/

Ingraham, R. W. (2006). *The Nagios 2.X event broker module API.* Retrieved from http://nagios.sourceforge.net/download/contrib/documentation/misc/NEB%202x%20Module%20API.pdf

Inlab Software GmbH. (2010). *Balance.* Retrieved from http://www.inlab.de/balance.html

Internet World Stats. (n. d.). *The Internet big picture, world Internet users and population stats.* Retrieved from http://www.internetworldstats.com/stats.htm

INTEROP. (2008). *Enterprise software customer survey.* Retrieved from http://www.interop.com/

ITRS. (2010). *International technology roadmap for semiconductors.* Retrieved from http://public.itrs.net

Jagdev, H. S., & Thoben, K. D. (2001). Anatomy of enterprise collaborations. *Journal of Production Planning and Control: The Management of Operations*, 12(5), 437–451. doi:10.1080/09537280110042675

Jensen, M., Schwenk, J., Gruschka, N., & Lo Iacono, L. (2009). On technical security issues in cloud computing. In *Proceedings of the IEEE International Conference on Cloud Computing*, Bangalore, India.

Jeyarani, R., Nagaveni, N., & Vasanth Ram, R. (2011). Self adaptive particle swarm optimization for efficient virtual machine provisioning in cloud. *International Journal of Intelligent Information Technologies*, 7(2), 25–44.

Jin, H., Tao, Y., Wu, S., & Shi, X. (2008). Scalable dht-based information service for large-scale grids. In *Proceedings of the 5th Conference on Computing Frontiers* (pp. 305-312). New York, NY: ACM Pres.

Justice.gov. (2003). *United States of America vs. H. Marc Watzman.* Retrieved from http://www.justice.gov/usao/iln/indict/2003/watzman.pdf

Kadam, A. W. (2007). Information security policy development and implementation. *Information Systems Security*, 16(5), 246–256. doi:10.1080/10658980701744861

Kandukuri, B. R., Ramakrishna, P. V., & Rakshit, A. (2009). Cloud security issues. In *Proceedings of the IEEE International Conference on Services Computing* (pp. 517-520).

Kannammal, A., & Iyengar, N. C. S. N. (2007). A model for mobile agent security in e-business applications. *International Journal of Business and Information*, 2(2), 185–198.

Kant, K. (2009). Data center evolution- A tutorial on state of the art, issues, and challenges. *Computer Networks*, 53(17), 2939–2965. doi:10.1016/j.comnet.2009.10.004

Kaplan, J. M., Forest, W., & Kindler, N. (2008). *Revolutionizing data centre energy efficiency.* Chicago, IL: McKinsey & Company.

Katharine, S., & David, B. (2010). Current state of play: Records management and the cloud. *Records Management Journal, 20*(2), 217–225. doi:10.1108/09565691011064340

Kaufman, L. M. (2009). Data security in the world of cloud computing. *IEEE Security & Privacy, 7*(4), 61–64. doi:10.1109/MSP.2009.87

Kaushal, V., & Anju, B. (2011). Autonomic fault tolerance using HAProxy in cloud environment. *International Journal of Advanced Engineering Sciences and Technologies, 7*(2), 222–227.

Kazman, R., Klein, M., Barbacci, M., Longstaff, T., Lipson, H., & Carriere, J. (1998). *The architecture tradeoff analysis method* (Tech. Rep. No. CMU/SEI-98-TR-008 ESC-TR-98-008). Pittsburgh, PA: Carnegie Mellon Software Engineering Institute.

Keffler, D. (1999). *Application and solution of heat equation in one and two dimensional systems using numerical methods*. Knoxville, TN: University of Tennessee.

Kerris, N., Dowling, S., & Beermann, T. (Eds.). (2005). *Apple to use Intel microprocessors beginning in 2006*. Retrieved from http://www.apple.com/pr/library/2005/jun/06intel.html

Kessler, G. (2011). *File signature table*. Retrieved from http://www.garykessler.net/library/file_sigs.html

Kessler, G. C. (2007). *Anti-forensics and the digital investigator*. Burlington, VT: Champlain College.

Khaund, B. (2009). *Cloud Architecture-SaaS*. Retrieved from http://www.codeproject.com/KB/webservices/CloudSaaS.aspx

Kim, K. H., Beloglazov, A., & Buyya, R. (2011). Power-aware provisioning of virtual machines for real-time cloud services. *Concurrency and Computation: Practice and Experience*.

Kim, S., Han, H., Jung, H., Eom, H., & Yeom, H. Y. (2010). Harnessing input redundancy in a MapReduce framework. In *Proceedings of the ACM Symposium on Applied Computing*, Sierre, Switzerland (pp. 362-366).

Kirkpatrick, M. (2007). *IBM unveils Blue Cloud - what data would you like to crunch?* Retrieved from http://www.readwriteweb.com/archives/ibm_unveils_blue_cloud_what_da.php

Korri, T. (2009, April). *Cloud computing: Utility computing over the Internet*. Paper presented at the Seminar on Internetworking, Helsinki, Finland.

Kushida, K. E., Breznitz, D., & Zysman, J. (2010). *Cutting through the fog: Understanding the competitive dynamics in cloud computing*. Berkeley, CA: The Berkely Round Table on the International Economy (BRIE).

Lakshman, A., & Malik, P. (2009). Cassandra: Structured storage system over a P2P network. In *Proceedings of the 28th ACM Symposium on Principles of Distributed Computing* (p. 5).

Lakshman, A., & Malik, P. (2010). Cassandra: A decentralized structured storage system. *ACM SIGOPS Operating Systems Review, 44*(2), 35–40. doi:10.1145/1773912.1773922

Lall, B., Guenin, B., & Molnar, R. (1995). Methodology for thermal evaluation of multichip modules. *IEEE Transactions on Components Packaging & Manufacturing Technology Part A, 18*(4), 758–764. doi:10.1109/95.477461

Lassing, N., Rijsenbrij, D., & Vliet, H. V. (1999). On software architecture analysis of flexibility, complexity of changes: Size isn't everything. In *Proceedings of the 2nd Nordic Software Architecture Workshop* (pp. 1103-1581).

Laureano, M., Maziero, C., & Jamhour, E. (2004). Intrusion detection in virtual machine environments. In *Proceedings of the 30th IEEE EUROMICRO Conference* (pp. 520-525). Washington, DC: IEEE Computer Society.

Lee, K.-W., Ko, B.-J., & Calo, S. (2004). Adaptive server selection for large scale interactive online games. In *Proceedings of the 14th International Workshop on Network and Operating Systems Support for Digital Audio and Video* (pp. 152-157). New York, NY: ACM Press.

Lee, Y. C., & Zomaya, A. Y. (2010). Energy efficient utilization of resources in cloud computing systems. *The Journal of Supercomputing, 53*, 1–13. doi:10.1007/s11227-010-0435-x

Lee, Y. C., & Zomaya, A. Y. (2010). Resource allocation for energy efficient large-scale distributed systems. *Information Systems. Technology and Management, 54*(1), 16–19.

Lefèvre, L., & Orgerie, A.-C. (2010). Designing and evaluating an energy efficient Cloud. *The Journal of Supercomputing, 51*(3), 352–373. doi:10.1007/s11227-010-0414-2

Le, H. Q., Starke, W. J., Fields, J. S., O'Connell, F. P., Nguyen, D. Q., & Ronchetti, B. J. (2007). IBM POWER6 microarchitecture. *IBM Journal of Research and Development, 51.*

Leverich, J., & Kozyrakis, C. (2009). On the energy (in)efficiency of Hadoop clusters. In *Proceedings of the Workshop on Power Aware Computing and Systems*, Big Sky, MT (pp.61-65).

Lewin, K. (1947). Frontiers in group dynamics: Concept, method, and reality in social science. *Human Relations, 1*(1), 5–42. doi:10.1177/001872674700100103

Lewis, C. (2008). *Databases in the cloud.* Retrieved from http://clouddb.blogspot.com/

Ley, K., Bianchini, R., Martonosi, M., & Nguyeny, T. D. (2009). *Cost- and energy-aware load distribution across data centers.* Paper presented at the 22nd ACM Symposium on Operating Systems Principles.

Li, B., Li, J., Huai, J., Wo, T., Li, Q., & Zhong, L. (2009). EnaCloud: An energy-saving application live placement approach for cloud computing environments. In *Proceedings of the IEEE International Conference on Cloud Computing* (pp. 17-24)

Li, C. S. (2010, July 5-10). Cloud computing in an outcome centric world. In *Proceedings of the IEEE International Conference on Cloud Computing*, Miami, FL.

Li, L. (2009). An optimistic differentiated service job scheduling system for cloud computing service users and providers. In *Proceedings of the 3rd International Conference on Multimedia and Ubiquitous Engineering* (pp. 295-299).

Lim, H. C., Babu, S., & Chase, J. S. (2010). Automated control for elastic storage. In *Proceedings of the International Conference on Autonomic Computing* (pp. 19-24). New York, NY: ACM Press.

Lim, H. C., Babu, S., Chase, J. S., & Parekh, S. S. (2009). Automated control in cloud computing: challenges and opportunities. In *Proceedings of the 1st Workshop on Automated Control for Datacenters and Clouds* (pp. 13-18). New York, NY: ACM Press.

Lin, Q., Neo, H. K., Zhang, L., Huang, G., & Gay, R. (2007). Grid-based large-scale web3d collaborative virtual environment. In *Proceedings of the Twelfth International Conference on 3D Web Technology* (pp. 123-132). New York, NY: ACM Press.

Linux Information Project. (2006). *Vendor lock-in definition.* Retrieved from http://www.linfo.org/vendor_lockin.html

Litty, L. (2005). *Hypervisor-based intrusion detection.* Unpublished doctoral dissertation, University of Toronto, ON, Canada.

Liu, L., Wang, H., Liu, X., Jin, X., He, W. B., Wang, Q. B., et al. (2009). GreenCloud: A new architecture for green data center. In *Proceedings of the Sixth International Conference on Autonomic Computing and Communications Industry Session*, Barcelona, Spain (pp. 29-38).

Longstaff, F. A., & Schwartz, E. S. (2001). Valuing American options by simulations: A simple least-squares approach. *Review of Financial Studies, 14*(1), 113–147. doi:10.1093/rfs/14.1.113

Lorch, J. R., & Smith, A. J. (2001). Improving dynamic voltage scaling algorithms with PACE. In *Proceedings of the ACM SIGMETRICS Conference on Measurement and Modeling of Computer Systems* (pp. 50-61).

Loreto, S., Mecklin, T., Opsenica, M., & Rissanen, H.-M. (2010). IMS service development API and testbed. *IEEE Communications Magazine, 48*(4), 26–32. doi:10.1109/MCOM.2010.5439073

Luis, M. V., Luis, R, Caceres, J., & Lindner, M. (2008). A break in the clouds: Towards a cloud definition. *SIGCOMM Computer Communication Review, 39*(1).

MacAskill, E. (2010). *WikiLeaks website pulled by Amazon after US political pressure.* Retrieved from http://www.guardian.co.uk/media/2010/dec/01/wikileaks-website-cables-servers-amazon

Maggiani, R. (2009). Cloud computing is changing how we communicate. In *Proceedings of the IEEE International Professional Communication Conference* (pp. 1-4).

Magoules, F., Pan, J., Tan, K.-A., & Kumar, A. (2009). *Introduction to computing, numerical analysis and scientific computation series*. Boca Raton, FL: CRC Press.

Marchany, R. (2010). *Cloud computing security issues: VA Tech IT security*. Retrieved from http://www.issa-centralva.org

Marcus, B. (2009). *How on the cloud computing can benefit small and medium enterprises (SMMEs)*. Retrieved from http://www.stumbleupon.com/url/www.helium.com/items/1624860-how-on-the-cloud-computing-can-benefit-small-and-medium-enterprises-smes

Marks, E. A., & Lozano, B. (2010). *Executive's guide to cloud computing*. New York, NY: John Wiley & Sons.

Mark-Shane, E. S. (2009). Cloud computing and collaboration. *Library Hi Tech News*, *26*(9), 10–13. doi:10.1108/07419050911010741

Marshall, P., Keahey, K., & Freeman, T. (2010). Elastic site: Using clouds to elastically extend site resources. In *Proceedings of the 10th IEEE/ACM International Symposium on Cluster, Cloud and Grid Computing*, Melbourne, Australia (pp. 43-52). Washington, DC: IEEE Computer Society.

Mather, T., Kumaraswamy, S., & Latif, S. (2009). *Cloud security and privacy*. Sebastopol, CA: O'Reilly Media.

Matthews, J., Garfinkel, T., Hoff, C., & Wheeler, J. (2009). Virtual machine contracts for datacenter and cloud computing environments. In *Proceedings of the 6th International Conference on Autonomic Computing and Communications* (pp. 25-30).

McEvoy, G. V., & Schulze, B. (2008). Using clouds to address grid limitations. In *Proceedings of the 6th International Workshop on Middleware for Grid Computing* (pp. 1-6).

Mckinsey & Company. (2010). *Highlights and features*. Retrieved from http://www.mckinsey.com/

McKinsey & Company. (2010). *Management consulting & advising*. Retrieved from http://www.mckinsey.com/

McLeod, S. (2005). *SMART anti-forensics*. Retrieved from http://www.forensicfocus.com/smart-anti-forensics

McMillan, R. (2009). *Researchers hack into Intel's vPro: Invisible things labs discovers a way to compromise the vPro architecture*. Retrieved from http://www.infoworld.com/d/security-central/researchers-hack-intels-vpro-604

Mello Ferreira, A. (2010). *An energy-aware approach for service performance evaluation*. Paper presented at the International Conference on Energy-Efficient Computing and Networking.

Mell, P., & Grance, T. (2009). *Effectively using the cloud computing paradigm*. Gaithersburg, MD: Nist Information Technology Laboratory.

Mell, P., & Grance, T. (2010). The NIST definition of cloud computing. *Communications of the ACM*, *53*(6), 50–50.

Menken, I., & Blokdijk, G. (2009). *Cloud computing specialist certification kit: Virtualization*. Brisbane, Australia: Emereo Publishing.

Milanović, M., Gašević, D., Wagner, G., & Devedžić, V. (2009, Novmeber 2-5). Modeling service orchestrations with a rule-enhanced business process language. In *Proceedings of the Conference of the Center for Advanced Studies on Collaborative Research*, Ontario, Canada (pp. 70-85). New York, NY: ACM Press.

Milenkovic, M., Castro-Leon, E., & Blakley, J. R. (2009). Power-aware management in cloud data centers. In M. G. Jaatun, G. Zhao, & C. Rong (Eds.), *Proceedings of the First International Conference on Cloud Computing* (LNCS 5931, pp. 668-673).

Millers, R. M. (2011). *Option valuation*. Niskayuna, NY: Miller Risk Advisor.

Mitre.org. (2011). *Common vulnerabilities and exposures (CVE) database*. Retrieved from http://cve.mitre.org/

Mohay, G., Anderson, A., Collie, B., & del Vel, O. (2003). *Computer and intrusion forensics* (p. 9). Boston, MA: Artech House.

Molla, A. (2008, December 3-5). *GITAM: A model for the adoption of green IT*. Paper presented at the 19th Australian Conference on Information Systems, Christchurch, New Zealand.

Moran, J. (2010). *ZScaler web security cloud for small business.* Retrieved from http://www.smallbusinesscomputing.com/webmaster/article.php/3918716/ZScaler-Web-Security-Cloud-for-Small-Business.htm

Moreno, M., & Navas, J. F. (2001). On the robustness of least-square Monte Carlo (LSM) for pricing American derivatives. *Journal of Economic Literature Classification.*

Morris, M. (n. d.). *Employee theft schemes.* Retrieved from http://www.cowangunteski.com/documents/EmployeeTheftSchemes_001.pdf

Mullen, K., & Ennis, D. M. (1991). A simple multivariate probabilistic model for preferential and triadic choices. *Journal of Psychometrika, 56*(1), 69–75. doi:10.1007/BF02294586

Mullen, K., Ennis, D. M., de Doncker, E., & Kapenga, J. A. (1988). Models for the duo-trio and triangular methods. *Journal of Bioethics, 44,* 1169–1175.

Murty, J. (2010). *Programming Amazon Web Services: S3, EC2, SQS, FPS, and SimpleDB.* Sebastopol, CA: O'Reilly Media.

Murugesan, S. (2008). Harnessing green IT: Principles and practices. *IT Professional, 10,* 24–33. doi:10.1109/MITP.2008.10

Naditz, A. (2008). Green IT 101: Technology helps businesses and colleges become enviro-friendly. *Sustainability: The Journal of Record, 1,* 315–318. doi:10.1089/SUS.2008.9931

Nagios. (2009). *The industry standard in IT infrastructure monitoring.* Retrieved from http://www.nagios.org/

Naraine, R. (2007). *100% undetectable malware' challenge.* Retrieved from http://zdnet.com/blog/security/rutkowska-faces-100-undetectable-malware-challenge/334

Nascimento, A. P., Boeres, C., & Rebello, V. E. F. (2008). Dynamic self-scheduling for parallel applications with task dependencies. In *Proceedings of the 6th International Workshop on Middleware for Grid Computing* (pp. 1-6). New York, NY: ACM Press.

Nathuji, R., Isci, C., & Gorbatov, E. (2007). Exploiting platform heterogeneity for power efficient data centers. In *Proceedings of the Fourth IEEE International Conference on Autonomic Computing* (p. 5).

National Institute of Standards and Technology. (2010). *NIST homepage.* Retrieved from http://www.nist.gov/

Nelson, B., Phillips, A., & Steuart, C. (2010). *Guide to computer forensics and investigations* (4th ed.). Cambridge, MA: Course Technology.

Nelson, M. R. (2009). The cloud, the crowd and public policy. *Issues in Science and Technology,* 71–76.

Netcraft. (2010). *November 2010 web server survey.* Retrieved from http://news.netcraft.com/archives/2010/11/05/november-2010-web-server-survey.html

Neugebauer, R., & McAuley, D. (2001) Energy is just another resource: Energy accounting and energy pricing in the nemesis OS. In *Proceedings of the 8th IEEE Workshop on Hot Topics in Operating Systems,* Schloss Elmau, Germany (pp. 59-64).

Newsham, T., Palmer, C., & Stamos, A. (2007). *Breaking forensics software: Weaknesses in critical evidence collection.* Retrieved from http://www.isecpartners.com

NIST. (2010). *Cloud computing forum and workshop.* Retrieved from http://csrc.nist.gov/groups/SNS/cloud-computing/

Niyato, D., Chaisiri, S., & Sung, L. B. (2009). Optimal power management for server farm to support green computing. In *Proceedings of the 9th IEEE/ACM International Symposium on Cluster Computing and the Grid* (pp. 84-91).

Nóbrega, A. D., Nyczyk, P., Retico, A., & Vicinanza, D. (2007). Global grid monitoring: The egee/wlcg case. In *Proceedings of the Workshop on Grid Monitoring* (pp. 9-16). New York, NY: ACM Press.

Nordman, B., & Christensen, K. (2010). Proxying: Next step in reducing IT energy use. *IEEE Computer, 43*(1), 91–93.

OpenStack. (2010). *OpenStack open source cloud computing software.* Retrieved from http://www.openstack.org/

Oppel, A. (2004). *Databases DeMYSTiFieD* (2nd ed.). New York, NY: McGraw-Hill.

Osipov, C., Goldszmidt, G., Taylor, M., & Poddar, I. (2009). *Develop and deploy multi-tenant web-delivered solutions using IBM middleware: Part 2: Approaches for enabling multi-tenancy.* Retrieved from http://www.ibm.com/developerworks/webservices/library/ws-multitenantpart2/index.html

Oxford Dictionaries. (2010a). *Adaptable.* Retrieved from http://oxforddictionaries.com/view/entry/m_en_gb0007570

Oxford Dictionaries. (2010b). *Elasticity.* Retrieved from http://oxforddictionaries.com/view/entry/m_en_gb0980800

Pajorova, E., & Hluchy, L. (2010, May 5-7). 3D visualization the results of complicated grid and cloud-based applications. In *Proceedings of the 14th International Conference on Intelligent Engineering Systems*, Las Palmas, Spain.

Pan, K., Turner, S. J., Cai, W., & Li, Z. (2009). Multi-user gaming on the grid using a service oriented HLA RTI. In *Proceedings of the 13th IEEE/ACM International Symposium on Distributed Simulation and Real Time Applications* (pp. 48-56). Washington, DC: IEEE Computer Society.

Papazoglou, M. P., & Heuvel, W. (2007). Service oriented architectures: Approaches, technologies and research issues. *The International Journal on Very Large Data Bases, 16*(3), 389–415. doi:10.1007/s00778-007-0044-3

Patel, C. D., & Shah, A. J. (2005). *Cost model for planning, development and operation of a data center.* Retrieved from http://www.hpl.hp.com/techreports/2005/HPL-2005-107R1.pdf

Patel, C. D., Bash, C. E., Belady, C., Stahl, L., & Sullivan, D. (2001, July 8-13). *Computational fluid dynamics modeling of high compute density data centers to assure system inlet air specifications.* Paper presented at the Pacific Rim/ASME International Electronic Packaging Technical Conference and Exhibition, Kauai, HI.

Patterson, D., Armbrust, M., Fox, A., Griffith, R., Jseph, A. D., Katz, R. H., et al. (2009). *Above the clouds: A Berkeley view of cloud computing* (Tech. Rep. No. UCB/EECS-2009-28). Berkeley, CA: University of California.

Paz, A., Perez-Sorosal, F., Patiño-Martínez, M., & Jiménez-Peris, R. (2010). Scalability evaluation of the replication support of JonAS, an industrial application server. In *Proceedings of the European Dependable Computing Conference* (pp. 55-60). Washington, DC: IEEE Computer Society.

Pearson, S. (2009). Taking account of privacy wehn designing cloud computing services. In *Proceedings of the International Workshop on Software Engineering Challenges of Cloud Computing.* Vancouver, BC, Canada.

Peltier, T. R. (2004). Developing an enterprisewide policy structure. *Information Systems Security, 13*(1), 44–50. doi:10.1201/1086/44119.13.1.20040301/80433.6

Perry, G. (2008). *How cloud & utility computing are different.* Retrieved from http://gigaom.com/2008/02/28/how-cloud-utility-computing-are-different/

Piazzesi, M. (2010). *Affine term structure models.* Amsterdam, The Netherlands: Elsevier.

Pinheiro, E., Bianchini, R., Carrera, E. V., & Heath, T. (2001). Load balancing and unbalancing for power and performance in cluster-based systems. In *Proceedings of the Workshop on Compilers and Operating Systems for Low Power* (pp. 182-195).

Plummer, D. C., Smith, D. M., Bittman, T. J., Cearley, D. W., Cappuccio, D. J., & Scott, D. (2009). *Five refining attributes of public and private cloud computing.* Stamford, CT: Gartner.

Popli, G. S., & Rao, D. N. (2009). *An empirical study of SMEs in electronics industry in India: Retrospect & prospects in post WTO era.* Retrieved from http://papers.ssrn.com/sol3/papers.cfm?abstract_id=1459809

Porter, G. (2010). Decoupling storage and computation in Hadoop with SuperDataNodes. *ACM SIGOPS Operating Systems Review, 44*(2), 41–46. doi:10.1145/1773912.1773923

Preda, S., Cuppens, F., & Cuppens-Boulahia, N. Alfaro. J., Toutain, L., & Elrakaiby, Y. (2009). Semantic context aware security policy deployment. In *Proceedings of the 4th International Symposium on Information, Computer, and Communications Security*, Sydney, Australia.

Provos, N., McNamee, D., Mavrommatis, P., Wang, K., & Modadugu, N. (2007). The ghost in the browser analysis of web-based malware. In *Proceedings of the RST Conference on First Workshop on Hot Topics in Understanding Botnets*, Berkeley, CA (p. 4).

Provos, N., Abu Rajab, M., & Mavrommatis, P. (2009). Cybercrime 2.0: When the cloud turns dark. *ACM Queue; Tomorrow's Computing Today*, 7(2), 46–47. doi:10.1145/1515964.1517412

Raftery, T. (2008). *How green is cloud computing?* Retrieved from http://greenmonk.net/how-green-is-cloud-computing/

Raghavendra, R., Ranganathan, P., Talwar, V., Wang, Z., & Zhu, X. (2008). No power struggles: Coordinated multi-level power management for the data center. In *Proceedings of the 13th International Conference on Architectural Support for Programming Languages and Operating Systems* (pp. 48-59).

Ramim, M., & Levy, Y. (2006). Securing e-learning systems: A case of insider cyber attacks and novice IT management in a small university. *Journal of Cases on Information Technology*, 8(4), 24–34. doi:10.4018/jcit.2006100103

Rao, M. L. N. (2010). *SaaS opportunity for Indian channels*. Santa Clara, CA: Jamcracker Inc.

Raza, A., & Abbas, H. (2008). *Security evaluation of software architectures using ATAM*. Paper presented at the IPID ICT4D PG Symposium, Joensuu, Finland.

Reservoir. (2010). *Reservoir fp7*. Retrieved from http://62.149.240.97/

Rewatkar, L. R., & Lanjewar, U. A. (2010). Data management in market-oriented cloud computing. *Advances in Computer Science and Technology*, 3(2), 217–222.

RightScale. (n. d.). *Cloud management platform*. Retrieved from http://www.rightscale.com/

Rittinghouse, J. W., & Ransome, J. F. (2009). *Cloud computing implementation, management and security*. Boca Raton, FL: Taylor & Francis Group/CRC Press.

Rodero-Merino, L., Vaquero, L. M., Gil, V., Galán, F., Fontán, J., & Montero, R. S. (2010). From infrastructure delivery to service management in clouds. *Future Generation Computer Systems*, 26(8), 1226–1240. doi:10.1016/j.future.2010.02.013

Roy, B., & Graham, T. C. (2008). *Methods for evaluating software architecture: A survey* (Tech. Rep. No. 2008-545). Kingston, ON, Canada: Queen's University.

Rusu, C., Ferreira, A., Scordino, C., Watson, A., Melhem, R., & Moss'e, D. (2006). Energy-efficient real-time heterogeneous server clusters. In *Proceedings of the 12th IEEE Real-Time and Embedded Technology and Applications Symposium*, San Jose, CA (pp. 418-428).

Rutkowska, J. (2007). *Security challenges in virtualaized environments*. Paper presented at the Nordic Virtualization Forum.

Saha, D., Sahu, S., & Shaikh, A. (2003). A service platform for on-line games. In *Proceedings of the 2nd Workshop on Network and System Support for Games* (pp. 180-184). New York, NY: ACM Press.

SANS. (2006). *Information security policy templates*. Reston, VA: SANS Institute.

Sasikala, P. (2011). Cloud computing in higher education: Opportunities and issues. *International Journal of Cloud Applications and Computing*, 1(2), 1–13.

Sasikala, P. (in press). Cloud computing: Present status and future implications. *International Journal of Cloud Computing*.

Sasikala, P. (in press). Research challenges and potential green technological applications in cloud computing. *International Journal on Cloud Computing*.

Schneider, E. (2010). *Go green, save green: The benefits of eco-friendly computing*. Retrieved from http://www.apcmedia.com/salestools/SLAT-7DCQ5J_R0_EN.pdf

Scholl, H. J. (2003). E-government: A special case of ICT-enabled business process change. In *Proceedings of the 36th Hawaii International Conference on System Sciences*.

Schooff, P. (2009). *The insider's guide to business and IT agility*. Retrieved from http://www.ebizq.net/soaregistercloudfutures.html

Schubert, L., Jeffery, K., & Neidecker-Lutz, B. (2010). *The future for cloud computing: Opportunities for European cloud computing beyond 2010 (public version 1.0)*. Retrieved from http://cordis.europa.eu/fp7/ict/ssai/docs/cloud-report-final.pdf

Schulz, W. (2009). *What is SaaS, Cloud Computing, PaaS and IaaS*. Retrieved from http://www.s-consult.com/2009/08/04/what-is-saas-cloud-computing-paas-and-iaas/

Secureworks.com. (2008). *Security 101: Botnets*. Retrieved from http://www.secureworks.com/research/newsletter/2008/05/

See, S. (2008). *Is there a pathway to a green grid??* Retrieved from http://www.ibergrid.eu/2008/presentations/Dia%2013/4.pdf

Seelam, S. R., & Teller, P. J. (2007). Virtual I/O scheduler: A scheduler of schedulers for performance virtualization. In *Proceedings of the 3rd International Conference on Virtual Execution Environments* (pp. 105-115).

Seifert, R. (1998). *Gigabit ethernet*. Reading, MA: Addison-Wesley.

Shachor, G. (2001). *Working with mod_jk*. Retrieved from http://tomcat.apache.org/tomcat-3.3-doc/mod_jk-howto.html

Sharma, P. (2009). *Is India ready for cloud computing?* Retrieved from http://dqchannels.ciol.com/content/space/109102902.asp

Shepler, S., Callaghan, B., Robinson, D., Thurlow, R., Beame, C., Eisler, M., et al. (2003). *Network File System (NFS) version 4 protocol (request for comments no. 3530): IETF*. Retrieved from http://www.ietf.org/rfc/rfc3530.txt

Silberstein, A., Cooper, B. F., Srivastava, U., Vee, E., Yerneni, R., & Ramakrishnan, R. (2008). Efficient bulk insertion into a distributed ordered table. In *Proceedings of the International Conference on Management of Data* (pp. 765-778).

Sinharoy, B., Kaal, R. N., Tendler, J. M., Eickerneyer, R. J., & Joyner, J. B. (2005). POWER 5 system micro architecture. *IBM Journal of Research and Development, 49*(4-5), 505–521. doi:10.1147/rd.494.0505

Smarr, L. (2010). Project GreenLight: Optimizing cyber-infraestructure for a carbon-constrained world. *IEEE Computer, 43*, 22–27.

Smith, J. E., & Nair, R. (2005). *Virtual machines- Versatile platforms for systems and processes*. San Francisco, CA: Morgan Kaufmann.

Snodgrass, R. T., Yao, S. S., & Collberg, C. (2004). Tamper detection in audit logs. In *Proceedings of the Thirtieth International Conference on Very Large Data Bases* (pp. 504-515).

SOA. (n. d.). *Service-oriented architecture definition*. Retrieved from http://www.service-architecture.com/web-services/articles/service-oriented_architecture_soa_definition.html

Sotomayor, B., Montero, R. S., Liorente, I. M., & Foster, I. (2008). Capacity leasing in cloud systems using OpenNebula engine. In *Proceedings of the Workshop on Cloud Computing and its Applications*, Chicago, IL.

Sotomayor, B., Montero, R. S., Liorente, I. M., & Foster, I. (2009). Virtual infrastructure management in private and hybrid clouds. *IEEE Internet Computing, 13*(5), 14–22. doi:10.1109/MIC.2009.119

Springboard Research. (2010a). *Cloud computing in Asia Pacific 2010*. Beijing, China: Springboard Research.

Springboard Research. (2010b). *IT market in India 2009-2013*. Beijing, China: Springboard Research.

Srikantaiah, S., Kansal, A., & Zhao, F. (2009). Energy aware consolidation for cloud computing. *Cluster Computing, 12*, 1–15.

Srinivasan, M. (2010). Cloud security for small businesses. In Proceedings of the *Allied Academies International Conference of the Academy of Information &. Management Science, 14*, 72–73.

Srinivasan, S. (2007). *Security and privacy vs. computer forensics capabilities*. ISACA Online Journal.

Stakhanova, N., Basu, S., & Wong, J. (2007). A taxonomy of intrusion response systems. *International Journal of Information Security, 1*(1-2), 169–184.

Standage, T. (2002). The weakest link. *The Economist*, 11-14.

Stanley, M. (2008). *Technology trends*. Retrieved from http://www.morganstanley.com/institutional/techresearch/pdfs/TechTrends062008.pdf

Stenberg, D. (2010). *libcurl - the multiprotocol file transfer library*. Retrieved from http://curl.haxx.se/libcurl/

Stoica, I., Morris, R., Karger, D., Kaashoek, M. F., & Balakrishnan, H. (2001). Chord: A scalable peer-to-peer lookup service for internet applications. In. *Proceedings of the Conference on Applications, Technologies, Architectures, and Protocols for Computer Communications, 29*, 149–160.

Strickland, J. (n. d.). *How cloud computing works*. Retrieved from http://communication.howstuffworks.com/cloud-computing1.htm

Sudha Sadhasivam, G., Jeyarani, R., Vasanth Ram, R., & Nagaveni, N. (2009). Design and implementation of an efficient two level scheduler for cloud computing environment. In *Proceedings of the International Conference on Advances in Recent Technologies in Communication and Computing*, Kottayam, India.

Sun Microsystems. (2009). *Introduction to cloud computing architecture*. Retrieved from http://webobjects.cdw.com/webobjects/media/pdf/Sun_CloudComputing.pdf

Sun, W., Zhang, X., Guo, C. J., Sun, P., & Su, H. (2008). Software as a service: Configuration and customization perspectives. In *Proceedings of the IEEE Congress on Services Part II* (pp. 18-25).

Swanson, M., & Guttman, B. (1996). *Generally accepted principles and practices for securing information technology systems*. Retrieved from http://csrc.nist.gov/publications/nistpubs/800-14/800-14.pdf

Takeda, T., Oki, E., Inoue, I., Shiomoto, K., Fujihara, K., & Kato, S. (2008). Implementation and experiments of path computation element based backbone network architecture. *IEICE Transactions on Communications, 91*(8), 2704–2706. doi:10.1093/ietcom/e91-b.8.2704

Ta-Shma, P., Laden, G., Ben-Yehuda, M., & Factor, M. (2008). Virtual machine time travel using continuous data protection and checkpointing. *ACM SIGOPS Operating Systems Review, 42*, 127–134. doi:10.1145/1341312.1341341

Taylor, M. (2010). *Enterprise architecture – architectural strategies for cloud computing: Oracle*. Retrieved from http://www.techrepublic.com/whitepapers/oracle-white-paper-in-enterprise-architecture-architecture-strategies-for-cloud-computing/2319999

Tekinerdogan, B. (2004). ASAAM: Aspectual software architecture analysis method. In *Proceedings of the Fourth Working IEEE/IFIP Conference on Software Architecture* (pp. 5-14). Washington, DC: IEEE Computer Society.

The Climate Group. (2008). *SMART 2020: Enabling the low carbon economy in the information age*. Retrieved from http://www.smart2020.org/_assets/files/02_Smart-2020Report.pdf

Thomson, R. (2009, February 24). Socitm: Cloud computing revolutionary to the public sector. *Computer Weekly*.

Thorsten. (2008). *Setting up a fault-tolerant site using Amazon's availability zones*. Retrieved from http://blog.rightscale.com/2008/03/26/setting-up-a-fault-tolerant-site-using-amazons-availability-zones/

Thuen, C. (2007). *Understanding counter-forensics to ensure a successful investigation*. Retrieved from http://citeseerx.ist.psu.edu/viewdoc/summary?doi=10.1.1.138.2196

Tian, C., Zhou, H., He, Y., & Zha, L. (2009). A dynamic MapReduce scheduler for heterogeneous workloads. In *Proceedings of the 8th International Conference on Grid and Cooperative Computing*, Lanzhou, China (pp. 218-224).

Trickyways.com. (2009). *How to change timestamp of a file in Windows*. Retrieved from http://www.trickyways.com/2009/08/how-to-change-timestamp-of-a-file-in-windows-file-created-modified-and-accessed/

Trusted Computing Group. (2010). *Cloud computing and security – a natural match*. Retrieved from http://www.infosec.co.uk/

Twitchell, D. P. (2006). Social engineering in information assurance curricula. In *Proceedings of the 3rd Annual Conference on Information Security Curriculum Development*, Kennesaw, GA.

Uptime Institute. (2009). *Clearing the air on cloud computing*. Retrieved from http://uptimeinstitute.org

US-Computer Emergency Readiness Team. C. (2008). *Computer F=Forensics.* Retrieved from http://www.us-cert.gov/reading_room/forensics.pdf

Vallee, G., Naughton, T., Ong, H., & Scott, S. (2006, October 17). Checkpoint/restart of virtual machines based on Xen. In *Proceedings of the High Availability and Performance Computing Workshop*, Santa Fe, NM.

Vaquero, L. M., RoderoMerino, L., Cáceres, J., & Lindner, M. (2009). A break in the clouds: Towards a cloud definition. *ACM SIGCOMM Computer Communications Review, 39*(1), 50-55.

Varley, I. T. (2009). *No relation: The mixed blessings of non-relational databases.* Austin, TX: University of Texas.

Velte, A. T., Velte, T. J., & Elsenpeter, R. (2010). Cloud computing basics. In *Cloud computing: A practical approach* (pp. 3–22). New York, NY: McGraw-Hill.

Venkatachalam, V., & Franz, M. (2005). Power reduction techniques for microprocessor systems. *ACM Computing Surveys, 37*(3), 195–237. doi:10.1145/1108956.1108957

Verizon Business. (2010). *2010 data breach investigations report.* Retrieved from http://www.verizonbusiness.com/resources/reports/rp_2010-data-breach-report_en_xg.pdf?&src=/worldwide/resources/index.xml&id=

Verma, A., Ahuja, A., & Neogi, A. (2008). pMapper: Power and migration cost aware application placement in virtualized systems. In *Proceedings of the ACM/IFIP/USENIX 9th International Middleware Conference* (pp. 243-264).

Viega, J. (2009). Cloud computing and the common man. *IEEE Computer, 42*(8), 106–108.

Vieira, K., Schulter, A., Westphall, C., & Westphall, C. (2009). Intrusion detection techniques in grid and cloud computing environment. *IT Professional, 12*(4), 38–43. doi:10.1109/MITP.2009.89

Vigfusson, Y., Silberstein, A., Cooper, B. F., & Fonseca, R. (2009). Adaptively parallelizing distributed range queries. *Very Large Data Base Endowment, 2*(1), 682–693.

Vining, J., & Di Maio, A. (2009). *Cloud computing for government is cloudy.* Stamford, CT: Gartner.

VMware. (2009). *VMware distributed power management concepts and use* (Tech. Rep. No. IN-073-PRD-01-01). Palo Alto, CA: VMware Inc.

Von Laszewski, G., Wang, L., Younge, A. J., & He, X. (2009). Power-aware scheduling of virtual machines in DVFS-enabled clusters. In *Proceedings of the IEEE International Conference on Cluster Computing*, New Orleans, LA (pp. 1-10).

Wagner, J. (2010). *Cloud computing: Preparing for the next 20 years of IT.* Retrieved from http://www.worldsystems-it.com/?p=209

Walcott, C. (2005). *Taking a load off: Load balancing with balance.* Retrieved from http://www.linux.com/archive/feature/46735

Wallis, P. (2008). *A brief history of cloud computing: Is the cloud there yet?* Retrieved from http://cloudcomputing.sys-con.com/node/581838

Wang, C. (2009). *How secure is your cloud: A close look at cloud computing security issues.* Retrieved from http://www.forrester.com/rb/Research/how_secure_is_cloud/q/id/45778/t/2

Wang, D. (2007). Meeting green computing challenges. In *Proceedings of the International Symposium on High-Density Packaging and Microsystem Integration* (pp. 1-4).

Wang, X., & Yu, H. (2005). How to break MD5 and other hash functions. In *Proceedings of the Annual International Conference on the Theory and Applications of Cryptographic Techniques* (pp. 19-35).

Wang, H., Zhang, Y., & Cao, J. (2005). Effective collaboration with information sharing in virtual universities. *IEEE Transactions, 21*(6), 840–853.

Ware, M., Rajamani, K., Floyd, M., Brock, B., Rubio, J. C., Rawson, F., & Carter, J. B. (2010). Architecting for power management: The IBM POWER 7 approach. In *Proceedings of the IEEE 16th International Symposium High Performance Computer Architecture* (pp. 1-11).

Warfield, A., Ross, R., Fraser, K., Limpach, C., & Hand, S. (2005, June 12-15). Parallax: Managing storage for a million machines. In *Proceedings of the 10th Workshop on Hot Topics in Operating Systems*, Santa Fe, NM (pp. 1-11).

Waters, D. (2008). *Quantitative methods for business* (4th ed.). Upper Saddle River, NJ: Prentice Hall.

Wei, J., Zhang, X., Ammons, G., Bala, V., & Ning, P. (2009). Managing security of virtual machine images in a cloud environment. In *Proceedings of the ACM Cloud Computing Security Workshop* (pp. 91-96).

Weinhardt, C., Anandasivam, A., Blau, B., Borissov, N., Meinl, T., & Michalk, W. (2009). Cloud computing – a classification, business models, and research directions. *Journal of Business and Information Systems Engineering, 1*(5), 391–399. doi:10.1007/s12599-009-0071-2

Weinhardt, C., Anandasivam, A., Blau, B., & Stober, J. (2009). Business models in the service world. *IEEE Computer, 11*(2), 28–33.

Weiss, A. (2007). Computing in the clouds. *netWorker, 11*(4), 16–25. doi:10.1145/1327512.1327513

Wei, Y., & Blake, B. M. (2010). Service-oriented computing and cloud computing: Challenges and opportunities. *IEEE Internet Computing, 14*(6), 72–75. doi:10.1109/MIC.2010.147

Whitman, M., & Mattord, H. (2009). *Principles of information security* (3rd ed.). Boston, MA: Course Technology.

Whitteker, M. (2008, 11). Anti-forensics: Breaking the forensics process. *ISSA Journal*.

Wikipedia. (2010). *Welcome to Wikipedia.* Retrieved from http://en.wikipedia.org/wiki/

Wikipedia. (n. d.). *Cloud computing.* Retrieved from http://en.wikipedia.org/wiki/Cloud_computing

Wikipedia. (n. d.). *Green computing.* Retrieved from http://en.wikipedia.org/wiki/Green_computing

Wikipedia. (n. d.). *Information lifecycle management.* Retrieved from http://en.wikipedia.org/wiki/Information_Lifecycle_Management

Wikipedia. (n. d.). *Service management facility.* Retrieved from http://en.wikipedia.org/wiki/Service_Management_Facility

Wikipedia. (n. d.). *Software as a service.* Retrieved from http://en.wikipedia.org/wiki/Software_as_a_service

Wittow, M. H., & Buller, D. J. (2010). Cloud computing: Emerging legal issues for access to data, anywhere, anytime. *Journal of Internet Law, 14*(1), 1–10.

Wojtczuk, R. (2008). *Track: Virtualization.* Retrieved from http://blackhat.com/html/bh-usa-08/bh-usa-08-speakers.html#Wojtczuk

Wolter, K. (2010). *Stochastic models for fault tolerance - Restart, rejuvenation and checkpointing.* New York, NY: Springer.

Wood, T., Shenoy, P., Gerber, A., Ramakrishnan, K. K., & Van der Merwe, J. (2009). *The case for enterprise-ready virtual private clouds.* Paper presented at the Workshop on Hot Topics in Cloud Computing.

Wu, Z. Z., & Chen, H. C. (2006). Design and implementation of TCP/IP offload engine system over gigabit Ethernet. In *Proceedings of the 15th International Conference on Computer Communications and Networks* (pp. 245-250).

Yang, H., Hu, S., & Guo, J. (2009). Cost-oriented task allocation and hardware redundancy policies in heterogeneous distributed computing systems considering software reliability. *Computers & Industrial Engineering, 56*(4), 1687–1696. doi:10.1016/j.cie.2008.11.001

Ylonen, T., & Lonvick, C. (2006). *The Secure Shell (SSH) protocol architecture (request for comments no. 4251):* IETF. Retrieved from http://www.ietf.org/rfc/rfc4251.txt

Yogesh, K. D., & Navonil, M. (2010). It's unwritten in the Cloud: The technology enablers for realising the promise of cloud computing. *Journal of Enterprise Information Management, 23*(6), 673–679. doi:10.1108/17410391011088583

Young, J. (2008). 3 ways that web-based computing will change colleges and challenge them. *The Chronicle of Higher Education, 55*(10), 16.

Youseff, L., Butrico, M., & Da Silva, D. (2008). *Towards a unified ontology of cloud computing.* Retrieved from http://www.cs.ucsb.edu/~lyouseff/CCOntology/Cloud-Ontology.pdf

Zaharia, M., Borthakur, D., Sarma, J. S., Elmeleegy, K., Shenker, S., & Stoica, I. (2010). Delay scheduling: A simple technique for achieving locality and fairness in cluster scheduling. In *Proceedings of the 5th European Conference on Computer Systems*, Paris, France (pp. 265-278).

Zaharia, M., Konwinski, A., Joseph, A. D., Katz, R. H., & Stoica, I. (2008). Improving MapReduce performance in heterogeneous environments. In *Proceedings of the 8th USENIX Symposium on Operating Systems Design and Implementation* (pp. 29-42).

Zahn, B. (1998, March). Steady state thermal characterization of multiple output devices using linear superposition theory and a non-linear matrix multiplier. In *Proceedings of the Fourteenth Annual Semiconductor Thermal Measurement and Management Symposium* (pp. 39-46).

Zdun, U., Hentrich, C., & Dustdar, S. (2007). Modeling process-driven and service-oriented architectures using patterns and pattern primitives. *ACM Transactions on the Web*, *1*(3), 14. doi:10.1145/1281480.1281484

Zhen, J. (2008). *Five key challenges of enterprise cloud computing.* Retrieved from http://cloudcomputing.syscon.com/node/659288

Zibin, Z., Zhou, T. C., Lyu, M. R., & King, I. (2010). FTCloud: A component ranking framework for fault-tolerant cloud applications. In *Proceedings of the 21th IEEE International Symposium on Software Reliability Engineering* (pp. 398-407).

Zikos, S., & Karatza, H. D. (2010). Performance and energy aware cluster-level scheduling of compute-intensive jobs with unknown service times. *Simulation Modelling Practice and Theory*, *19*(1), 239–250. doi:10.1016/j.simpat.2010.06.009

Zou, Y. Q., Liu, J., Wang, S. C., Zha, L., & Xu, Z. W. (2010). CCIndex: A complemental clustering index on distributed ordered tables for multi-dimensional range queries. In *Proceedings of the 7th IFIP International Conference on Network and Parallel Computing*, Zhengzhou, China (pp. 247-261).

Zscaler. (2010). *Zscaler web security cloud for small business.* Retrieved from http://www.zscaler.com/pdf/brochures/ds_zscalerforsmb.pdf

Index